RENDERING LIFE MOLECULAR

EXPERIMENTAL FUTURES
Technological Lives, Scientific Arts, Anthropological Voices
A series edited by Michael M. J. Fischer and Joseph Dumit

RENDERING LIFE MOLECULAR

MODELS, MODELERS, AND EXCITABLE MATTER

Natasha Myers

DUKE UNIVERSITY PRESS
Durham and London | 2015

Designed and typeset by Tseng Information Systems, Inc. in
Minion DTP and Hypatia Sans

Library of Congress Cataloging-in-Publication Data
Myers, Natasha, [date]
Rendering life molecular : models, modelers, and excitable matter /
Natasha Myers.
pages cm—(Experimental futures : technological lives, scientific arts,
anthropological voices)
Includes bibliographical references and index.
ISBN 978-0-8223-5866-4 (hardcover : alk. paper)
ISBN 978-0-8223-5878-7 (pbk. : alk. paper)
ISBN 978-0-8223-7563-0 (e-book)
1. Proteins—Structure. 2. Molecules—Models. 3. X-ray crystallography.
4. Molecular biologists. I. Title. II. Series: Experimental futures.
QP551.M99 2015
572′.33—dc23
2015005600

Duke University Press gratefully acknowledges the support of the Faculty of
Liberal Arts & Professional Studies Book Publication Subvention Program, York
University, Canada, which provided funds toward the publication of this book.

Cover art: "The Inner Life of the Cell," © 2006–2014 President and Fellows
of Harvard College. Created by Alain Viel, PhD, and Robert Lue, PhD, in
collaboration with XVIVO, LLC, and John Liebler, lead animator. Made possible
through the generous support of the Howard Hughes Medical Institution's
Undergraduate Science Education Program.

This book is dedicated to the memory of my grandparents:

    Ruth and Irv Mudrick
      and
    Sadie and Ken Myers

I can still see your smiling eyes.

CONTENTS

What are you made of? Look at your hands. Draw one palm across the other. Feel the density of your tissues, the bones, musculature, and sinuous ligaments. What gives your tissues substance and form? You have probably been told that your body is composed of trillions of living cells. But what are your cells made of? What is the stuff of life?

Those of us hailed by contemporary technoscience are likely familiar with the notion that proteins are the molecular building blocks of life. But what is a protein?[1] And why do we care so much about the molecular constitution of our bodies? Proteins are remarkably prominent actors in our everyday explanations of life and health. We so often hear that "you are what you eat." When you are hungry, what is it you are looking for? You don't have to be an athlete, nutritionist, or dieter to know that protein-rich foods satisfy your hunger. If you live in the relative privilege of North America or Europe you are unlikely to escape a consumer culture that finds you constantly on the lookout for ways to improve your body and your health. Maybe you compulsively read the nutrition labels on packaged food at the supermarket, those little charts that fractionate foods according to chemical composition, listing the amounts of fat, salt, sugar, cholesterol, and protein you are about to consume. How many grams of protein are in that package of tofu? Or that slice of processed cheese?

Proteins, we are told, provide sustenance. Yet scientists also figure proteins as the enzymes that catalyze life-sustaining chemistry; the substances that transduce signals within and between cells; the molecules that transcribe, translate, and rewrite DNA to produce more proteins; and the materials that provide the dynamic—continually cycling, growing, and retracting—architectural support for cellular life. The evolutionary stories we tell ourselves about the origins of life and the adaptations of organisms rely on knowledge of heritable changes in DNA that affect the molecular constitution of proteins.

Pharmaceutical advertisements animate the specific chemistry of drugs to show you how they bind to proteins and intervene in your physiological processes at the molecular scale.[2] In the twenty-first century, life and living bodies have been rendered thoroughly molecular, and proteins now share the stage with DNA and other molecules as pivotal actors in our stories about cellular life.[3]

This propensity to parse the world into molecular components has a long history that extends back to ancient philosophers who postulated that the worldly stuff we could see and feel had an unseen "inner constitution"; matter, it was thought, was made up of subvisible atoms or particles.[4] For centuries scientists in a range of disciplines have expended great effort refining techniques to make matter visible, tangible, and workable at the molecular scale. The vindication of atomic theories of matter and the allure of mechanistic explanations of life in the nineteenth and twentieth centuries have swept researchers up in efforts to see through the obscuring density of cells and tissues. Researchers seek chemical and physical explanations to deepen their mechanistic understanding of living processes.[5] It is through their concerted efforts that the stuff of life has *come to matter* at the molecular scale.[6] This book is about the practitioners who *render life molecular.*

Who knows about proteins? Who can tell us what they look like and what they are up to in a cell? Life science researchers in the multiple and overlapping disciplines of cell biology, molecular genetics, and genomics care a great deal about the activities of proteins and other molecules in the cell, but they generally do not have the tools or techniques to make protein structures visible at the atomic scale. One group of researchers who call themselves "protein crystallographers" are especially interested in resolving the precise atomic configuration of protein molecules. They want to be able to see proteins and intervene in cellular processes at the molecular scale. In order to make proteins visible, tangible, and workable in their laboratories, these researchers build three-dimensional models in a wide range of media. This book is a study of the remarkable techniques and media forms that crystallographers and other protein modelers have developed to make visible otherwise imperceptible molecular phenomena. It examines protein crystallography as a practice and a culture, tracking how these practitioners gather data, conduct experiments, test hypotheses, and demonstrate results.

Protein crystallographers face significant challenges in their efforts to build sound models of things they cannot otherwise see. What remains for them unseen, intangible, and unimaginable? And how do they fill in the gaps between a perceptible world of aggregate substances (like cells, tissues, and bodies) and the largely imperceptible realm of molecular phenomena? This book explores the remarkable ways that they confront the limits of what they can see and what they can know. It documents how they hone their sensibilities and intuitions, and the creative ways that they engage their bodies in the production and dissemination of visual facts. And, by tracking both expert researchers and their more novice students, it inquires into the ways that modelers teach their students what they see, feel, imagine, and know about molecular life.

These modelers craft the models that are defining the material substructure of living bodies today. Their models come to stand as authoritative descriptions of the molecular world. The facts these practitioners craft in their laboratories are also the facts of life that many others come to live by.[7] These are the facts that experts rely on to tell us what our bodies are made of, what makes us tick, what makes us sick, and what might make us better. The explanations of life many of us have come to rely on, including the evolutionary stories we tell ourselves about the origins of life, hinge on knowledge of the molecular structures of proteins.[8] Scholars in feminist theory and science studies have shown that it is crucial to pay close attention to the techniques and practices that shape matter and materiality.[9] Protein modelers are the scientists to watch in order to see what forms of life and what materialities are coming to matter in the twenty-first-century life sciences.

And yet, the forms of life coming to matter in the hands of these practitioners are perhaps not what we might expect from a discipline grounded in mechanistic approaches to life. This book examines a range of phenomena that are not easy to see, especially if one just reads the texts scientists write. It amplifies forms of life in the laboratory that are muted in other accounts. It pays close attention to what modelers see, say, feel, imagine, and know about molecules, and in the process, reveals a marvelous range of ways that proteinacious life is rendered in their laboratories. Modelers, it turns out, cultivate intimate relationships with their molecules as they get themselves caught up in the involving work of molecular visualization. It is in the energetically and emotionally charged space of the laboratory that protein modelers confront living matter as an unruly, wily substance. This account tunes in to the *affective entanglements of inquiry* in the laboratory to document the extraordinary ways that modelers' lively renderings of molecular life disrupt conventional, mechanistic accounts. This book argues that these modelers are shifting the contours of the contemporary biological imagination and reconfiguring the nature of living substance. Throughout, it asks: What is molecular life becoming in their hands?

ACKNOWLEDGMENTS

This book took a long time to write. I am forever grateful to all those who helped keep me engaged in this project over the course of such a long journey. I feel blessed to be surrounded by the most brilliant and generous mentors, teachers, collaborators, students, colleagues, and friends one could hope for. My deepest gratitude goes to Joseph Dumit, Stefan Helmreich, and Donna Haraway, whose loving care, inspiration, and guidance made this all possible. Joe welcomed me into the History | Anthropology | STS graduate program at MIT when I began my PhD. He worked with me closely and patiently for hours every week on each essay, thought, and concept I puzzled through over the course of my graduate degree. Stefan Helmreich, who arrived at MIT a few years later, inspired me with the depth of his insights into all things anthropological, and with his intellectual acuity and incredible commitment to mentorship. Stefan and Joe continue to be my most cherished mentors and collaborators. Partway through my degree, I had the opportunity to work with Donna Haraway as a visiting student in the History of Consciousness Department at the University of California, Santa Cruz. The time I spent in lectures, in seminars, and in meetings with her was transformative, and I continue to be buoyed by the sharpness, depth, and power of her teaching and thinking.

This book owes its existence to the generosity of many researchers, graduate students, postdocs, educators, and laboratory directors, including protein crystallographers, biological engineers, and many others who welcomed me into their laboratories and classrooms, sat with me in interviews for countless hours, and shared their remarkable stories. I wish I could thank them all by name. Many of the students I worked with requested anonymity, so the names of all participants have been changed and the location of their laboratories and institutions has been left unnamed.

As a graduate student I had the privilege of working closely with David

Kaiser, Michael Fischer, Susan Silbey, and Sheila Jasanoff. I learned a great deal from each of them. As a National Science Foundation (NSF) doctoral fellow, I had the opportunity to develop my research in collaboration with Sherry Turkle, Joe Dumit, Susan Silbey, Hugh Gusterson, David Mindell, Yanni Loukissas, Rachel Prentice, and others on a multidisciplinary research initiative. Susan's support throughout the duration of this project was invaluable. I am especially grateful to my graduate student cohort Candis Callison, Anita Chan, Richa Kumar, Jamie Pietruska, and Will Taggart. Many thanks also to Etienne Benson, Nate Greenslit, Shane Hamilton, Eden Medina, Esra Ozkan, Rachel Prentice, Anne Pollock, Aslihan Sanal, Jenny Smith, Livia Wick, Rebecca Woods, and Anya Zilberstein for their camaraderie and for making Boston more than just a pretty town. Many thanks to the members of Biogroop, including Stefan Helmreich, Sophia Roosth, Michael Rossi, Sara Wiley, and Rufus Helmreich for generating such a rich space for exploring the history and anthropology of the life sciences. Thanks also to Clementine Cummer for all the learning that came out of our movement and performance collaboration.

In Santa Cruz I found a lively community of brilliant thinkers. There I met some of my dearest collaborators and friends, including Natalie Loveless, Harlan Weaver, Maria Puig de la Bellacasa, and Astrid Schrader. I can no longer think without them. Many thanks also to Sarah Bracke, Lindsay Kelly, Sandra Koelle, Sha LaBare, and Lisette Oliveres for inspiring conversations. I spent the last months of my graduate degree at teaching UC Davis and working alongside Joe Dumit. I am grateful to Jim Griesemer, Colin Milburn, Moon Duchin, Andrés Barragan, Michelle Stewart, Chris Kortright, Nicholas D'Avella, Fabiana Li, and Vivian Choi for their warm welcome.

I returned to York University as a faculty member in the Department of Anthropology six years after completing my master's degree there in environmental studies. I feel blessed to have such an open and inviting place to do interdisciplinary work, and am grateful to my colleagues for their collegiality, friendship, and collaborations. My thanks especially to Naomi Adelson, Kathryn Denning, Leesa Fawcett, Shubhra Guruani, Zulfikar Hirji, Teresa Holmes, Edward Jones-Imhotep, Bernie Lightman, Ken Little, Maggie MacDonald, Aryn Martin, Nadya Martin, Carlota McAlister, Cate Sandilands, Albert Schrauwers, Joan Steigerwald, Penny van Esterik, Ana Viseu, Walter Whitely, and Daphne Winland. My students at York are amazing, and have taught me a great deal. It has been such a joy to see their fascinating research projects flourish. Many thanks to members of META Lab (Laboratory for

Modes of Embodiment in Technoscience and Anthropology) and the Plant Studies Collaboratory, including Melissa Atkinson-Graham, Laurie Baker, Heather Barnick, Jessica Caporusso, Lisa Cockburn, Heather Cruickshank, Bretton Fosbrook, Kelly Fritsch, Kristin Hardy, Peter Hobbs, Carla Hustak, Duygu Kasdogan, Kelly Ladd, Cameron Murray, Lisa Richardson, Andrew Schuldt, Emily Simmonds, and Annabel van Barren. Many thanks also to the graduate students in the Rendering Life Itself seminar for their generous readings of an earlier draft of this book. Above all, I am indebted to Michelle Murphy for her inspiration and brilliance and for her friendship over the many years we have collaborated to build a feminist science studies community around the Technoscience Salon here in Toronto.

This book has benefited from conversations and critical engagements at many conferences and meetings, including invited talks at the Joint Workshop of the Science and Technology Studies Programs at Harvard and MIT; the Performing Science Workshop at the Annenberg School for Communication, University of Pennsylvania; the Workshop on Scientific Collaboration, Interdisciplinary Pedagogies and the Knowledge Economy at Oxford University; the Knowledge/Value Conference at the University of Chicago; the Cosmopolitics Round Table at CUNY's Graduate Center; the Science, A Moving Image Symposium at Harvey Mudd College; the conference on Biomedicine and Aesthetics in a Museum Context at the Medical Museion in Copenhagen; and Seminar Series talks at the New School's Department of Anthropology, McGill University's Social Studies of Medicine, Sarah Lawrence College's Departments of Dance and Environmental Studies, and the Institute for History and Philosophy of Science at the University of Toronto. Many thoughtful scholars have engaged this work closely, and for their generous readings I want to thank Mark Auslander, Karen Barad, Lisa Cartwright, Hasok Chang, Tim Choy, Gail Davies, Richard Doyle, Martha Flemming, Jack Halberstam, Dehlia Hannah, Orit Halpern, Sarah Kember, Martha Kenney, Eben Kirksey, Hannah Landecker, Susan Leigh Star, Michael Lynch, Annmarie Mol, Iwan Morus, Ruth Müller, Susan Oyama, Anand Pandian, Heather Paxson, Hugh Raffles, Kaushik Sunder Rajan, Dave Richardson, Jessica Rosenberg, Dimitrina Spencer, Sergio Sismondo, Isabelle Stengers, Lucy Suchman, Charis Thompson, Christien Tompkins, Jennifer Tucker, Nina Wakeford, Catherine Waldby, Sha Xin Wei, Richard Wingate, Katherine Youssoff, and Charles Zerner.

At Duke University Press, I am indebted to Ken Wissoker for his incredible patience and care as I labored over this manuscript. Many thanks also to Leigh Barnwell and Elizabeth Ault for their support with the manuscript at critical stages. For their close and careful readings of earlier drafts of this manuscript I especially want to thank Melissa Atkinson-Graham, Joe Dumit, Orit Halpern, Stefan Helmreich, Nadine Levin, Michelle Murphy, Katja Pettinen, Sophia Roosth, Dorion Sagan, Astrid Schrader, Lucy Suchman, Alma Steingert, and Charis Thompson.

My dearest friends and family provided all the love and support I needed to see this project through. I extend huge gratitude to Rutvica Andrijasevic, Suzanne Bradley-Siskind, Leah Cowen, Ana Francisca de la Mora, Joseph Johnson-Cami, Ayelen Liberona, Natalie Loveless, Lynn Margulis, Alorani Martin, Shawn and Gloria Maximo, Saara Nafici, Maria Puig de la Bellacasa, Dorion Sagan, Robin Shulman, Geoff Siskind, Emma Somers, Inge Tamm, Evan Thompson, Rebecca Todd, and Karin von Ompteda. I am especially grateful to Debra Bluth for the depth of what she has taught me about movement over the past decade, and to Katherine Duncanson for her inspired teachings. To my parents, Sheila and Martin; to my sister and sister-in-law, Stephanie and Gillian; to my step-moms, Maggie, Kathie, and Virginia; and to Alice-the-cat: thank you for everything. You have kept me buoyed through it all.

This research was generously funded by a four-year Social Sciences and Humanities Research Council of Canada (SSHRC) Doctoral Fellowship (Award No. 752-2002-0301) and by the National Science Foundation through both a Predoctoral Fellowship (Grant No. 0220347) and a Dissertation Improvement Grant (Award No. SES-0646267), as well as a series of grants from the Faculty of Liberal Arts and Professional Studies at York University.

Material in this book has appeared in earlier publications, including: "Animating Mechanism: Animation and the Propagation of Affect in the Lively Arts of Protein Modeling," *Science Studies* 19, no. 2 (2006): 6–30; "Molecular Embodiments and the Body-work of Modeling in Protein Crystallography," *Social Studies of Science* 38, no. 2 (2008): 163–99; "Performing the Protein Fold," in *Simulation and its Discontents*, ed. Sherry Turkle, Cambridge, MA: MIT Press, 2009; "Pedagogy and Performativity: Rendering Laboratory Lives in the Documentary Naturally Obsessed: The Making of a Scientist," *Isis* 101, no. 4 (2010): 817–28; "Dance Your PhD: Embodied Animations, Body Experiments, and the Affective Entanglements of Life Science Research," *Body & Society* 18, no. 1 (2014): 151–89; and "Rendering Machinic Life," *Representation in Scientific Practice Revisited*, ed. Catelijne Coopmans, et al., MIT Press, 2014.

The cartoon shown in figure 1.1 features two male scientists in a genetics research laboratory. One is seated at the lab bench busy at work with his instruments and test tubes. The other scientist is twisting his body into the shape of a double helix. The seated scientist looks over his shoulder, chiding the contorted one: "Very good, Michaels—you're a DNA molecule. Now, get back to work.[1]

It would be easy to laugh with the scientist at the bench who derides his colleague's playful contortions as distractions from more important work. But what if the joke were on him? He thinks that he is getting his work done hunched over at the bench. This book argues that it is perhaps the helically wound-up scientist who is doing the important experiment. He uses his body to reason through the molecular structure of a complex biological molecule. It is by conducting a *body experiment*, an embodied twist on the well-known thought experiment, that he *figures out* the specificity of molecular form.

Those resembling the curmudgeonly scientist sitting at the bench are scarce among practitioners in the diverse disciplinary fields that converge around the task of protein modeling. Protein modelers engage their bodies actively in their work. They learn how to feel through molecular structures by experimenting with the forces and tensions in their own bodies. They get entangled—*kinesthetically and affectively*—in their modeling efforts. The term "kinesthetics" as I use it here describes the visceral sensibilities, movements, and muscular knowledge that modelers bring to their body experiments.[2] The term "affect," on the other hand, indexes the energetics, intensities, and emotions that propagate through modelers' efforts.[3] Both the kinesthetic and the affective dimensions of modelers' practices converge in the familiar realm of what we tend to call "feeling."[4]

Rendering Life Molecular documents the multifarious modes of body-work and play that are integral to protein modelers' research and teaching practices.

"Very good, Michaels – you're a DNA molecule. Now, get back to work."

I.1. A body experiment. © 1992 by Nick Downes; from *Big Science.* Courtesy of the artist.

These modalities are striking in that they challenge assumptions about the kinds of labor required to do scientific research, and the ways that scientists are supposed to stand in relation to their objects of inquiry. Scientific objectivity is conventionally understood as a neutral, rational, and so disembodied practice. Scientists are expected to dissociate their cognitive activities from their bodies' complicating passions and proclivities. Michaels's body experiment, however, challenges these assumptions as it makes explicit the ways that seeing, feeling, and knowing are entangled in laboratory research. Michaels demonstrates well the dense thicket of kinesthetic and affective entanglements involved in model building. Like Michaels, the practitioners documented in this book reveal that life science research is a full-bodied practice.

An example from my ethnographic fieldwork in a protein crystallography laboratory at a research university on the East Coast of the United States is instructive. I spent several days alongside Edward, a postdoctoral researcher from the UK, as he conducted routine work in the lab.[5] On one of these days we were sitting in the computer room looking at his data and at the computer graphic models he had been working with on screen. He walked me through the steps he had taken in order to build an atomic resolution model of a protein structure (for an example of a crystallographic structure of a protein molecule, see plate 1; readers not yet familiar with the basics of protein science or protein crystallography may want to consult the appendix to this book for a brief introduction). He showed me a computer program he had been using to help solve a recalcitrant problem with his model. This was software developed to facilitate pharmaceutical research. Programs like this can be useful for researchers who try to design drugs to perform specific functions in cells and tissues. Their designs build on a mechanistic model of biochemical interactions first proposed by German chemist Emil Fischer in 1885. Fischer suggested that proteins and their substrates, the molecules they interact with, "fit together like a lock and key."[6] Once researchers know the structure of a protein they want to target, they can design molecules that fit into the "active site" of a protein. By "docking" or binding to a chemically reactive crevice in the protein, a drug can disrupt or amplify the protein's biochemical activity.

In the course of our conversation, Edward became frustrated. "I used the automated docking programs, but they were giving me garbage." This was, it seemed, one more in a long series of challenges he faced trying to coax workable data out of his computer. There were so many ways, it seemed, that his computer programs failed him. He explained where the snag was. He gestured at the screen to show me how to see what he was saying:[7] "They were putting the [molecule] there, which is just not right. I thought screw it. I'll just *look at it* because often common sense is just as good as a software program." I was struck by his use of the term "common sense." What I had been learning from him and his colleagues was how much their work relies on carefully honed expert judgment to evaluate the volumes of data that are generated by their computer programs.[8] When automated software fails him, Edward builds his computer graphic models and interprets molecular interactions by eye and by hand. To do this he draws on a kind of molecular intuition he has built up over the long process of his training. The "common sense" that he invoked was perhaps common only among his teachers and colleagues. This was an expertise and a sensibility that he had cultivated over time through intensive training.

Why is it is so difficult to use off-the-shelf software to predict how proteins might bind, or dock with one another? Edward explained that these tools don't work well for large molecules like proteins because "proteins are breathing entities." I interrupted him. I didn't think I had heard him right. "Did you say proteins are . . . breathing?" "Yes. Breathing entities," he responded, adding, "I don't know. Sounds a bit romantic, doesn't it." Where the model on-screen remained static, he relayed the qualities of his breathing molecule by wrapping his hands around an invisible, pulsing sphere. According to all measures, Edward is a well-trained crystallographer. He tells me that he takes a "mechanistic approach" to protein function. He is clearly wary of enchantments that animate matter with mysterious forces. He certainly doesn't want to be seen anthropomorphizing molecules. Yet, the breath-like quality of proteins that he demonstrated for me was distinct from the kinds of random, Brownian motions that molecules are subject to inside cells. His animated gestures performed his conviction that molecules actively move and change in their watery, subcellular milieu.

His close study of chemical laws and the physical properties of proteins have certainly honed his "common sense." And yet, this sense of things has also been contoured by a kinesthetic and affective sensibility that he did not learn from books. He is particularly critical of the static data forms that are published in scientific papers. Two-dimensional images depict molecules as rigid bodies. He told me that these static images pose serious problems for those without the expertise to interpret the data. Sound interpretation of protein structures is an acquired skill, and according to him, many practitioners in the life sciences don't have the know-how to make sense of the data. In the space of our conversation he made some clear distinctions between different kinds of life scientists. Crystallographers are, for him, distinct from those practitioners who analyze cellular processes by manipulating genetic codes. He referred to them as "molecular biologists," and admonished them for being "notorious" for misinterpreting structures: "The main criticism crystallographers have about molecular biologists is that they don't think about the structure as a *breathing entity*. [For them] it's just a rigid body."

I understood well what he meant by this. I was trained as a molecular biologist and had started a PhD in 1997 to study the molecular genetic processes involved in plant and flower development. Over the course of my undergraduate and graduate training, I had never been introduced to protein structures. This was an era when genetic sequence data held sway and captivated researchers' attentions. At that time the precise atomic structures of the proteins encoded by the genetic sequences I worked with in the lab were unknown, and I had no way of making the leap between the one-dimensional genetic codes I was manipulating and the complex three-dimensional cellular structures and tissues that took shape over the course of development. Edward was right: just looking at the structure on the screen, I had no idea how proteins moved or how they participated in cellular activities. I did not yet have a feel for the dynamic physical and chemical properties of these biological molecules. With all my training in the life sciences, I was still a novice in this field. The static two-dimensional images and three-dimensional models of proteins he showed me just did not convey the kinetic dimensions of molecular form.

This encounter with Edward crystallizes the central themes in this book. What kind of model is Edward building on his computer screen? How did he learn how to make proteins visible, tangible, and workable in this way? What are the skills and dexterities that distinguish protein crystallographers from other life scientists? Why can't he rely on computer programs to automate his modeling efforts? Moreover, what is significant about the tension between his mechanistic approach to protein modeling and his intuitions about molecules as breathing entities? And why is it that he is moved to articulate the forces and movements of this breathing molecule with his own body?

Part I of this book shows how practitioners like Edward cultivate intuitions about how proteins move and breathe through the time-consuming and laborious process of building models in the laboratory. Part II takes a close look at just what these molecular models stand for, and how members of this research community adjudicate the truth status of these models. Just as Edward leaned into the space between us to *effect* the *affects* of a lively body, part III of this book documents the analogies and anthropomorphisms that modelers use to animate their protein models. It examines how both lively and mechanistic articulations shape modelers' molecular imaginaries and their renderings of life. Throughout, this book asks: What is life becoming in protein modelers' hands?

This anthropological study pays close attention to scientists' modes of embodiment in the construction and propagation of visual facts.[9] It argues that the visual cultures of science must be understood simultaneously as performance cultures. Throughout, it shows how protein modelers' moving bodies and their moving stories are integral to scientific inquiry. As a sensory ethnography of scientific pedagogy and training it pays close attention to how protein modelers hone their intuitions and cultivate the kinesthetic and affective dexterities to construct and adjudicate crystallographic models and data. It observes how practitioners get entangled with their molecules, models, and machines in the course of their experiments. By homing in on the *affective entanglements of inquiry*, this ethnography challenges conventional assumptions about the practice of objectivity.[10] What is more, this book explores how protein modelers *do mechanism* in what at first might seem surprising ways.[11] This study reveals moments when these practitioners do not abide by the deanimated, mechanistic theories of life they are supposed to avow. By opening up gaps and fissures in mechanistic reasoning, this book documents practitioners' failure to mobilize mechanism in a way that would fully disenchant the life sciences. In the fields I document here, mechanism does not acquire the hegemonic status many would assume it has achieved by the twenty-first century. Indeed, it is by paying attention to the affective entanglements of scientific inquiry that this book is able to amplify protein modelers' otherwise muted views about the *affectivity of matter*.[12] In their hands, protein molecules "breathe." Expert practitioners can tell when proteins are happy, stressed, in pain, under strain, or relaxed; when they are behaving and when they are misbehaving. If living substance is for them at least partially reducible to mechanical principles, it is also simultaneously lively, wily, and unruly. It is by observing how modelers perform their molecular knowledge through gestures, stories, and animate renderings that the mechanistic theories of matter they adhere to in their scientific texts can be seen to give way to livelier ontologies. This book thus reveals forms of animacy immanent to mechanistic logics. Modelers' invocations of the excitable life of matter offer up novel views of the sciences of life, with the promise that the biosciences today are more and other than what we may have long anticipated.

AN ANTHROPOLOGIST AMONG PROTEIN MODELERS

This study builds on a relatively recent tradition of ethnographic research in and around scientific laboratories. Anthropologists and their allies in science studies and history of science have, since the 1980s, turned their attentions to science as a culture and a practice. By extending the ethnographic methods of long-term fieldwork and participant-observation, these researchers have studied an array of sites inside and outside laboratories that allow them to examine how scientific facts are made, how they are made to circulate, and how these facts participate in larger economies of power and knowledge.[13] Anthropologists ask not only how facts are made, but also how these facts are "lived" and what these lived facts come to mean for scientists and their broader publics.[14] Moreover, a wide array of studies have shown that what counts as knowledge and as scientific method can vary widely between communities.[15] This book builds on insights from studies that have examined scientists' forms of life and their livelihoods, their practices of objectivity, and their status as brokers of knowledge and arbiters of truth. It is concerned with core anthropological issues such as the material, visual, and performance cultures of science, and human relations with nonhuman realms. It examines how transformations in techniques, technologies, and scientific "thought styles" participate in the formation of new ways of knowing and forms of expertise.[16]

This account of protein modeling is based on five years of anthropological fieldwork, between the years 2003 and 2008. I conducted this study among several communities of structural biologists and biological engineers working in academic laboratories in the United States. My primary field site was a private research university on the East Coast. There, and at nearby institutions, I observed laboratory practice and conducted multiple in-depth interviews with protein crystallographers and other protein modelers working on projects in the varied fields of biology, chemistry, physics, synthetic biology, biological engineering, computer science, mathematics, and mechanical engineering. Some of these practitioners focused on protein folding and molecular dynamics, and some used different techniques such as electron microscopy to model their molecules. The participants in this study were at various stages in their careers, and they included principal investigators, research coordinators, course directors, postdoctoral researchers, graduate students, teaching assistants, and undergraduate students. My training in the biological sciences gave me the opportunity to engage my interlocutors in conversations on matters they cared deeply about, and at the same time our conversations gave them space to think through issues and express ideas they did not otherwise have the opportunity to voice. And while I was fluent in the language and laboratory techniques of molecular genetics, the practices of protein crystallographers and biological engineers were strange to me. Little was self-evident to me about their practices and ways of knowing.

In order to understand how protein modelers acquire their skills and intuitions, I observed semester-long graduate and undergraduate courses, including courses on macromolecular crystallography, biomolecular kinetics and cellular dynamics, protein folding, basic biology, and biological engineering, as well as a hands-on laboratory course for biological engineering majors. To document how experts in this field communicate their knowledge of protein structure, I observed numerous public lectures on structural biology, protein crystallography, and other modalities of biological visualization and interviewed a number of protein crystallographers and structural biologists working at other institutions on the East and West Coasts of the United States. I learned about the ways visual facts in this field circulate by attending several professional conferences and meetings, including a weeklong interdisciplinary workshop on protein folding dynamics attended by mathematicians, protein crystallographers, biophysicists, mechanical engineers, and computer scientists in 2008. That year I also tracked the history of protein models in the Archives of the Laboratory of Molecular Biology in Cambridge, UK, and conducted interviews with long-term members of that institution.

In addition to reading scientific papers, I searched the Internet for news sources, blogs, and videos that would help me stay abreast of ongoing events in the field. I spent considerable time exploring the online archive of protein structures, downloading data sets and manipulating molecular models on-screen. I also tuned in to the public life of protein science by examining a range of pedagogical materials, videos, and films that were circulating widely on YouTube and other web-based media platforms. As a life-long dancer, my attentions were especially attuned to the relationship between movement and forms of knowing in science. And so I watched with delight when beginning in 2008 a science journalist teamed up with the American Academy for the Advancement of Science (AAAS) and *Science Magazine* to mount what has become an annual dance competition that invites scientists to stage their findings in choreographic form. These sites generated significant ethnographic insight into the modes of embodiment and performance cultures of science.

It is crucial that I situate myself in this ethnography. As will become clear in this book, I am no neutral observer of science. I care a great deal about the life sciences and what life is becoming in laboratories today.[17] The account I offer here feeds on all sorts of concerns and anxieties about what the life sciences are up to and desires for how they could be otherwise. This is an aspirational account: in response to descriptions that tend to flatten both scientists' practices and the stuff of life, this ethnography attempts to render life and science in ways that might change what we think science is and what it could become. My intervention works by amplifying a range of practices that are otherwise muted, overlooked or even disavowed. These are practices that remain tacit among scientists, or are otherwise not readily perceptible to observers of science. In this sense, my account remains partial, and necessarily occludes as much as it reveals. Its findings are not meant to be decisive or complete. Rather, the aim is to supplement current work in the anthropology of science and science and technology studies by shifting perceptions of scientific practice in a way that may change the questions we ask about life, matter, and forms of knowing.

TANGIBLE BIOLOGY

Today the world is messages, codes, and information. Tomorrow what analysis will break down our objects to reconstitute them in a new space? What new Russian doll will emerge?

—François Jacob, *The Logic of Life*, 1973

The distinction Edward made between protein crystallographers and molecular biologists—those who work with protein structures and those who work with genetic codes—raises important questions about the status of protein crystallography in the life sciences today. This is especially so in light of the prominence of molecular genetics and genomic approaches, and at a moment when biology is being lauded as an information science.[18] How are the three-dimensional data forms generated by protein modelers reconfiguring biological explanations and imaginaries? Are informatic models of life giving way to a new kind of tangible biology?

Proteins had a particularly rich history in the life sciences during the late nineteenth and early twentieth centuries. In the 1860s, for example, British scientist Thomas Henry Huxley popularized a "protoplasmic theory of life." He proposed that the unifying basis of all life—the material that united the plant and animal kingdoms—was the proteinaceous substance of the cell that he named the "protoplasm."[19] It was the irritable and contractile capacity of the protoplasm that demonstrated for him the vital powers of the cell. Protoplasm was a veritably "excitable" substance.[20] Yet, Huxley was no vitalist: his theory proposed a mechanistic view of life in which the "vital forces" of the cell could be reduced to mechanical, "molecular forces."[21] By the late nineteenth century, the protoplasm was already figured as a molecular substance that adhered to mechanical laws. As historian Lily Kay has shown, by the 1930s Huxley's protoplasmic theory had given way to a widespread view that proteins were the "principal substances" of life. Indeed, through the 1930s and 1940s, and up until the determination of the structure of DNA in 1953, proteins were thought to be the material basis of heredity, and intensive effort was invested in determining their elemental composition, chemical specificities, and cellular activity.[22]

Efforts to visualize protein structures were first initiated in the UK in the 1930s. In contrast to Edward's twenty-first century nomenclature that distinguished molecular biologists from protein crystallographers, "molecular biology" was the term mathematical physicist Warren Weaver coined in 1938 to circumscribe research into the structural properties of biological molecules. Molecular biology promised to bring physics to the study of life at the molecular scale.[23] W. T. Astbury, a biophysicist and member of the growing "protein community" in the UK, was among the first to popularize the field.[24] In 1951, Astbury insisted that "molecular biology" was to be understood as the "predominantly three-dimensional and structural" study of the biophysical and chemical properties of molecules.[25] By 1967, however, the definition of molecular biology was already changing. In his widely cited lecture "That Was the Molecular Biology That Was," biologist Gunther Stent forecasted the decline of the structural school of molecular biology. Stent defended the structural school's "down-to-earth," "physical" approach, which promoted the "idea that the physiological function of the cell" could be understood "only in terms of the three-dimensional configuration of its elements." And yet, at that time Stent did not see how these contributions could be "revolutionary to general biology."[26] After all, by that time, it had taken over twenty years to determine the structures of just two proteins: hemoglobin and myoglobin.[27] The revolution was, according to Stent, going to be led by the "one-dimensional" or "informational school," whose "intellectual origin" in the emerging computational cultures of cybernetics and cryptography in the 1950s and 1960s was "diametrically opposite" to the physical understandings of molecules championed by the structural school.[28]

While "structural biology" did not disappear, the contributions of this field did lose traction during the sequencing craze of the molecular genetics and genomics revolutions.[29] During the 1980s and 1990s, in particular, a kind of "genetic fetishism" swept over the life sciences.[30] This new "molecular vision of life" that took root in the wake of the determination of the genetic code was a vision that flattened life into thin threads of genetic "information." However, since the late 1990s, with the completion of the genomes of humans and other organisms, and the ramping up of postgenomic investigations, the terrain is shifting again. Researchers, funding bodies, and venture capitalists are launching a range of "omic" initiatives, such as proteomics and metabolomics, which aim to document and analyze all the molecular species or metabolic processes in a given organism at a particular stage of development or over the course of its lifetime. In the process, these projects are revealing the limitations of genetic sequence data for accessing the multidimensional problems that biology poses.[31] Today, as major journals such as *Science* and *Nature* are publishing newly determined protein structures almost weekly, life scientists can be seen turning from matters of code to matters of substance—that is, from spelling out linear gene sequences to inquiring after the multidimensional materiality of the protein molecules that give body to cells.

This recent and dramatic rise of structural biology can be mapped through the history of its primary data archive, the Protein Data Bank (PDB). This essential research tool enables practitioners to share their data and to access the atomic coordinates of molecules determined by X-ray crystallography and other structure determination techniques.[32] A data archive was first conceived in the year 1970 when crystallographic techniques were just beginning to be refined. By that time, practitioners were already facing challenges archiving and sharing their data. The issue was that protein structures generated massive data sets. Some proteins are made up of several thousands of atoms, and crystallographic data sets describe the three-dimensional coordinates of each atom in the molecule. The computational capacity available in the 1960s and 1970s was limited to punch card computers. In order to share data, stacks of punched cards would have to be sent through the mail. Given that "each atom was represented by a single card," the exchange of data for a crystallographic structure of a molecule like myoglobin "required more than 1000 cards."[33] Transferring the data set for hemoglobin, a molecule that is four times larger than myoglobin, would have required over four thousand cards. Data sharing was severely hindered by the labor required to prepare data sets for transfer, and so at that time the "coordinates for individual entries had only been exchanged among a few research laboratories."[34]

American crystallographer Helen Berman, the director of the archive in its current form, recounts that she and her colleagues first came up with the concept of a "central repository for coordinate data" at a 1970 meeting of the American Crystallographic Association in Ottawa, Canada.[35] The following year, Cold Spring Harbor Laboratories hosted "Structure and Function of Proteins at the Three Dimensional Level," a symposium that was described as "a coming of age" for structural biology.[36] At that meeting, British crystallographer Max Perutz, who would go on to win a Nobel Prize for solving the structure of hemoglobin, convened an "informal" gathering of protein crystallographers to explore "how best to collect and distribute data." By October 1971, crystallographers at the Brookhaven National Laboratories on Long Island in New York, and the Cambridge Crystallographic Data Centre in the UK had come together to establish the Protein Data Bank to facilitate the electronic exchange of data. According to Berman, when the first newsletter for the PDB was issued in 1974, "thirteen structures were ready for distribution and four were pending." Just two years later the PDB had archived a total of twenty-three structures, and in 1976 alone, "375 data sets had been distributed to 31 laboratories."[37] Ten years later, the PDB issued a press release announcing that fifty thousand data sets had been uploaded (see figure 1.2).[38]

As of December 2014 the coordinates of 105,839 protein structures have been deposited online in the RCSB PDB,[39] and contributions continue to grow exponentially from laboratories around the world. In 2013 alone, the PDB logged a total of 3,850,473 unique visitors from about 190 countries worldwide. These queries accessed approximately 19,479 GB of data.[40] The RCSB considers itself a "global" resource.[41] It now hosts the WWPDB, a "worldwide"

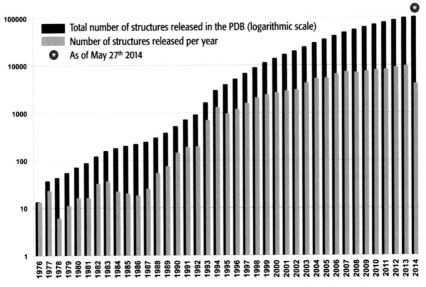

I.2. "Yearly Growth of Structures Released in the PDB Archive," visualized on a logarithmic scale. RCSB Protein Data Bank, 2013 Annual Report.

data bank that integrates the RCSB's PDB with the Macromolecular Structure Database at the European Bioinformatics Institute (MSD) and PDB Japan (PDBj) at the Institute for Protein Research at Osaka University, Japan.[42]

As the data bank extends its geographic reach, it is also expanding the diversity of its collection of proteinaceous forms.[43] The archive includes data on protein structures derived from a vast menagerie of organisms, including species of microbes, viruses, animals, and plants. *Escherichia coli* and *Saccharomyces cerevisiae*, two of the most common microorganisms in biology labs, account for 18.4 percent and 9.1 percent of all the proteins in the database, respectively. Proteins sourced from humans (7.06 percent), mice (3.68 percent), chickens (2.23 percent), wild boar (1.47 percent), wheat (0.40 percent), and fruit flies (0.34 percent) are among the best represented in the database.[44] The largest proportion, a remarkable 27 percent of all proteins archived, are derived from one relatively rare microorganism, *Thermus thermophilus*, a thermophilic, or heat-loving, extremophile. Since the publication of its genome in 2004 this microbe has become a major model organism for researchers in the growing field known alternately as structural genomics or proteomics. Researchers in these fields are developing high-throughput technologies for protein structure determination.[45] As they make rapid contributions to the PDB, they are reconfiguring the distribution of data on model organisms, and with this, which bodies are coming to matter in biomedical research.

For centuries scientists have relied on illustrated atlases to document the remarkable diversity of natural phenomena. These "atlases of observables" enabled experts and novices to train their visual sensibilities and hone their expert judgments on a panoply of natural forms.[46] The Protein Data Bank can be thought of as an extension of this tradition of the scientific atlas. It offers a collection of data sets that can be visualized on a platform that allows modelers to compare and contrast protein structures derived from distinct species and distinct experimental contexts, as well as structures that have been synthesized *de novo*. And yet this atlas does not just train researchers' eyes: the protein structure data contained in the PDB is made tangible in the form of three-dimensional computer graphic models that allow users to manipulate the structures on screen. As an "atlas of manipulables" it offers an interactive medium through which structural biologists can entrain their sensibilities to a wide range of protein folds and forms. The PDB is thus making visible and tangible the structural properties of a vast array of life's molecular possibilities.

EMBODIED VISION

The trials and tribulations Edward encountered working with automated software offer a glimpse into the challenges crystallographers face confronting the indirect nature of molecular vision. He draws attention to the fact that there is no automated computer program or assay that can detect the chemical structures of a protein and provide a readout of the coordinates of its atoms. Crystallographic modeling techniques require the active participation of the modeler at every stage of the process. This makes model building painstakingly slow and laborious. One of the most remarkable and time-consuming features of this technique is that modelers must turn protein molecules, the very objects of their inquiry, into devices that are integral to the apparatus used to make them visible. Chapter 1 of this book documents closely how crystallographers transform proteins into visualization technologies by working with these molecules in crystalline form. Protein crystals are different from the sometimes-colorful mineral crystals that are familiar from museum collections and shops that sell New Age paraphernalia. While those often grow under pressure in the dark recesses of the earth, proteins can be coaxed to grow into microscopically sized crystals in vitro in the laboratory. Well-formed crystals are highly organized materials: their growing forms incorporate thousands of molecules into regular, repeating arrays. Crystals have fascinating properties, and one is their ability to scatter high-energy radiation, like X-rays. Modelers use protein crystals as the inverse of a microscope lens: rather than focusing light into an image, as a lens does, their crystals diffract X-ray radiation into irregular patterns that show up as spots on a detector. They expend great effort trying to decipher these patterns of scattered spots. With the help of computer power and carefully honed intuitions, they render their data into maps and models that indicate the probable three-dimensional configuration of hopefully most of the atoms in a given molecule.

To appreciate the peculiarities of protein crystallography, this technique must be understood in the context of other modalities of scientific visualization.[47] Microscopy offers a generative counterpose. In the seventeenth century, British scientist Robert Hooke recognized the prosthetic nature of his microscope as a tool that could "enlarge" his senses, drawing "the hitherto inaccessible, impenetrable, and imperceptible" into view.[48] Hooke sought a means to bolster what he perceived to be the "inherent fallibilities" of perception and so used his "Instruments" as "artificial Organs" in an attempt to remedy "the 'infirmities' of the 'human senses.'"[49] Like microscopists, protein crystallographers must get themselves fully entangled with their instruments to facilitate a prosthetic extension of their senses into the molecular realm.[50] The diffractive optics they engage in their work certainly reconfigure their senses and sensibilities. Historians of science Steven Shapin and Simon Schaffer astutely observe that scientific instruments like microscopes "imposed both a correction and a discipline upon the senses."[51] In the context of protein crystallography it is, however, necessary to understand the terms "discipline" and "correction" in the most generative sense; that is, in a way that appreciates how modelers' senses are enlisted, honed, cultivated, and trained, rather than merely controlled or constrained. Indeed, what were once considered the "fallibilities" and "infirmities" of the senses are now, in the context of protein crystallography, deemed essential resources. The indirect vision generated through diffractive optics and the hands-on labor involved in building models require that practitioners get actively involved in the work of making molecules visible. In addition to their well-trained intuitions and dexterities, modelers must engage their entire sensorium in the work of building models. Modelers' subjective perceptions are not so much a corruptive force to be disciplined; rather, in this context the "capacities" and "productivity of the observer" are celebrated.[52]

The phenomenological tradition in philosophy offers a generative response to the long-standing illegitimacy and suspect status of the human senses. Maurice Merleau-Ponty's phenomenology challenges the disembodied claims of Cartesian objectivity.[53] His approach emphasizes a body's full participation in acts of perception. Whereas eyes have long been conceived as disembodied instruments of vision, for Merleau-Ponty sight and seeing become the capacities of a lively sensorium tethered to a lively world. He proposes a "crossing over" between the "visible" and the "tangible," such that our visual exchanges with other bodies and objects engage us viscerally in physical encounters with the world.[54] From this vantage point, our eyes and hands become extensions of one another, such that looking becomes a way of feeling out the world, and seeing, a way of being touched by others.[55] Phenomenological studies of scientific vision insist that rather than corrupting "objective" methods, bodily knowledge is integral to science, including practices such as theory making and reasoning, which are conventionally understood as the domain of the mind.[56]

Similarly, feminist science studies scholar Donna Haraway draws attention to the embodied nature of scientific vision to offer an antidote to the moralizing discourses of distance and neutrality that condition conventional accounts of objectivity.[57] Her insistence on embodied vision resists the alluring myth of objectivity conceived as a form of disembodied and omniscient vision. She outs this view as a "god trick" that pretends to see "everything from nowhere."[58] Her approach to vision challenges the "myth of body-lessness" that is perpetuated in standard accounts of scientific practice and objectivity.[59] By grounding vision in the peculiarities and specificities of technologically mediated bodies, she articulates a feminist epistemology that insists all claims to truth must be located. Where Cartesian approaches to objectivity hinge on a disavowal of one's complicity, feminist objectivity requires accounting for the limits, contingencies, and partiality of what we can and cannot see and know about the world.[60] It is by training ethnographic attention on the remarkably multisensory, affective dimensions of model building that this book makes palpable how corporeal knowledge shapes the facts of molecular life.[61] Protein modelers' practices offer a refreshing counterpose to conventional norms of objectivity. Theirs is a "situated knowledge practice" in Haraway's sense of the term. Indeed, these modelers are insistent about locating their contributions to the work of crafting molecular facts.

CRAFTING MODELS

If it is possible to make a generalization about scientific models, it is that they are indeterminate objects that are hard to pin down. Philosophers and historians of science have made major contributions to the literature on a wide array of scientific models, including among others, two-dimensional diagrams, flow charts, and analogical models.[62] This work builds on movements in the history, philosophy, and sociology of science that aim to reorient long entrenched assumptions about science as a theory-driven activity.[63] Until recently, historians have largely regarded three-dimensional models as "mere" "memory tools" or "mneumotechnical devices" that aid in teaching and learning, with little to offer scientific research.[64] In an attempt to recuperate a lost history, numerous scholars have endeavored to bring attention to models as tools in research contexts.[65] Studies of the experimental lives of models have shown that modeling practices are intimately entangled in the tacit knowledges, social negotiations, moral economies, work cultures, and figural vocabularies that shape laboratory life.[66] Philosopher of science Ian Hacking suggests "models are doubly models": they are both representations of theories and of phenomena.[67] Philosophers Mary Morgan and Margaret Morrison describe this dual function of models as their capacity to set up a "relation" to their two referents, both theories and worldly phenomena.[68] A model can be a theoretical elaboration, an empirically informed abstraction, a figment of the imagination, or all of these at the same time. Science studies scholar Sergio Sismondo suggests that models occupy a "messy category," one that we should not try to clean up.[69] He insists that models spread out across a "continuum" of possible forms and functions and "cut across boundaries of pure categories": they are "monsters necessary to mediate between worlds that cannot stand on their own, or that are unmanageable."[70]

Three-dimensional models are a unique species along this continuum. Scientists who build and use three-dimensional models explicitly demonstrate the embodied, multisensory nature of scientific visualization.[71] Researchers rely on three-dimensional models as objects-to-think-with: they are recursively made and remade in attempts to conceptualize and actualize new hypotheses and new modes of inquiry. Chapter 2 of this book examines three-dimensional models as handcrafted objects that disrupt assumed binaries between the intellectual and physical labor of research.[72] More than visual traces, marks, or inscriptions, three-dimensional models explicitly blur the boundaries between automated machinic productions and the skilled work of scientists. In this sense, they do not conform to the "immutable mobiles" that Bruno Latour has described in his studies of molecular biology labs. Automated "inscription devices" are tools that are meant to ensure fidelity in recording the signatures of nature. One familiar example might be electrophoresis gels so common in molecular genetics laboratories. By applying electrical current to a sample of a substance embedded in a thin rectangle of agar gel, these devices can separate substances by molecular weight and so be used to indicate the presence or absence of a particular molecule in a sample. Once stained or visualized under UV light, such gels can be "read" like a two dimensional graph or chart. For Latour, such flat inscriptions promise the efficient movement of facts through scientific networks, and deftness in mobilizing allies to resolve arguments over scientific claims.[73] Yet, historians have argued that the wide circulation of three-dimensional facts through models demonstrates that the "visual worlds of science" aren't so "flat."[74] In order to appreciate the import of three-dimensional models in research contexts it is necessary to understand that they are not legible in the same ways as flat inscriptions. Historians of the life sciences Nick Hopwood and Eric Francoeur insist that building and using three-dimensional models cannot be reduced to textual practices of reading and writing.[75] Rather, model building is a full-bodied practice: making three-dimensional facts visible and legible demands ongoing corporeal engagement with tangible media, whether these are physical materials or virtual tools.

ENACTING MODELS

This study brings attention to models as they are built and used. Reflecting on the nature and use of three-dimensional models in the history of the life sciences, philosopher and historian of biology James Griesemer insists that accounts of modeling practices must include a history of the "gestural knowledge" that shapes how models are made and how they are made to circulate. Studies must take into account the *enactment* of models, which includes "gestural as well as symbolic knowledge and the variety of means and modes of making, experiencing, and using models."[76] Where historians are often faced with the challenge of reconstructing this gestural knowledge from the wear and tear on objects and instruments, and textual documents and images stored in archives, ethnographers can observe this ongoing gestic choreography in practice.

Modelers get physically entangled in their modeling efforts. To do their work well they depend on a synesthetic tangle of sensory perceptions. Crystallographers do not just see with their eyes, they practice a kind of "haptic vision"; that is, their ability to perceive molecular worlds is intimately coupled with the visceral modalities that make up the sense we commonly call touch. Modelers' moving bodies and their curious hands are informed through the senses of kinesthesia, a kind of muscular sensibility, and proprioception, an awareness of their bodies in space.[77] These synesthetic modalities do not just inform modelers' bodily tissues; they simultaneously inflect their ways of thinking. As such this form of "haptic vision" is coupled to a kind of "haptic creativity" that extends modelers' intuitions, memories, and imaginations as they engage their bodies and various forms of tangible media to play through hypothetical permutations in protein form.[78] Such forms of haptic creativity are vividly demonstrated in Michaels's body experiment and Edward's performance of his breathing molecule. Indeed, modelers engage their entire bodies, including their hands, arms, shoulders, heads, necks, torsos, and even legs in model building. Chapter 8 of this book shows how they practice a kind of "molecular calisthenics" as they figure out (for themselves) and relay (for others) their intimate knowledge of molecular forms and movements. While this is a practice that both enables and constrains how they imagine molecular worlds, their bodies provide a pliable, readily available medium for reasoning through the specificities of protein structure and sharing their insights with others. In this way, modelers not only "do things with words," in the sense that philosopher J. L. Austin defined "the performative," they craft expert modes of communication and reasoning by *folding semiosis into sensation* and propagating an affectively charged repertoire of gestures and movements.[79] These enactments can serve to enroll new generations of life scientists by tacitly and explicitly entraining them to these subtle ways of knowing. Protein crystallographers demonstrate well how the visual facts of science are both *performed* and *performative*.[80]

RENDERING LIFE MOLECULAR

This book develops the concept of rendering to account for both the performance of molecular models as they are made and used, and the performativity of molecular facts. In other words it pays close attention to how particular enactments *rend the world as molecular*, changing meanings and material realities for practitioners, their students, and wider publics. Consider the indirect nature of crystallographers' molecular vision. They do not "see" molecules or produce "images" of biological phenomenon; rather, they *make* models to *render* the molecular world visible, tangible, and workable.

Protein models are renderings in that they are representations of molecules.[81] Like a translation, a work of art, or a detailed architectural drawing, these models show their users and viewers what molecules "look like." However, the verb "to render"—which also means "to make"—draws our attention explicitly to the craft of model building. Models are things *made*. Ian Hacking has suggested that scientific representation hinges on modes of intervention.[82] For example, one simply cannot just look through a microscope and see the details of a cellular world. Rather, in order to make a translucent smear on a microscope slide visible and legible as an aggregate of individual cells, a microscopist must intervene directly in the optical system by applying dyes that bind to specific regions of the cell or by modifying the light sources. In microscopy, such interventions assume some contiguity between the thing on the microscope slide and the image that is perceived through the objective lens: the image changes visibly as the microscopist manipulates the material on the microscope slide. Crystallographic vision is, however, much more indirect. There is no material or optical contiguity between the diffraction pattern generated by a protein crystal and the three-dimensional model that rotates on a modeler's computer screen. The resulting model is in this sense a fabrication.[83] Modelers make molecules visible and palpable by crafting proxies that can stand in as analogs for molecular configurations. The concept of rendering is particularly salient in contexts like this where the gap between the representation and its referent is so wide. For crystallographers this gap between model and molecule is traversed by an elaborate choreography involving living substance, instruments, X-ray sources, computers, and an array of modeling materials. Additionally, this experimental configuration must be supplemented by the modeler's creativity, intuitions, sensibilities, and kinesthetic dexterities. As things made, these models are also thus partly "made up";[84] this concatenation of fact and fabulation is a key feature of renderings.

The concept of rendering can account for the creative ways that practitioners confront the limits of molecular vision. Consider one meaning of the term, where a rendering is understood as an artistic performance, as in the rendering of a play or musical score. A rendering carries the mark of the artist: different singers, for example, will inflect the same musical score with unique tones, textures, and affects. Crystallographic renderings are more than empirical descriptions of molecular structures. They are simultaneously creative elaborations, shaped by modelers' intuitions and imaginations. Renderings are not just performances; they are also *performative*. What this means is that they not only represent the molecular realm, they also make the world molecular and so sediment particular ways of seeing and knowing. As parts II and III of this book show, renderings are simultaneously material and semiotic.[85] Both models and modelers act as proxies, speaking as and for molecules; in so doing, they participate in a kind of molecular storytelling that renders salient some aspects of molecular life, while obscuring other dimensions.

As a performative approach to representation, the idiom of rendering insists on paying attention to how models *rend* the world. What forms of life come to matter in the hands of protein modelers, and which do not?[86]

MODELING PROTEINS, MAKING SCIENTISTS

Edward demonstrates how crystallographers must grapple directly with their data in order to make protein structures visible. His reliance on what he calls "common sense" raises crucial questions about classroom pedagogy and laboratory training, and especially what counts as "proper" training in this field. How is it that Edward can tell the difference between correct structures and the "garbage" that his automated software was spewing out? How do scientists-in-training learn to discriminate between good and bad models? How do practitioners teach their students how to see, feel, and know the difference?

Laboratories are not just factories for the production of scientific facts: they are sites for the making of new scientists.[87] Laboratory life, especially in academic institutions, revolves around the training of novice scientists.[88] These are also sites where expert practitioners continually retrain as they experiment with new techniques to approach new research questions. Educators and researchers in this rapidly expanding field are struggling to find ways to introduce protein structures into classrooms and innovate molecular modeling techniques in the laboratory. They are charged with training a new generation of scientists whose forms of knowing must simultaneously be attuned to the subtlest chemical affinities, physical forces, and molecular movements, and keyed to the tangible logic and rhetoric of a mechanistic vision of life.

How do expert practitioners model proper scientific practice? Pedagogy can itself be approached as a modeling practice. Experts model techniques and practices through the face-to-face apprenticeships that take shape in laboratories. These include both formal mentorships between students and their professors, and the more informal, day-to-day peer mentorships through which students are continually called on to share their know-how with lab mates.[89] It is in the space of apprenticeship that an expert models their tacit knowledge for a novice. Tacit knowledges include the skills that cannot be read about in textbooks and the kind of know-how that one picks up through experiences of trial and error and observing others in practice.[90] In a field where experts must often rely on their well-trained intuitions to know the difference between a good and bad model, or between good and bad data, tacit knowledge is highly valued. As chapter 3 of this book shows, in protein crystallography the act of model building itself is a crucial pedagogical event. This book documents how, as students submit themselves to the challenges of model building, they learn the physical and chemical forms and affinities that hold a protein together. Edward's well-honed "common sense" is a reminder that, over time, model building refines a student's sensibilities, intuitions, and judgments. It turns out that making sense of these otherwise imperceptible phenomena demands intensive sensory engagement from modelers.

Art historian Caroline Jones defines the term "sensorium" as "the subject's way of coordinating all of the body's perceptual and proprioceptive signals as well as the changing sensory envelope of the self."[91] The sensorium is for her, "at any historical moment shifting, contingent, dynamic, and *alive*."[92] The senses, as I explore them here, are not innate, biological functions that can be divided up into the five familiar modalities of sight, touch, taste, hearing, and smell.[93] Sensing bodies are not separate from the worlds they reach toward in acts of sense making. Sensing takes shape between bodies and worlds. Moreover, broader social, cultural, and economic forces condition both the contours of a person's sensorium and what Jacques Rancière identifies as "distribution of the sensible." Such distributions are the effects of sensory regimes, themselves shaped by processes such as capitalism, colonialism, and biopower.[94] These regimes of the sensible powerfully shape what we can and cannot see, say, feel, or know, and often remain so self-evident that they are as imperceptible as the air we breathe.

This book focuses on the distribution of the sensible and the making of expert sensoria in the context of protein modelers' efforts to make sense of life at the molecular level. Bruno Latour's marvelous account of how perfumiers train their noses to distinguish the finest notes in a fragrance offers a generative approach to the aliveness and plasticity of a sensorium-in-training. He proposes the concept of "articulation" to account for the ways that practitioners train their senses to distinguish—and so articulate—finer and finer differences in a phenomenon. Perfumiers train their noses through "odor kits" that present smells in a particular sequence. Before training, students are "inarticulate": odors would waft over them "without making them act, without making them speak, without rendering them attentive, without arousing them in precise ways: any group of odors would have produced the same general undifferentiated effect or affect on the pupil. After the session, it is not in vain that odors are different, and every atomic interpolation generates differences in the pupil who is slowly becoming a 'nose.'"[95]

For Latour, a "nose" describes a practitioner who has become sensitive to the finest "atomic" propositions of matter. For him, "it is not by accident that the person is called 'a nose' as if, through practice, she had acquired an organ that defined her ability to detect chemical and other differences."[96] A nose is someone who has become articulate, who has learned how "to be affected" by the subtlest of differences in odors.[97] Becoming articulate demands "passionate" interest in the phenomenon, and a willingness to be "put into motion by new entities" and so "register" their differences "in new and unexpected ways."[98] In this formulation, Latour produces a performative, nonrepresentationalist theory of the senses. That is, rather than assuming that there is a preexisting world "out there" that impinges on the mechanics of an observer's sensory physiology, he insists that sensing is an active process that relies on the passionate involvement of the observer in an ever more interesting world. It is a practitioner's willingness to expand sensory dexterities that allows them to begin to learn how to articulate the world's remarkable multiplicity of propositions. In this view, both subject and world become increasingly interesting and articulate, in the fullest senses of these terms.

In the context of protein modeling, novices must subject themselves to training in order to articulate their entire sensorium. Only this will allow them to articulate the remarkable diversity of protein forms. Here the dual meaning of Latour's concept becomes clear: articulation is both the discrimination of difference and the naming of that difference. In the place of an odor kit, it is in the very process of building their first models that students learn to resolve and report finer and finer differences in a molecule's chemical configuration. The concept of articulation propagates throughout this book to describe the myriad of ways that practitioners acquire the kinesthetic and affective dexterities that allow them to distinguish not only where each amino acid is located in the molecule, but also between good and bad models, skills they need in order to learn how to adjudicate these complex visual facts. It is thus in the process of articulating protein models that articulate modelers are made.

To conduct this research I too have had to articulate my sensorium. Passionately interested in the relation between movement and knowing, I have learned how to entrain my ethnographic attention to modelers' movements, gestures, and affects. In order to "tune in" and parse their nuanced gestural vocabularies, I have relied on an affinity for movement built up over twenty-five years of training, first in classical ballet, and later in contemporary dance. As an ethnographer, I have had to experiment with ways to detect, recall, and relay modelers' subtle bodily affects.[99] In the process I have learned how to remember and document the nuanced cadences, tempos, rhythms, and tones that inflect their models, animations, and animated forms of body-work.

In 2009 a documentary film about graduate student life in a protein crystallography laboratory was released. This film challenged me to think carefully about how scientists and observers of science tell stories about laboratory life. *Naturally Obsessed: The Making of a Scientist* is set in Larry Shapiro's laboratory at Columbia University. It documents the harsh realities of graduate student life in this challenging field. The filmmakers, Richard Rifkind, professor emeritus and former chair of the Sloan-Kettering Institute for Cancer Research, and his wife, Carole Rifkind, a filmmaker and educator, spent three years filming in the lab and a full year in the editing studio. They are frank about their struggle to stitch together a good story from the hundreds of hours of footage they generated in their extensive study.[100] The final cut is moving. It is tuned to a musical score that lifts emotions to a fever pitch. Their edit dramatizes recurring scenes of dismal failure and desperation; and then, finally, when one of the students publishes his crystal structure, triumph. Stylistically it wavers ambiguously between the genres of documentary and competitive reality TV shows. It has received widespread attention, critique, and commentary and has been viewed in laboratories, university lecture theaters, and high schools across North America and Europe.[101] This rendering of laboratory lives tells some salient stories, if it also simultaneously occludes others. Given its widespread distribution and the wide-ranging responses of its audiences, this documentary provides an exceptional archive of ethnographic material to consider in this book.

The filmmakers track the lives of a few select members of Larry Shapiro's protein crystallography laboratory. In addition to Larry, the documentary features three graduate students at various stages of their training. Kil, Gabe, and Rob are in the thick of things. Stitched-together scenes document the students' mundane, daily activities in the lab and show them fumbling with machines, instruments, and recalcitrant materials. Viewers listen in as the students reflect on their sometimes-wavering goals, desires, and drives. The cameras follow the lead characters out of the lab and into the streets of New York City. The audience is invited into their cramped apartments, where we meet their partners and their pets, and observe how their training interferes with their relationships, tests their resolve, and forces each of them to second-guess their dreams of becoming a scientist. The documentary makes palpable how graduate students must contort and reorient their lives around the task of crafting molecular facts. The film documents how they submit themselves to training in order to fashion themselves into scientists who can persevere under the most challenging circumstances.

As students in the film struggle to master the experimental techniques proper to protein crystallographers, they not only learn how to calibrate their instruments, they also learn how to adjust their attitudes, postures, and sensibilities to a research culture contoured by intense competition. The film makes palpable the kinds of "institutional gazes, bodies, gestures, architectures, routines, incitements, examinations, and punishments" that condition the bodies and minds of scientists-in-training.[102] We get to see close up what Michel Foucault has called the productivity of power and the "positive economy" of training and discipline—that is, how regimes of "institutionalized training" shape students' modes of embodiment, norms, mores, and values, as well as the very grounds of knowledge.[103]

Larry Shapiro sets ambitious goals. His students work on high-impact molecules whose mechanisms are thought to have major biomedical significance. These are the kinds of protein structures that will get graduate students a publication in a top-tier journal like *Science* or *Nature*. One of the proteins that students in his lab work on is AMP protein kinase, an enzyme involved in regulating the metabolism of fat. This protein structure promises to be a "goldmine" for drug development. Figured as a mechanical "metabolic switch," it is identified as a prime target for therapeutic drug design in obesity and diabetes research. Already figured as a form of "promissory capital," this protein structure anticipates serious returns for pharmaceutical companies seeking to develop drugs that can control metabolism at the molecular scale.[104]

Getting a grip on the molecular mechanisms that "drive" vital processes is depicted as a high-stakes game in *Naturally Obsessed*.[105] Tensions run high as the students in this film compete with one another and with other laboratories to be the first to solve the structure of this valuable molecule. The rivalry becomes intense when the students recognize that their futures hang in the balance. At one point in the film Rob turns to his lab mate Kil, who has just made headway on his project, and says, "The race is on, and I'm behind." In the style of a competitive game show, this documentary's scenes are charged with frustration, jealousy, and despair, glimmers of hope and moments of sheer elation.

Observers of the life sciences attuned to the coproduction of power and knowledge, and wary of the ways that scientific knowledge about life can so readily be used to control and govern life, will see *Naturally Obsessed* as a vindication of their critiques of biocapital and the biopolitical economy of science. If a biopolitical economy is one geared toward bringing "life and its mechanisms" into the realm of "explicit calculation," then *Naturally Obsessed* offers a perfect portrait of one that is doubly fecund.[106] Larry's laboratory is not only home to competitive, entrepreneurial scientists who gear their labors toward the capture and control of life's vital mechanisms; it is also a site for the production of new scientists whose technical and affective dexterities are in the process of being finely tuned to this task. Able critics will find in *Naturally Obsessed* all the elements of a smoothly functioning biopolitical machine: compulsively laboring scientists, fetishized facts, life mechanized and captured, capital incentives, and exploitative values. And if the documentary continues to circulate widely and does its work as a pedagogical device, it will serve to recruit would-be scientists to commit their lives to the labor of extracting even more capital from life.

LIFE ITSELF, CAPTURED?

Larry Shapiro's efforts to model proteins whose structures may contribute valuable data to the development of drugs that can treat obesity and diabetes is one example of the broader practical horizon of protein crystallographers' labors. As these practitioners transform the forms of data that circulate through life science laboratories, they are also shifting research agendas. Protein structures are becoming objects of multidisciplinary interest and investment. With the promise of novel insights into basic biological processes, biomedical research, drug development, biofuel engineering, and environmental remediation, biologists, chemists, physicists, engineers, mathematicians, and computer scientists are accessing the coordinates of the vast array of structures housed in the Protein Data Bank. Value can be extracted from this data in the form of patentable designs and innovations.[107] Protein structures are especially valuable to practitioners working in the fields of biological engineering. Once frustrated by the opacity and recalcitrance of the gooey, proteinaceous substances that constitute living systems, biological engineers can now engage proteins as concrete, physical objects that they can measure, manipulate, and redesign. As I show in chapter 6 of this book, in their hands proteins are rendered as the "machinery of life," and this machinery can be reengineered, repurposed, and "enterprised up."[108] "Life itself," it seems, has been captured and put to work in the form of a streamlined assemblage of molecular machines that hum productively on the factory floor of our cells.[109]

Protein structures also have military applications, such as in the design of biological weapons and molecular defenses that can "preempt" enemy attacks.[110] A sobering reminder of such applications of protein structure research came in 2007, when I received an invitation from DARPA (Defense Advanced Research Projects Agency), a research unit funded by the U.S. military. I was asked to participate in a biological weapons design workshop at their headquarters in Arlington, Virginia. Their invitation suggested they had "tantalizing evidence" of the "role of shape, rather than chemical composition, in biological systems." They were hosting a one-day event "to explore innovative methods to produce synthetic macromolecular assemblies for applications in biological control." Their research team was working with the hypothesis "that analogues" that "mimic the shape—but not the composition—of their natural counterparts, may be fabricated and used for exquisite control of biological processes in a number of denied physiological applications." I found the invitation chilling—I could not comprehend how I got on this list of invitees, which included a large group of leading practitioners in biological engineering and protein modeling. I never did get the chance to attend the meeting. As soon as the organizers discovered that I was not the artist they had hoped would inspire them to think in new ways, but an anthropologist, and, what was even worse, a Canadian, I was promptly uninvited.[111] Apparently there was no way I could get security clearance in time to attend the meeting.

The militarization of protein structures is just one reminder of how inquiry in the life sciences is never innocent;[112] knowledge of the structure of biological molecules can always be applied to bring "life and its mechanisms" into the realm of "explicit calculation." In this sense, protein crystallographers and their collaborators participate in a kind of *molecular biopolitics*,[113] where knowledge of the stuff of life is intimately entangled in efforts to govern human and nonhuman lives and worlds.[114] Scholars in science studies and anthropology of science have generated crucial critiques of the recent ramping up of efforts to capture life in commodity form. They have closely examined the ways that scientists and their knowledge projects are complicit in global regimes of capitalism, war, racism, colonialism, and neoliberalism.[115] These critiques offer essential tools for grappling with the complex formations of power and knowledge in the life sciences.

At the same time, some of these critiques also participate in sedimenting stereotypes about science and scientists, as if all participants in technoscience were bent on the capture of "life itself" for capital gain. One effect is that these critiques can further constrain assumptions about the ways that scientists relate to the living phenomena they study in laboratories. While I am convinced these analyses are necessary and generative, I can't help but wonder whether there are other analytic frames and other ways of telling stories about the sciences and lives in science.[116]

Take the example of *Naturally Obsessed*. When viewed through the lens of these critiques it becomes clear that the documentary propagates the very caricatures and stereotypes about science and scientists that give critical analyses of biocapital and biopolitics their traction. This book seeks to disrupt the temptation to read this documentary, or interpret my field notes, through a frame that can see only the negative, controlling dimensions of biopower and biocapital. While this study builds on insights from the film, it does not take this documentary as a complete description of life science practice. Its story line is too constrained by what have become all-too-convenient scripts and conventions. Rather than taking these scripts and norms literally, it is necessary to make their self-evidence strange. To follow the scripts, without recognizing the specificities and constraints of their form, would be to generate an impoverished rendering of laboratory life. This book asserts that there are other ways to observe and tell stories about the life sciences, ways that don't foreclose what this practice is and what these practitioners are up to in their laboratories. Throughout I examine a number of the scenes in the film, as well as outtakes that are posted online, to explore contexts where slippages and deviations from such constraining scripts and "thought styles" begin to surface. I show that close ethnographic attention to the entanglements among modelers, their molecules, models, and machines can amplify the subtle ambivalences, contradictions, and diverse subjectivities that are integral to laboratory life.

This book thus offers a supplement to critical accounts of biocapital and biopolitics by bringing ethnographic attention back to the laboratory to trouble pervasive assumptions about the life sciences today. This study renders lives in science in ways that aim to keep open what it is possible to see, say, feel, and know about both scientific practice and the stuff of life. It offers what Beatriz da Costa and Kavita Philip have called a "tactical biopolitics," an approach that recognizes that the productivity of biopower can generate unexpected forms of life, and that power can move in unpredictable ways.[117] In so doing, this account gestures toward the possibility of an "affirmative" or "life affirming" biopolitics that holds out hope that there are forms of life and subjectivities taking shape in these laboratories, ones that are perhaps not so readily captured.[118]

EXCITABLE ONTOLOGIES

Fernando is a fifth-year PhD student working in the same lab as Edward. In the course of one of our many conversations, we talked about the nature of molecular life. He insisted that proteins are not alive and tried hard to stick to mechanical analogies to describe living processes. He likened the cell to the factory floor of a Ford car plant (see chapter 6). And yet, his mechanistic description of living substance wavered and in several instances veered toward livelier articulations. For him, the stuff of life is not inert. "We have," he explained, "a physical, very *motional* relationship to the world": "Things are always in change. Okay. When things reach stasis, equilibrium, they die. That's basically death. You have reached equilibrium. Nothing comes in nothing comes out. So how can you not describe molecules and life as a set of motions, as tensions? Okay. Someone described this to me once as: 'Life is a constant struggle between the hydrophobic and the hydrophilic.'"

"Hydrophobic" and "hydrophilic" describe two of many possible affective states of matter. It is the water-loving and water-hating properties of the different amino acids that make up protein molecules that shape how a protein folds and unfolds in its watery milieu. Fernando's insistence that it is a love/hate (-philic and -phobic) relationship that sets life in motion caught my attention. Everything, he assured me, including our relationships to the world, is "motional." It is kept in motion through the chemical affinities and repulsions of molecules that either "love" or "hate" water. For Fernando, when this "tension" is lost, "things reach stasis," and they die. In his essay, "How to Talk about the Body?," Bruno Latour recalls his response to the provocative question, "What is the opposite of a body?" Channeling Spinoza and Deleuze, he suggests that, if "a body" is defined by its capacity to affect and be affected, then the opposite of "a body" is the loss of this capacity to be moved by another. The opposite of a body for Latour amounts to death.[119] Fernando's definition concurs. *Affectivity* in Fernando's formulation is a life-giving property of matter. The liveliness of matter is its capacity to affect and be affected by other bodies. Matter, in other words, is *excitable*.

Fernando demonstrates how, in spite of his efforts to deanimate matter and provide mechanistic explanations, the stuff of life keeps coming alive his hands. Like Fernando, Edward's breathing molecule is not just a "romantic" holdover from some more enchanted moment in the history of the life sciences. His enactment is hooked into a widespread phenomenon among protein modelers, who are continually confronted by living phenomena that evade capture in their experimental apparatuses. Protein modelers render their objects in *ambi-valent* registers: in their hands proteins waver continuously between deterministic machines and lively, wily bodies. These simultaneously machinic and lively renderings come alive in modelers' conversations with one another at the laboratory bench, in group meetings, in conference talks, and during classroom lectures. These phenomena would be imperceptible if one's attention were solely focused on the texts that scientists write and disseminate. While scientific publications offer excellent resources for examining the prominence of mechanism in conventions of scientific writing, ethnographic observation of modelers in their laboratories and classrooms makes palpable the slippages and deviations from this script. Part III of this book tunes in to ask how these practitioners "do mechanism," amplifying the otherwise muted registers in which modelers articulate their molecular knowledge. What it finds is that in practice, mechanism fails to cohere as a singular, hegemonic discourse. Rather, it is always contaminated by livelier stories. This study shows that forms of animism are immanent to mechanistic logics.[120] The ongoing oscillation between lively and mechanistic renderings produces a new discourse and way of knowing among protein modelers. Theirs is a *lively mechanism* that conjures an *excitable* world of wily molecules. This is a significant shift given the foundational logics of neo-Darwinism that undergird most concepts in the life sciences today. Neo-Darwinism relies on a firmly mechanistic approach to biological processes. Part III and the conclusion to this book explore the implications of protein modelers' *excitable ontologies* for the evolutionary stories we tell ourselves about the origins of life and the adaptations of organisms.[121]

To be clear, the account I offer here is not an enchantment or reenchantment of the life sciences. It is an account of practitioners' failure to comply with a mechanism that would fully disenchant their practice or deanimate their objects.[122] It is by hitching a ride on the stories these modelers tell through their animated renderings that I have learned that proteins are "up to stuff" in cells. Indeed, these modelers generate a lively account of what might best be called the *molecular practices of cells*. These are stories that work athwart the deterministic logics of mechanistic descriptions of molecular and cellular life, those renderings that can so easily get swept up by the pharmaceutical industries and military interests. The practitioners documented in this book thus model forms of life that may as yet escape complete capture.

RENDERING LABORATORY LIFE

render, v. . . . to give in exchange, to give back, to produce, to yield . . . to let go, give up, to surrender . . . to pay (service) to . . . to bring or put (someone) into a particular state or condition, to make . . . to emit, give out, give off . . . to bring up, to deliver, take, lead, . . . to repeat, report, to deliver or give up (a person) to a religious life . . . to surrender, transfer, to pronounce (sentence, judgment, etc.) . . . to translate word

for word, to represent, to reproduce, to portray . . . to restore . . . to cause to re-
appear, to expel, throw up, to throw back (an image), reflect, to echo, to reproduce
in speech, repeat, to utter in reply . . . to ascribe, attribute, to hand over . . . to give an
account of, to set forth, to give (judgments) . . . to bring forth, to bring about, to cause
to be or become . . . to perform . . . to fulfill, to carry out.

—Oxford English Dictionary

The verb "to render" is multivalent, and its many meanings propagate through-out this book. The concept itself serves as a guide to the chapters that follow. In order to introduce the techniques involved in protein modeling, part I of the book is situated in protein crystallography laboratories. The first chapter, "Crystallographic Renderings," provides a guide to these techniques with a focus on the effort and skill that are required to purify proteins and coax them to form crystals. These practices make palpable other meanings of the term "rendering," which includes "to separate," as in the rendering of fat from bone when making chicken soup. This sense of the term is a reminder that "to rend" is also to tear or rip things apart, an especially salient connotation in light of the visceral and sometimes violent work involved in extracting, separating, and purifying proteins from once-living bodies.[123] Another dimension of the concept comes to the fore in chapter 1, which documents how students submit themselves to training. In this sense they "render themselves up," giving themselves over to the labor of model making in ways that evoke early uses of the term to mean "to surrender (one's life, soul)."[124] Indeed, these efforts to render life molecular get modelers entangled with their molecules, models, and machines.

Chapter 2, "Tangible Media," examines the materials and media crystallographers use to build physical and computer graphic maps and models. It pays close attention to the material cultures of modeling and the improvisational, experimental forms of haptic creativity that are integral to crafting tangible maps and models.[125] In the field of computer modeling, a rendering is "the processing of an outline image using color and shading to make it appear solid and three-dimensional."[126] Where in the past, models were built with ready-to-hand physical materials, protein modelers now make extensive use of computer graphic processing to elaborate, add to, and augment their data to render it in three-dimensional form.

To render also means "to decipher." This is an apt description for the work involved in rendering crystallographic data into the form of maps and models. This labor is frequently likened to "puzzle solving." Chapter 3, "Molecular Embodiments," examines how crystallographers must grapple directly with their data in order make it legible and workable. That chapter also explores other nuances of the verb "to render," including its meaning "to give over to" and "to give birth to." It documents how modelers give their bodies over to the work of model making in a practice that articulates their senses and sensibilities. This chapter attends to the intimacies that take shape among modelers and their models, relationships that shift conventional assumptions about the imagination, scientific reasoning, and even intellectual property.

Part II of the book raises epistemological and ontological questions by exploring how modelers know what they know, and examining closely how their models stand in relation to molecular phenomena. Chapter 4, "Rending Representation," documents protein crystallographers grappling with the limits of their vision and their ability to represent otherwise imperceptible phenomena. It examines experts' anxieties about the ways that static models can misrepresent dynamic molecular phenomena. To render means to represent in the sense of "to recite," "to echo," and "reflect," but it also means "to create." This chapter draws attention to the performative dimensions of rendering. It explores how protein models do not just represent the molecular world; they also materialize a world that is constituted by molecules.

Another meaning of the verb "to render" is "to give" or pay "homage," or "allegiance." This meaning is salient in the ways that crystallographers' models must stay true to molecular form; in this sense their models must maintain an "allegiance" to actual chemical configurations. Only robust facts should stand up for review and evaluation. To render is also "to put forward for consideration, scrutiny, or approval," and "to hand over or submit" (as in "to render up" a verdict or a document). Crystallographers must upload their structures to the Protein Data Bank in order to make them available to their colleagues for evaluation and further investigation. This raises the question of how practitioners adjudicate crystallographic models. Chapter 5, "Remodeling Objectivity," documents responses to three recent events that spurred serious controversies in the protein structure community. These events included the high-profile retractions of numerous protein structures published by two different laboratories, and the simultaneous publication of two different models of the same molecular assemblage. These events shed light on the peculiar culture objectivity that has taken shape among these practitioners. Eschewing myths of objectivity as a neutral, disembodied practice, modelers appear to advocate a modest, situated objectivity grounded in partial truths.

Part III of the book picks up on rendering as a practice that "inflects" representations. It draws attention to rendering as the action of "infusing a quality into a thing," "describing as being of a certain character," and "portraying, or depicting artistically." This part of the book considers the forms of molecular life that are coming to matter in the hands of protein modelers. Chapter 6, "Machinic Life," homes in on biological engineers' renderings of proteins as molecular machines, while chapter 7, "Lively Machines," examines how these machinic renderings are inflected with a new range of affects when they are animated in time-based media. Chapter 8, "Molecular Calisthenics," turns attention to the ways modelers use their own bodies as animating media to communicate the fine details of protein models in their laboratories, in classrooms, and at conferences. Drawing ethnographic attention to these ephemeral practices, modelers can be seen *moving with and being moved by* molecular phenomena. Indeed, another meaning of the verb to render is "to surrender," "to give in exchange," "to yield to," and "to utter in reply." This chapter explores how modelers give their bodies over to a practice of *mimetic emulation*. In the process, it shows how modelers *transduce* and so propagate a range of molecular affects.

Ethnography is also a rendering practice. The concept of rendering is especially salient to the methods I engage here to document forms of life in structural biology. Modelers work to *amplify* otherwise imperceptible phenomena and *animate* the peculiar qualities of their proteins. In turn, my rendering amplifies and animates modelers' practices. In so doing, I render up a range of practices that are otherwise hard to see. I tune in to amplify the otherwise muted registers in which scientists articulate their intimate knowledge of protein molecules, and I animate a host of subjectivities, sentiments, and values that are not otherwise readily visible in twenty-first-century laboratories. Rendering is an inherently performative practice: in other words, a rendering not only depicts a phenomenon; it conjures and so materializes some aspects of that phenomenon, to the exclusion of others. This ethnography animates some forms of life in the laboratory and not others. My aim here is to keep open rather than foreclose what it is possible to see, say, feel, and know about scientific practice and the living world. And as supplement to other accounts of science, this ethnography aims to render life science *otherwise*.

PART ONE | **LABORATORY ENTANGLEMENTS**

Diane is the head of a protein crystallography laboratory. She is a tenured faculty member based in the departments of chemistry and biology at research institute on the East Coast of the United States.[1] Her laboratory hums with activity. It takes up a number of rooms on the fourth floor of a large building that also houses chemistry laboratories and several other academic units, including the anthropology department. The groups' dozen graduate students and postdocs occupy a large open space lined with laboratory benches. Stacks of used petri dishes, test tubes, Styrofoam containers, used pipette tips, and beakers are scattered across the surfaces of their black laboratory benches. Some benches have apparatuses to purify proteins or automated PCR (polymerase chain reaction) machines for amplifying specific strands of DNA. Microscopes, water baths, hot plates, and shakers take up the remaining surfaces. Shelves lining the walls are stacked high with darkened glass jars full of chemicals, fresh supplies wrapped in plastic, and gleaming glassware. The fume hoods that line the walls are filled with instruments that must be carefully calibrated. Personal workspaces are decorated with pictures of family, friends, comic strips and cartoons, and one or two of the desks sport glamour shots of celebrities.

The machines scattered through the lab are part of a larger assemblage of attendant refrigerators, incubators, and centrifuges that take up space in the halls and adjoining rooms. Vapors billow forth from liquid nitrogen canisters and −80°C freezers.[2] A massive X-ray diffraction machine is housed in its own set of rooms. A protective glass wall separates this machine and its radiation from users and the computers they use to process diffraction data. Down the hall a darkened computer room equipped with several workstations feels like a quiet sanctuary. It is here that lab members render their long-sought-after crystallographic data into three-dimensional computer graphic models.

Just down the hall from the computer lab is Diane's office. When I arrived for our first interview, her aging dog, Max, greeted me with sweet, mournful eyes. His arthritic gait told me he wasn't long for this world. He was Diane's constant companion in the lab, and her grad students would come by throughout the day to take him out for walks. His presence sometimes made the lab seem like a family, especially when he came to the weekly lab meetings. These group meetings gave Diane and her students opportunities to present their progress on current research projects, as well as seek support to navigate difficult problems. At the same time this was also their opportunity to negotiate who was going to take on routine chores, tasks that involved cleaning communal workspaces and maintaining equipment. If some took up these tasks with a sense of collegial duty, others had to mute their disdain for washing glassware or cleaning out the refrigerators.

In our first interview, I learned a lot from Diane about what motivated her to pursue a career in the field of protein crystallography. Diane studied chemistry as an undergraduate student. One of the questions that fascinated her was how enzymes and their substrates interacted to catalyze biochemical processes. She explained: "I really liked the idea of trying to understand enzymes. What did they do? How did they catalyze this reaction? What was the detailed mechanism involved?" She wanted to be able to "see" biochemical reactions unfold: "I figured out that I couldn't think of working on a project if I didn't understand *what it looked like*. I needed that first. . . . You have to have some kind of concrete thing to start with, even if it's just one picture. At least there is something more tangible involved there. It was just that it seemed to me that that was the starting place for science. The first thing you ask is, '*What does it look like?*'" This desire to see what proteins "look like" is evocative of the efforts of nineteenth-century natural historians' efforts to describe "nature's panopoly" through drawings, images, and models.[3] It is only once they could see what proteins "look like," that modelers could begin to compare and contrast these remarkably diverse forms. What Diane was proposing seemed at first like a high-tech, high-resolution natural history of enzymes.

Protein crystallography was the technique that could give her visual access to these unseen dimensions of cellular life. Yet, this was more than a descriptive project. It was only with a three-dimensional model of the precise atomic configuration of a molecule that she would be able to discern the fine details of a protein's "active site," the area in the molecule that reacts chemically with other molecules. This was crucial knowledge if she was going to be able to figure out how that protein interacted with other molecules and how it catalyzed chemical reactions in the cell. Crystallographic models were "tangible" objects that could give her "concrete" access to these molecular interactions. The models she built would enable her to design experiments to intervene in an enzymatic reaction, and these interventions would help her interpret what particular proteins were up to inside the cell.

When Diane was ready to begin her graduate work in the 1990s, however, many of her mentors saw protein crystallography as a dead-end discipline. This was a time when rapid advancements in molecular genetic techniques and recombinant DNA technologies were shaping the direction of research questions and funding. Researchers in the biological sciences had already diverted their interest and investments from atomic-scale models of proteins to the automation of genetic sequence analysis. Informatic models of the genetic determinants of life held sway and molecular genetic studies dominated the life sciences.[4] Diane was warned not to pursue training in protein crystallography. In some ways, she was getting sound advice. It was hard to build models of protein structure. It took Nobel Prize laureate Max Perutz twenty-two years to solve an atomic resolution crystallographic structure of just one protein molecule. His crystallographic model of hemoglobin, published in 1967, was at that time only the second high-resolution protein structure to be determined. Twenty years later faster computers and better equipment helped to speed things up. But by 1990 the Protein Data Bank housed just 486 protein structures determined by this technique.[5] The future of the field did not look bright.

Yet, Diane's desire to gain visual access to the molecular realm was so strong that she ignored the sage advice, and began graduate work in protein crystallography. Her perseverance, it seems, paid off. By the time she began her tenure-track academic position in 1999, techniques had improved so much that contributions to the field had grown exponentially. That year alone nearly two thousand novel structures were determined by crystallographic methods, bringing the total to over nine thousand unique crystal structures archived in the Protein Data Bank. By 2004 she was tenured and head of a thriving lab at a prestigious research university.

This chapter offers a detailed examination of the techniques and practices protein crystallographers must learn in the course of their training. While it focuses primarily on the experiences of graduate students in Diane's laboratory, it also explores how graduate student life is rendered in the documentary *Naturally Obsessed*. In ways similar to the documentary, this chapter describes the practical and conceptual hurdles graduate students, postdocs, and their mentors encounter while attempting to build their first models. Alongside stories of scientists-in-training struggling to figure out how to comport themselves in the lab, this chapter traces the early history of X-ray diffraction and offers an overview of the suite of techniques that make up crystallographic vision. It foregrounds the limitations of X-ray diffraction methods and pays special attention to the contributions modelers must make at every step in the process to fill in the gaps between what is perceptible and what remains unseen. And as a step-by-step guide to help readers understand the technical challenges crystallographers encounter, it details the peculiar sequence of crystallographic techniques, from protein purification and modification methods to techniques for protein crystallization and diffraction. It offers an account of both the pleasures and the perils students face in their attempts to simultaneously master crystallographic techniques and learn to fashion themselves as scientists.

Naturally Obsessed is particularly generative to think within the context of this study, as it helps to keep in view the broader social, political, and economic forces that shape scientific lives. This chapter takes a close look at how audiences responded to *Naturally Obsessed* in various discussion forums. The documentary has spurred numerous conversations and debates on such issues as the naturalization of science as a competitive game, what counts as success, and who succeeds in science. These issues are crucial to keep in the foreground as this chapter hones attention on forms of affective labor that shape modelers' intimate relationships with the molecules, materials, and machines that populate their laboratories. It is by turning close ethnographic attention to the affective entanglements of laboratory practice, without losing sight of the political economy in which laboratory labor gains its traction, that it is possible to hear practitioners' accounts of their encounters with the affectivity of matter. This chapter shows how, in their hands, the stuff of life comes to matter as a lively, wily, and excitable substance. What becomes clear is that the forms of life that materialize in the intimate spaces of these laboratories are not so readily captured by any technique or industrial enterprise.

ENROLLING NEW SCIENTISTS

In the fall of 2006, during one of her lab's weekly group meetings, Diane practiced a talk she was going to give at an upcoming symposium that featured researchers at the forefront of the field. A group of graduate students, undergraduates, and postdocs were seated around the long table. Standing, she gestured across the room to a half-finished PowerPoint slide. As she walked through the points she would make in her talk, she laid out her approach to the "incredible" chemical structures that "nature has tailored" in living cells.[6] Undeterred by the serious technical challenges protein crystallographers like her face in visualizing molecular forms, she told members of her lab that she wanted to "think bigger about what we can accomplish": "Instead of solving the structure of one enzyme from a [biochemical] pathway, we [want to] solve the structures of all the enzymes in the pathway, alone and in complexes. Instead of solving the structure of one protein from a superfamily, [we want to] solve structures of multiple members of that family—because in the comparison you can often figure out what is truly important in those molecules. . . . [We want to] solve many structures of one enzyme and capture states as it proceeds through its reaction cycle."

What Diane wanted her audience to consider were the dynamic properties of proteins: proteins move and change shape as they participate in chemical reactions in the cell. For the most part, protein crystallography is a static visualization technology, producing three-dimensional structures. Diane's ambitious proposal reimagines protein crystallography as a time-lapse imaging technology for playing through the sequence of structural changes that a protein undergoes as it encounters other molecules in a given biochemical reaction pathway.[7]

Protein crystallography requires "resolving" differences in each region of the molecule; a crystallographer must be able distinguish between each atom at high resolution. Yet, note Diane's use of the verb "to solve" here. Where natural historians claim to "discover" new species in nature, Diane relates to protein crystallography as a puzzle-solving exercise, or what she describes as "detective story." Recall that one of the meanings of the verb "to render" is "to decipher" (see introduction). This is an apt description of the kind of work involved in making crystallographic data visible in the form of molecular models. If solving a single protein structure requires arduous labor, generating multiple models of a molecule at each step in its reaction cycle would be a monumental task. Diane's lab has been exceptionally productive, and she has built up an international reputation for solving the structures of significant proteins, yet this vision still seemed ambitious.

As she and the other practitioners documented in this book have taught me, the determination of a protein structure is a remarkable achievement.[8] It takes serious work to model the atomic configuration of a protein. Practitioners require intensive training to cultivate the technical expertise they will need to accomplish this feat. According to Diane, to succeed as a protein crystallographer you must simultaneously be "a molecular biologist and a protein biochemist." Additionally, "you have to be a little bit of a physicist, you have to be a computer jock, *and* you have to be an artist." The wide range of dexterities and competencies bound up together in this "bricolage of expertise" means there is a steep learning curve for students in training.[9] Many students arrive in Diane's lab with an undergraduate degree in either biochemistry, inorganic chemistry, or molecular biology. Some have never worked directly with computer codes; others have never worked with proteins or living organisms; most have no experience doing laboratory research.

The field is changing with innovations in technology and new investments, and as researchers like Diane refine their techniques they are taking on more challenging questions. Faster computers, automated laboratory techniques, powerful software, interactive computer graphics interfaces, and synchrotron X-ray sources now make it possible to solve protein structures on the time-scale of a PhD student's tenure in the lab. Some labs are involved in large-scale collaborations to build models of massive protein complexes. Yet, crystallo-graphic techniques today are far from foolproof, and almost every project is plagued by setbacks, failures, and detours. Novice modelers need to cultivate patience to weather the slow, painstaking work required to isolate and crystal-lize proteins, collect their data, and build models. Each step in the process is time-consuming, and physically, intellectually, and emotionally demanding. Even today it might take five years to determine a high-resolution structure of a large molecular assemblage. This is often too long for students who must complete a structure in time to graduate. Some students get lucky, or have the skills to make quick progress. When I first met Dehlia, a third-year student in Diane's lab, she was just putting the final touches on a paper forthcoming in *Nature*. Some projects present too many obstacles. Things were not moving so quickly for Amy. Already in her fifth year, she was still struggling to get usable crystallographic data.

In this sense, Diane's message wasn't just geared to impress the audience at the following week's symposium; her practice talk also served as a rallying call to her graduate students and postdocs. Indeed, the weight of these ambi-tious goals would fall on their shoulders. They would each spend at least five years apprenticing with her and learning from the other students and postdocs in the lab. They wouldn't graduate until they had successfully modeled their own proteins.[10] In order to produce such fine-tuned, high-resolution visions of molecular life, Diane would have to invest her energies in training this new generation of scientists. Thus, while her laboratory is dedicated to the work of producing scientific facts, it is also a site geared to the task of making new scientists. So while lab directors like Diane are keen to solve protein structures and publish their results, they are also explicitly invested the training and continual retraining of graduate students and postdoctoral researchers. It is in the process of learning how to craft molecular facts that novices begin to fashion themselves as scientists. But what is the model for a good scientist? What does success in science look like?

Academic laboratories, such as the one that Diane runs, are sites where a student's *ethos* and *habitus* are still in formation. The concept of "ethos" is used here to evoke the tangle of affects, values, attitudes, sentiments, styles, and sensibilities that shape practice and laboratory culture.[11] For example, modelers must learn what counts as good or bad data, when to trust or mistrust their experimental results, and how to care for their objects, materials, and equipment. Laboratory training attunes them not only to proper attitudes and conduct, but also to proper techniques. "Habitus," a term first introduced by the French anthropologist Marcel Mauss, has come to stand for the "unselfconscious orchestration of practices" that condition a practitioner's sensorium, posture, comportment, and bodily know-how.[12] At each stage of their training novices can be found fumbling with their instruments and materials until they begin to get the gist of things. In other words, it is in training laboratories like Diane's that graduate students slowly acquire the kinesthetic and affective dexterities they need to build crystallographic models.

Naturally Obsessed documents well how graduate students cultivate the ethos and habitus they need to succeed in science. It offers a remarkable view into the ways they give their bodies and minds over to training, how they fashion themselves into scientists and learn how to persist under the most challenging circumstances. It makes palpable how students must contort, reorganize, and reorient their lives around the task of molecular modeling. And yet, it is crucial to ask: Which ethos? Whose habitus? And, what forms of life come to matter in this particular rendering of laboratory life?

Naturally Obsessed is a dramatization of laboratory lives. The filmmakers Richard and Carol Rifkind admit that they faced a challenge in crafting a salient story whose characters and dramatic structure could captivate its audiences. Cutting and splicing hundreds of hours of footage, the story that the filmmakers come to tell hinges on the familiar trope of science as a zero-sum game where the winner takes all. Winning the game in this rendering requires graduate students to solve the structure of a prized protein molecule before others beat them to it. The documentary has circulated widely since it was first released, and has been discussed at length in public forums as well as in laboratory meetings and high school screenings. Some video recordings of these screenings and discussions are posted on the documentary's website. The metaphor of science as a game filtered out into the spaces in which the film was discussed and evaluated. High school students came away from a screening of an early version of the film with the impression that a doctoral degree must be "won" in a competition. In the question period after the screening, one student stood up to ask the directors, "I just want to know, what are the qualifications to compete in a PhD and in the other sciences?" Another posed the question this way: "Is this like a Disney movie thing, where everyone gets a PhD? Or only Rob wins? Do some people win and others don't?"[13] These questions from high school students raise others: How does the trope of science as a game inflect and orient the attention of the filmmakers? How does this affect how participants in the documentary are rendered? How does it affect how they perform, as well as fail to perform, for the camera? Feminist science studies scholars Donna Haraway and Emily Martin have long encouraged observers of science to question the tropes they mobilize to tell stories about science.[14] They question the politics of mobilizing metaphors that describe science as a competitive, agonistic playing field in which success is defined as winning at a high-stakes game. Are there other stories to tell about science and perhaps other kinds of games to play, especially ones that are not so thoroughly configured around the thrill of the chase and the triumphant win?

The Rifkinds are open about the great effort they put in shaping the dramatic structure of the film.[15] At the same time they also affirm the "authenticity" of their film as an accurate and true-to-life representation of what they call "the culture of science." Richard Rifkind insists on the objectivity of their rendering: "Our camera observed the students absorb the values inherent to the culture of science."[16] In an online interview, Carol Rifkind offered this: "They [the audience] liked the authenticity of it . . . that this is the way X-ray crystallography is, this is the culture of a science lab, these are the relationships that make science go forward. I guess generally speaking, the culture of science is what our focus was, and what the scientists feel very keenly. I guess all audiences feel that that's what they've experienced."[17] Commentators in a variety of discussion forums, however, have questioned whether the film correctly depicts the values "inherent" to science or whether the film generates and propagates one particular set of values.

Naturally Obsessed and the conversations it has spurred offer generative ethnographic material for exploring what kinds of stories about "a life in science" are salient today. What does this documentary model as the ethos and habitus of a scientist who can succeed on the competitive playing field that it calls science? In other words, what are the values, norms, mores, desires, comportments, and postures of a successful scientist in this film? How does this story of failure and triumph model the kind of person who can *make it* in the economies that shape the life sciences today? Which facts and which values prevail when what counts as success in the lab is figured as the heroic and timely capture of life in the form of a valuable protein structure?

| BORN OR MADE? FASHIONING A SCIENTIFIC SELF

> Science is hard. Right. You got to push, and push, and push, and push. You gotta really want it. You either gotta be driven from behind or be pulled from the front. Right. So. Isn't this what obsessive [compulsive disorder is]? . . . Like . . . You have this focus. [Hands lift to face and form blinkers around eyes, then draw downward, directing his gaze] You have this thing. You just want it.
>
> —Rob Townley, *Naturally Obsessed*

The epigraph above is a quote from Rob taken from the *Naturally Obsessed* website which features a series of short interview clips as mini portraits of the lead characters in the film. Rob's animated face and hands are framed against a black backdrop. He speaks directly into the camera. The film's trailer features carefully spliced scenes of laboratory labor coupled with a soundtrack with a driving beat and voiceovers from the lead characters that reiterate Rob's compulsion to work:

ROB: There is a certain species of scientist that I refer to as being naturally obsessed. There's this itch or anxiety you have that can only be supplied by knowing.

GABE: What gets me to work every day is that maybe today I will have the answer to this question. And then I'll have more questions that I'll be able to ask.

KIL: It's like a challenge every day. You want to know why this works or how this works. You are just driven to find out.

LARRY: One of the best things that you can do as a scientist is suffer from obsessive-compulsive disorder. So that you become obsessed with a problem and can't stop working on it until you get . . . to your answer.

The documentary is saturated with invocations of obsession, compulsion, and unrelenting drive. What viewers learn is that good scientists are "obsessive compulsives" whose attentions and drives are geared unwaveringly to solving scientific problems. It is easy to get swept up by these impassioned invocations of dedication to science. But first, consider the full title of the documentary, *Naturally Obsessed: The Making of a Scientist*. While "Naturally Obsessed" figures obsession as the naturally endowed, innate disposition of a successful scientist, the second part of the title emphasizes that *scientists are made*. This ambivalence in the title is a potent reminder about the labor it takes to fashion such a disposition and make it seem natural. What is the effect of naturalizing obsession in this way? One effect is to bolster the conventional assumption that laboratories are sites where exceptionally gifted people gather to give their labor over selflessly in service of science. This, however, obscures the larger political and economic forces that shape how and why people commit themselves to scientific work. At the heart of this discussion is not only the question of whether scientists are born or made, it is about how the vocation of science is figured for would-be scientists, how the labor of scientific work is justified and valued in larger economic frames, and precisely who does and does not end up "making it" as a scientist.[18]

In March 2009, the City University of New York (CUNY) Graduate Center convened an event entitled "Our Future Scientists."[19] Fashioned as a town hall meeting, it aimed to wrestle with the fallout of the current "crisis" in science education in the United States. The flyer for the event read as follows: "The U.S. is falling behind in the production of new science PhDs. Is there a crisis looming? A discussion with laboratory scientists that follows the screening will allow an exchange of ideas on what's needed to maintain an ample pipeline of future scientists." This documentary, which raises so many questions about the challenges of scientific training, was to be the grounds for a public conversation about the state of the political economy of science in the United States. With its focus on bolstering the number of science PhDs to feed the nation's "pipeline" of researchers, this event brought to the fore the role of science training in maintaining the nation-state's scientific and technological prowess. Indeed, this context for the discussion makes palpable the neoliberal formations of capital that propel students to make such intensive commitments to laboring in life science laboratories.[20]

After the film was screened, National Public Radio (NPR) host Robert Krulwich moderated a discussion among three local scientists: Joy Hirsch from Columbia University, Ben Ortiz from Hunter College, and Susan Zolla-Pazner from New York University. The discussion quickly homed in on the question of whether one is born or made a scientist. What is the key to success for a scientist? The conversation turned to the title of the documentary and the theme of obsessive compulsion. The debate turned around the question of whether obsession really was what makes a real scientist tick. Is obsession innate? Is it something a scientist is born with? Joy Hirsch had this to offer: "You often times say that scientists are born, *not made*. And those of us that are in the business of teaching scientists to be scientists, that's a little bit frustrating because either they come to you as a scientist, or they don't. There's not much we can do as professors to make them scientists, I don't think."

Ben Ortiz took a different tack for his contributions to the conversation. He made a number of interventions that challenged core assumptions of privilege that are embedded in the idea that a successful scientist is one who is naturally endowed with his talents and unwavering drives. In each of his contributions to the conversation Ortiz emphasized the "diversity of approaches to science": "I think there is a quite large range of acceptable approaches to doing good science besides being obsessed, overwhelmed, and submerged all the time." In 2005 Ortiz was featured in a special edition of *The Scientist* focused on a group of researchers identified as members of "underrepresented groups." Ortiz had grown up in a diverse, low-income community in Brooklyn, New York. The article explained how he felt "ruled by practical considerations" and was reluctant to take risks: "When a professor talked to him about the possibility of graduate school in science," Ortiz responded by saying, "I don't have that kind of time. I have to go out and get a job."[21] And so, in the CUNY town hall meeting, Ortiz's response to the question of whether people are born to be scientists hit a chord: "I'm not sure people are born [to be scientists]. I don't think I was born to be a scientist. I was born to be a bank teller." Ortiz had enjoyed his job at the bank and had to be convinced to leave such a secure job in order to pursue an uncharted career path: "So, no, I'm not sure we are all born [to be this way]. I think we have to really be shown what it's about. And I grew up with no concept of what a scientist was. At all."

Ortiz understands full well the barriers that keep people out of elite institutions like science. For him, science is not a zone restricted to geniuses and obsessive compulsives. According to him, a scientist can be made: "I think the most important thing is just plain old clear thinking about basic facts. I always say that I wasn't born a genius, but I think clearly and that helps me very much. And so part of what I think you can do to help a student who has never been exposed to science or the scientific method . . . the biggest task is to clarify their thinking. Make their thinking less cluttered. And you can actually see that."

Ortiz's comments challenge those who benefit by trading on the notion that they are "naturally obsessed" and therefore innately gifted. Science historian Rebecca Herzig calls attention to self-fashioning and subjectivication as performative processes through which scientists actively cultivate their identities.[22] She recognizes that the affects of obsession and compulsion have to be cultivated; they are dispositions around which scientists have long fashioned their identities. She documents a history of scientists whose compulsive drives pushed them beyond the limits of pain and fatigue, scientists who felt compelled to gain knowledge at any cost. And yet she does not take these stories of heroism and sacrifice at face value. She shows how much work it takes to fashion a successful scientific self.

So, how do students participate in fashioning their identities as scientists? How do they work on themselves to ensure that they *effect* the *affects* appropriate to their vocation? Anthropologist Saba Mahmood's analysis of the cultivation of piety among a group of Muslim women in Egypt offers insight into processes through which a person can work on themselves to activate desired affective states.[23] Mahmood revises Foucault's analysis of the disciplining of "docile bodies" in order to foreground the intensive labor involved in submitting oneself to training. She reinterprets Aristotle's concept of "docility" as an actively cultivated disposition; the docile student is one who actively gives themselves over to training. A good student is one who *renders* him- or herself up to receive teachings from a master, and Mahmood foregrounds the novice's agency in cultivating this posture of submission.

Herzig and Mahmood help us to see what is obscured when "obsession" and "compulsion" are naturalized as innate dispositions. Once naturalized, it becomes more difficult to see how these very postures participate in the broader neoliberal formations of capital that shape the political economy of science today. Indeed, the figure of the obsessive-compulsive scientist is one whose ethos and habitus are perfectly tuned in to a neoliberal economy: this is an economy that depends on a person's will to improve their lot by entraining themselves and their skills to market demands. Unwavering dedication is precisely what is required for a "scientific entrepreneurialism" that is successful in producing the commodities and marketable products that can fuel industries organized around extracting value from biological processes.[24]

The naturalization of these dispositions also makes it easy to dismiss a student's lack of success in a lab by suggesting that they just don't have what it takes. Such a move makes it particularly difficult to recognize other forces that hold back students, including forms of exclusion that make it hard to secure strong mentorship relationships. Gabe, for example, the one woman featured in the film, struggles to get her experiments to work. The documentary makes it seem that Gabe doesn't have what it takes to handle the ups and downs of life in the lab. This issue came up in one of the high school screenings of the film that is documented on *Naturally Obsessed*'s website. One African American high school student stood up during the discussion after a screening at her school to say this: "My question is: How do women make out in this field?" The CUNY panel addressed this thorny issue:

KRULWICH: Did all three of you know that Gabe was headed for trouble? What did you see?

HIRSCH: [Long, awkward pause] That's hard to express. But yes. I sensed that probably she was the one that wasn't going to . . . That the thrill of the chase wasn't going to make her happy. That there were other things that were going to make her happier. The risk of being a scientist is sometimes very painful.

Hirsch's awkward response barely concealed some of the fraught relations of power that shape gendered notions of success in the film. Ortiz pushed back against Hirsch and challenged the film's depiction of Gabe as a failure:

I think perhaps she might have been better matched to a different type of project. She's not a failure by any stretch of the imagination. She certainly has a career now and that's fine. But, I wonder if she were matched to a project where it wasn't so all or none. Where she could have seen the incremental progress from time to time . . . She might have had a different experience. This gets back to the idea that there is really a diversity of approaches to science. And when I am working with a graduate student, you really need a feel for them and [you need to] figure out what they would like, what they would be good at and what they would need in order to be successful.

In this remark, Ortiz suggests a kind of neglect on the part of Gabe's mentor. And indeed, where the filmmakers feature many scenes of Larry actively working with Rob and Kil, we are given just one or two glimpses of Larry speaking with Gabe. At one point in the documentary Gabe acknowledges this lack of mentorship: "I've been aggressive in asking for help on how to design experiments and what the proper controls are and things like that. What I haven't been aggressive enough about, I think, is getting help from the boss."

Survival in this game apparently demands that students be "aggressive" about getting attention from their advisors. At one point in the film, Larry points to Kil and tells him, "It's a tough world out there. You gotta be a tough guy." Kil responds by saying, "I thought I was tough enough. And then I came here." Viewers of the film are left with the impression that Gabe's decision to leave the lab to become a research scientist in a biotech firm is a failure; she lost the game. To say that she wasn't tough enough or didn't have the innate disposition of a successful scientist would be to overlook the possibility that her project wasn't a good fit, that perhaps her mentor wasn't doing his job, or that the competitive and sometimes macho culture of this laboratory didn't leave space for her to flourish.[25]

By hailing an obsessive drive as the key to success for scientists, the documentary also seems to give scientists permission to block out distractions so that they can stay focused on the scientific problems in the lab. The panel was convened to address current anxiety in the United States about lowered productivity and problems with training in the sciences; a naturally obsessed disposition would seem to be just what was needed to keep scientists laboring in the lab without distraction. A self-described obsessive compulsive, Hirsch seems happy enough to keep the rest of the world at bay. When asked if she did science to make contributions to society, she responded by saying, "I don't care. I just want to solve the problems." By contrast, Ortiz, who works on the molecular mechanisms of HIV, is passionately committed to the larger social, political, and economic implications of his work. In response to Hirsch's refusal to care, he insisted, "You may know people with HIV who need the drugs!"

Conversations taking shape around *Naturally Obsessed* are potent reminders that academic teaching laboratories are not cordoned off from broader social, cultural, economic, or political forces. Students, postdocs, research associates, technicians, and principal investigators are enmeshed in complex relations of power, and their lives and labors inside and outside the lab are not immune to the systemic violences, injustices, and exclusions that are contoured by capitalism, colonialism, and by intersections of race, class, and gender. Christien Tompkins, a PhD student in the Department of Anthropology at the University of Chicago offered an astute set of readings on the silences and absences in *Naturally Obsessed*'s rendering of laboratory life.[26] When he was an undergraduate student at Columbia University, Tompkins was a community activist committed to challenging the university's expansion into the low-income African American and Latino communities of Washington Heights and Harlem. He also helped organize a hunger strike and participated in negotiations on behalf of strikers in response to the "razing and development" of the Audubon Ballroom where Malcolm X was assassinated in 1965. This historic site was "just a block east of Larry Shapiro's lab." Tompkins points out that diabetes and obesity, the very diseases that Shapiro's laboratory investigates, "are severe problems in the Washington Heights and Harlem communities" just beyond the walls of the laboratory. Even as the "Columbia University administration touts the disease-curing potential of its research as a community benefit," these potential benefits are "elided" in *Naturally Obsessed*'s "accounting of scientific practice."

As a recruitment tool for high school students, this documentary is itself a pedagogical device that can shape what its viewers come to think a life in science *should* be like. Audiences watching the documentary learn a number of important lessons. Viewers learn what counts as success in science and who can succeed in science; they learn the postures, attitudes, and relations scientists should cultivate with one another; they learn what submitting oneself to a life in science should look and feel like; moreover, they learn to distinguish the proper objects of scientific study and the proper stance they should take in relation to these objects. In this way, the documentary becomes a mirror in whose reflection would-be-scientists, scientists-in-training, and even accomplished researchers can fashion their identities. It is a rendering that they cite and recite, even if many viewers resist its norms.

AFFECTIVE LABOR

The account I offer is distinct from the one presented by *Naturally Obsessed*. For one thing, I am less interested in composing an account whose dramatic structure conforms to culturally salient scripts. And yet, for all its problems, this documentary is certainly generative. Indeed, viewers' critiques of this documentary help to frame laboratory labor in a broader sociopolitical and economic context. These forces must be kept in view, even as I turn attention in this chapter to the specificities of the techniques and practices students must master in protein crystallography laboratories. In what follows, I focus on one dimension of laboratory labor that is so often overlooked in accounts of the political economy of science. This is an attention to the affective labor involved in scientific research and training, a form of labor that is crucial to the work of producing and circulating valuable scientific facts. Affective labor is a concept developed by feminist scholars and political theorists to analyze forms of labor, such as nurture and care, which have historically been undervalued or otherwise made invisible.[27] Although such forms of caring labor have systematically been overlooked, according to political theorist Michael Hardt, affective labor has "never been entirely outside of capitalist production."[28] Affective labor is a crucial element, for example, in service industries, where the labors of maids, nannies, and other domestic workers are commoditized and monetized. Indeed, such forms of labor have been "incorporated and exalted" as "one of the highest value-producing forms of labor" in contemporary information and knowledge economies.[29] Affective labor demands attention as it is in the "production and reproduction of affects" that "collective subjectivities are produced and sociality is produced—even if those subjectivities and that sociality are directly exploitable by capital."[30] As Hardt suggests however, affective labor may also be a site for resisting and subverting capitalist modes of production.

The concept as I use it here invokes the incredible effort and energy modelers expend to care for all elements of their experimental configuration. Thinking with Herzig's performative account of scientists' dedication to their work, such forms of care must not be mistaken as evidence of a student's devotion to knowledge for knowledge's sake, as if they give their labor over selflessly in the name of the "greater good" of science. This affective labor must be understood as integral to the work they do to cultivate skills and expertise required for them to succeed in a highly competitive context that is contoured by neoliberal formations of capital. This investment they make in their training is part of a biopolitical imperative that Foucault called the "care of the self."[31] And yet, I argue here, it is in this very process of working on themselves to finely tune their ethos and habitus that novice scientists also come to care a great deal for their instruments, their experiments, and their objects. In other words, with this "care of the self" they also cultivate a kind of "care of the molecule," as if the very objects of their inquiry were extensions of themselves.[32] In what follows, I document how practitioners must go to extraordinary lengths to keep their proteins "happy" and to nurture the perfect conditions for their proteins to crystallize. A skilled practitioner is one who has become responsive to the demands of her often-finicky materials and one who has cultivated intimate knowledge of the machines she uses in experiments. For these practitioners, securing a fact in the form of an atomic-scale protein model is not just a matter of adhering to the technical requirements of a system; their task is lived as an evolving relation of obligation to their molecules, models, and machines.[33] Remarkable forms of affective labor take shape over the course of the long hours students spend doing what would otherwise seem monotonous or grueling work. What becomes clear in this account is that a crystallographic model is not just "matter of fact"; it is also what science studies scholar Maria Puig de la Bellacasa has called a "matter of care."[34] Molecular vision hinges on modelers sustaining intimate relationships with their molecules, models, and machines. And it is in the space of these very intimacies that practitioners come to tell remarkable stories about their encounters with the affectivity of matter; it is through their affective labors that they animate the wily excitability of molecular life. So while modelers' labors are geared toward the production of facts in a way that will allow them to graduate, get a job, and contribute to the economy, the effects of their labors produce forms of life that are perhaps less readily captured by capital.

DIFFRACTIVE OPTICS

In 1892 John W. Caldwell, a chemist at Tulane University in New Orleans, published an article in the journal *Science* entitled "Some Analogies between Molecules and Crystals." In the article, Caldwell offered a chemist's response to the alluring suggestion that molecules may have distinctive configurations. Fewer than twenty years earlier, Dutch chemist Jacobus H. van't Hoff had proposed that the atoms making up molecules were arrayed in three dimensions. In Caldwell's day, however, there had as yet been no visual proof. According to Caldwell, "the subject of molecular configuration is comparatively new; still we are becoming familiarized with diagrams and models intended to represent such relations. Many of us may have been at first indisposed to accept these views as anything more than visionary and fantastic; but the more we have pondered them, the more we have been impressed with their significance and beauty. Shape, form and volume must be attributed to molecule as well as to mass; the only trouble has been in regard to the former, the apparent audacity and hopelessness of any attempt to penetrate matter to such depths."[35]

For Caldwell and his readers, molecules and their atomic configurations remained imperceptible and elusive. The "depths" of matter—its three-dimensional "molecular configuration"—was hopelessly "impenetrable." It remained an "audacious" fantasy to hope to behold the "shape, form and volume" of matter at the scale of atoms. Too small to be seen with the instruments available to Caldwell, the molecular configuration of matter remained beyond grasp.

In 1895, just three years after Caldwell's publication, German physicist W. C. Röntgen discovered X-rays. This remarkable form of radiation made the molecular structure of matter accessible in profoundly new ways. A high-energy form of electromagnetic radiation, X-rays have significantly smaller wavelengths than visible light. It was in 1912 that physicists realized that crystals, made up of regular arrays of molecules, were able to interact with X-rays. Whereas X-rays could pass through large obstructions such as skin and muscle, in contrast, crystals placed in front of an X-ray source caused these rays to scatter, or diffract.

Diffraction is a phenomenon produced when any wave, such as a sound wave or wave of radiation passes through a regular array of objects. When waves encounter obstructions, they get deflected and change course, often forming patterns as one wave overlaps and interferes with another. Diffraction patterns, also known as interference patterns, map moments where waves come together with varying effects: sometimes the waves build on one another, and at other times they cancel one another out. Familiar diffraction patterns include those made in water, when, for example, waves pass through gaps in a breakwater or when pebbles are dropped on still water and the spreading waves overlap and interfere with one another.[36] The patterns that are generated are an effect of the entire configuration of elements, including the properties of the wave (for example, its wavelength), the size of the obstacles, and the size of the spaces between them. If the spaces between a set of objects are of the same order of magnitude as the length of the wave, an array of objects can act as a diffraction grating; that is, it can be used deliberately to scatter the wave. If the properties of the wave are known in advance, diffraction patterns can be deciphered to determine the size and configuration of the objects in the array.

Crystals are made up of repeating arrays of identical molecules and the spaces between the molecules are on the same order of magnitude as X-rays. It was British physicist William H. Bragg, who, building on the experiments of German physicist Max von Laue, discovered that crystals could be used to diffract X-rays. In 1913 he turned this X-ray diffracting property of crystals into a technology for investigating the very atomic configuration of molecules that were packed inside crystals. By 1914 Bragg and his son William L. Bragg were developing the first X-ray diffraction experiments with inorganic crystals, and the first structure determined was sodium chloride (table salt).

One of the remarkable features of this technology is that the crystal—the very object whose structure scientists want to describe—is transformed into a technology that is integral to the visualization apparatus. Figure 1.1 offers a schematic overview of the configuration of an X-ray diffraction experiment. A crystal placed in front of an X-ray source scatters the rays and produces a diffraction pattern on X-ray sensitive film. The pattern is made up of small dots of varying intensity, which are called reflections. The larger dots are sites

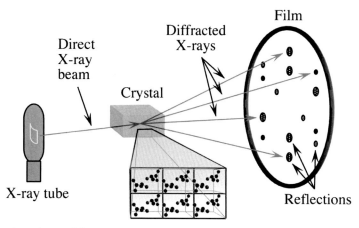

1.1. A simple X-ray diffraction experiment: "The crystal diffracts the source beam into many discrete beams, each of which produces a distinct spot (reflection) on the film. The positions and intensities of these reflections contain the information needed to determine molecular structures." From Rhodes, *Crystallography Made Crystal Clear: A Guide for Users of Macromolecular Models*, 11. Courtesy of Gale Rhodes.

where X-rays in the same phase have intersected, producing waves of higher intensity; the smaller dots occur when waves with different phases cancel one another out. Each spot in the diffraction pattern is an effect of the interaction of X-rays with every atom in the molecule, and every molecule in the crystal. The Braggs studied these diffraction patterns and determined mathematical proofs that would allow them to calculate the configuration of atoms in each molecule within the crystalline array.

W. H. Bragg once recalled that X-ray diffraction "increased the keenness" of his vision "over ten thousand times," making it possible to "'see' the individual atoms and molecules."[37] The Braggs, it seems, were successful in devising a form of molecular vision that Caldwell had at one time deemed so hopeless and audacious. While the Braggs worked on inorganic materials, British biochemist Dorothy Hodgkin, a pioneer in protein crystallography, took up the challenge to develop X-ray diffraction techniques that could be used to visualize complex biological molecules. Like the Braggs, she also wanted to be able to "see" molecules. She explains how, while working at Oxford and Cambridge Universities in the 1930s, she was "captivated by the edifices chemists had raised through experiment and imagination." And yet, she "still had a lurking question": "Would it not be better if one could really 'see' whether molecules as complicated as the sterols or strychnine were just as experiment suggested? The process of 'seeing' with X-rays was clearly more difficult to apply to such systems than my early reading of Bragg had suggested; it was with some hesitation that I began my first piece of research work."[38] Hodgkin went on to collaborate with crystallographer John D. Bernal, a socialist, public intellectual, and former student of W. H. Bragg. Together they determined the crystal structure of pepsin. Later she solved the structures of cholesterol and insulin, and won the 1964 Nobel Prize in Chemistry for her elucidation of the chemical structures of penicillin and vitamin B_{12}. In spite of the indirect nature of diffractive optics, Hodgkin achieved remarkably keen molecular vision.

And yet, Hodgkin's initial hesitation was well founded. She knew how challenging the work would be. Indeed, extending X-ray crystallographic techniques to large molecules made up of thousands of atoms was no small feat. The diffraction patterns generated by simple molecules are already hard to interpret, and large molecules produce more complex interference patterns as X-rays scatter and intersect. Learning to read those diffraction patterns continues to pose serious challenges for practitioners working in the field today. Even with major technological innovations since Hodgkin's time, including automated methods to measure the intensity of each reflection, molecular vision is still incredibly indirect.

What is it that the Braggs, Hodgkin, and Diane have been able to "see" with the diffractive optics of X-ray crystallography? What is the nature of their molecular vision? A brief overview of the techniques crystallographers use today will help to open up these questions for a deeper appreciation of the limits of vision in protein crystallography.

THE HUMAN-COMPUTER LENS AND THE LIMITS OF MOLECULAR VISION

> Diffraction does not produce "the same" displaced, as reflection and refraction do. Diffraction is a mapping of interference, not of replication, reflection, or reproduction. A diffraction pattern does not map where differences appear, but rather maps where the effects of difference appear.
>
> —Donna Haraway, "The Promises of Monsters"

I sat in on Diane's graduate course on macromolecular protein crystallography to gain deeper insight into the practice. The course took students through the steps involved in each stage of the model-building process and included laboratories where I had a chance to try my hand at some of the techniques. I learned early on that protein crystallographers often approach their technique as an optical system, that is, as a technology for visualizing molecules. To do this they compare and contrast crystallographic techniques to the techniques

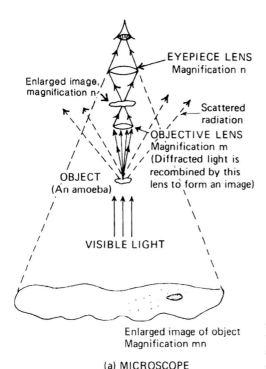

EYEPIECE LENS
Magnification n

Enlarged image,
magnification n

Scattered
radiation

OBJECTIVE LENS
Magnification m
(Diffracted light is
recombined by this
lens to form an image)

OBJECT
(An amoeba)

VISIBLE LIGHT

Enlarged image of object
Magnification mn

(a) MICROSCOPE

1.2. Microscopy. From Glusker and Trueblood, *Crystal Structure Analysis* (1985), figure 1.1 a. By permission of Oxford University Press.

of light microscopy to demonstrate the specificities and challenges of X-ray diffraction methods.[39] Diane adapted this canonical analogy in her introductory lecture on the first day of class. She presented a hand-drawn schematic on an overhead that distilled its insights from a diagram developed by crystallographers Jenny Glusker and Kenneth Trueblood (see figures 1.2 and 1.3) in their introductory textbook on crystallographic techniques.

Take a close look at the original figures generated by Glusker and Trueblood. The first image is a diagram of light microscopy. It shows how the objective lens of a microscope is used to recombine the visible light that is refracted from an illuminated amoeba. It conveys how the entire apparatus is organized to magnify the object and produce an enlarged image for the observer.[40] The second diagram offers a schematic overview of the steps involved in protein crystallography. Rather than a single apparatus, we see a complex configuration of inputs (X-rays), outputs (molecular models), practitioners, and technologies coming together to do the work of making models of protein structure.

Why not just apply a high-powered microscope to the task of molecular

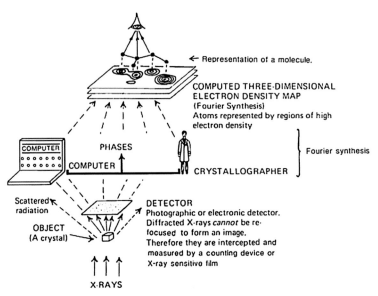

← Representation of a molecule.

COMPUTED THREE-DIMENSIONAL
ELECTRON DENSITY MAP
(Fourier Synthesis)
Atoms represented by regions of high
electron density

COMPUTER

PHASES

COMPUTER

CRYSTALLOGRAPHER

Fourier synthesis

Scattered radiation

OBJECT
(A crystal)

DETECTOR
Photographic or electronic detector.
Diffracted X-rays *cannot* be re-
focused to form an image.
Therefore they are intercepted and
measured by a counting device or
X-ray sensitive film

X-RAYS

1.3. Protein crystallography. From Glusker and Trueblood, *Crystal Structure Analysis* (1985), figure 1.1 b. By permission of Oxford University Press.

visualization? The juxtaposition of these diagrams demonstrates some crucial distinctions between these two visualization techniques. The key differences are the size of the objects being visualized and the wavelengths of light that each technique uses. While cells and their internal structures are large enough to interact with and refract the long wavelengths of light that compose the visible spectrum, protein molecules are too small. Visible light ranges from 380 to 740 nanometers, and protein molecules are so tiny that long waves of visible light pass right by them without registering their presence. With wavelengths on the order of 0.1 to 10 nanometers X-rays are ideal for diffraction because they are on the same order of magnitude as the spaces between molecules arrayed in crystalline form.

The problem that these diagrams emphasize is that diffracted X-rays cannot be focused with the aid of a simple instrument like a microscope lens. In place of the objective lens of the microscope, the authors have diagrammed an intricate system composed of a crystallographer and an assemblage of computer technologies and mathematical functions. Crystallographers must engage computer power to run a series of algorithms in order to visualize molecular structure. Their aim is to transform their diffraction data into three-dimensional electron density maps that can indicate the approximate positions of atoms within the molecule. These maps are read almost like three-dimensional topographical maps, where "peaks" of electron density mark the approximate positions of atoms.

Bragg and Hodgkin were right to qualify what they meant by "seeing" molecular structures. It is not actually possible to "see" molecules using X-ray diffraction techniques.[41] Molecular vision is extraordinarily indirect. As Donna Haraway suggests in the epigraph above, diffraction does not produce mirror-image reflections of phenomena. The interference patterns generated when X-rays are diffracted through a crystal are not immediately legible. A diffraction pattern does not map the molecule, rather it maps the effects of X-rays scattering as they pass through the complex configuration of amino acids that constitute each molecule among the thousands arrayed in a crystal. Once resolved, a diffraction pattern can provide at best a map of probabilities: the calculated electron density map with its peaks and valleys can only orient the modeler to sites where an electron is likely to be found, and it is up to the modeler to figure out to which of the potentially thousands of amino acids in the molecule this electron belongs. In a sense, crystallographers practice a kind of shadow vision: they work in the near-ghostly, dancing shadows of diverging and converging waves, never fully grasping an image of their coveted objects directly.

It is the "human-computer lens" diagrammed by Glusker and Trueblood that is critical in the work of interpreting these cryptic diffraction patterns. Until very recently, few steps of the model-building process had been automated. Though many computer scientists and mathematicians are trying to devise algorithms that can take on more of the labor of protein structure determination, they still cannot fully solve (that is, decipher, render) a protein structure from the amino acid sequence, or perfectly fit a protein model to its electron density map. The crystallographer remains an essential component of this visualization technology. In addition to their skill working with data and algorithms, their well-trained intuitions form an integral part of the technological "lens" that draws proteins into view.

Glusker and Trueblood's paired diagrams thus have further implications: if, in their diagrams, microscopists appear to rely on the "mechanical objectivity" of their technical apparatus to produce true-to-life, mirror image reflections of cells, crystallographers explicitly enact an entwined human-technological agency to navigate the shadowy realm of diffractive optics.[42] As the active concept "rendering" reminds us, modelers must get fully involved in the effort to decipher diffraction patterns and resolve visible models of protein structure.

The subtle vision of X-ray crystallography is challenging both to teach and to learn. Each stage in the long series of steps demands intensive energy, attention, and skill, and each poses significant challenges for novice students attempting to model their first molecules. It is to the laboratory and the very first steps involved in protein purification, modification, and crystallization that this account now turns.

CRYSTALLOGRAPHIC TECHNIQUES | PURIFICATION

With whose blood were my eyes crafted?

—Donna Haraway, "Situated Knowledges"

Living organisms make proteins. One of the first tasks involved in protein crystallography is purifying desired protein molecules from the mass of other substances that constitute living bodies. Recall that "to render" also means "to separate," or "extract," as in rendering of fat from animal carcasses. This is a potent reminder that proteins must be extracted from once-living bodies.[43] A practitioner must be able to collect large amounts of the protein of interest. In the early days of protein crystallography it was necessary to purify proteins directly from animal and plant sources. During the Second World War, British crystallographers Max Perutz and John Kendrew were working on the structures of myoglobin (a muscle protein) and hemoglobin (a blood protein) at the Laboratory of Molecular Biology (LMB) at Cambridge University.[44] Their experiments relied on abundant sources of fresh animal tissues. At the time, slaughterhouses provided some of the cheapest and most abundant supplies. Perutz harvested hemoglobin from several sources, including adult and fetal sheep, and Kendrew harvested myoglobin from horse heart. When horse heart myoglobin failed to produce crystals large enough for diffraction experiments, Kendrew turned to sperm whale tissue and apparently received a large cut of whale meat from Peru.[45]

Mike Fuller, now a partly retired member of the LMB, began working with Perutz and Kendrew when he was just fifteen years old. I met Mike during a visit to the LMB archives, and he recounted his early years in the lab. Apparently, on his first day on the job the first question the lab directors asked was: "Have you got a bike?" Once that was confirmed he was promptly sent off to the slaughterhouse to get blood from a freshly killed horse. "They actually killed the horse in front of you. And you had to stand there and reach out with the flask to catch the blood." He recounted the story with amazement, as if he were still horrified at the task that he was required to perform. The work was stomach churning. It was also quite messy. Others recounted stories of botched experiments that saw blenders bursting and spewing liquefied liver tissue all over the lab. The violence of these rendering practices ingathered affect with intensity. Indeed, Mike's experiences shed light on that peculiar mixture of love and violence that has long conditioned experimental work in the life sciences.[46]

Crystallographers continued harvesting protein from fresh animal sources through the 1970s. Sir Alan Fersht, a Cambridge University protein chemist who was knighted in 2003 for his work in protein engineering, was completing his doctorate at Cambridge in 1968. He saw firsthand how studies that relied on fresh tissues slowed "momentum" in the field.[47] The challenges of working with animal sources motivated him to develop new approaches, and he applied emerging recombinant DNA techniques to the problem of protein crystallography. In a retrospective on the field he notes how "refinements" in DNA sequencing and genetic engineering "profoundly altered the course of biology and protein science."[48] Once the DNA sequence of a protein was determined, that specific nucleotide sequence could be spliced into the genome of a viable host microbe, such as E. coli bacteria. Large quantities of these peptides could then be synthesized in the lab by harnessing the growth potential of bacteria. One of the most important features of this suite of tools was that it provided an abundant "source of experimentally tractable proteins," including "previously rare and unknown proteins" and genetically modified proteins that were designed for easy handling.[49] Today, the pungent smells of dead animal tissues have been replaced by nearly imperceptible sequences of DNA suspended in tiny vials of colorless liquid. Labs are now saturated with the sour aroma of E. coli bacteria, sloshing around in flasks of culture media while they are kept warm on mechanical shakers. These acrid smells fill the refrigerators where plastic petri dishes coated with agar and plated with bacteria sit stacked, labeled, and stored, ready for future use.

| MODIFICATION

"Experimentally tractable" proteins are hard to come by. Zeynep is a postdoctoral researcher working in a protein crystallography lab at a private research institute not far from Diane's laboratory. Her doctoral training was in molecular genetics, and so she was still getting oriented to techniques in protein crystallography. At the time of our interview, she was struggling with the work of engineering experimentally tractable proteins. She explained to me the challenges that she faced, given that the molecule she cares about is very large and composed of several different subunits of smaller peptides. It is often too difficult to crystallize a large protein made up of multiple peptides. To circumvent this problem, she has had to consider experimenting with molecules from different species of plant, animal, or microbe. Realizing that this strategy is not viable for her project, she has had to go to great lengths to modify the protein to make easier to work with. She is using genetic engineering techniques to splice and rework the original DNA sequence to see if she can craft peptide fragments that behave like the original, or what is often called "native" protein. The modifications she makes to this protein can't be too severe. She must test these engineered peptides to ensure that they retain the original protein's known biochemical functions. Biochemical assays can be used to ensure that the molecule still binds to known cofactors or participates in the same reaction pathways. Yet, there is active and ongoing debate as to whether such modifications produce sound data that reflect the characteristics and behavior of the molecule of interest. She explains that when crystallographers rely on data from modified proteins they make a major leap in assuming the structures that these peptides acquire in vitro and in crystalline lattices are the same as those that take their form inside living cells.

Zeynep expressed some concern as to whether the structures that her proteins acquire in their crystalline forms are adequate proxies for their in vivo molecular forms. The question that plagued her at this stage of her experiments was whether she could trust that the modified peptides she had engineered were representative of her target protein. She explained the reason for her hesitation by taking us into the crystal:

> You are looking at a protein that is being forced to sit in a very ordered array next to another protein and millions of them are sitting close to each other to form a crystal. That is not what is happening biologically, right? You always have to keep in mind [the question]: Is this the biologically relevant structure? . . . Is that what is really happening in real life? You have to use other methods, and there are plenty of methods I can use to test these things. . . . And again it is like the whole concept of science; I think you always have to double-check, triple-check, see if different things, different methods, even different approaches, give you the same result. And that's when you believe it.

Zeynep grapples with the imperceptibility of molecules and with the fabricated, made-up dimensions of protein engineering. She cannot see the proteins directly and so must test and retest them over and over again using biochemical assays that help her build up a kind of trust in her methods and data.

This is not blind faith, but a belief secured through the layering of confirmations reached through multiple forms of experimental demonstration.[50] Protein crystallographers are required to go to great lengths to get familiar with the qualities of each experimental peptide to ensure that they can trust that the substances and structures they are working with are "biologically relevant." If caution and doubt are crucial postures for scientists, well-founded belief, such as this trust that Zeynep builds up over the course of her assays, is also essential. Bruno Latour recognizes the role that belief plays in science. For him, "if there is such a thing as belief at all, it is the most complex, sophisticated, critical, subtle, reflective activity there is. But this subtlety can never unfold if one first attempts to break it down into cause-objects, source-subjects, and representations. To take away the ontology of belief, on the pretext that it occurs inside the subject, is to misunderstand objects and human actors alike."[51] Belief, in this sense, is not a subjective affect or attitude that inheres in the scientist. Rather, it is distributed among all elements of an experimental configuration. In this sense, Zeynep learns to move from doubt to trust through what science studies scholar Charis Thompson might call a careful "ontological choreography" with these fabricated and imperceptible objects.[52] Here belief takes shape as an affective entanglement that tethers modelers to their molecules, materials, and machines. Zeynep's situation makes palpable how practitioners-in-training struggle to acquire the ethos and habitus that will help them distinguish when their trust is warranted, and when disbelief and doubt should hold sway.

| CRYSTALLIZATION

LARRY: You know, all students, by the end of this, have had many dreams about crystals.

ROB: Oh, yeah, I see them in my sleep.[53]

Protein crystals are integral elements of the X-ray diffraction apparatus. A protein crystal is both the object crystallographers hope to describe, and the instrument they will need in order to visualize the molecules that make up the crystal.[54] For this reason students can't progress to the model building stage unless they can "grow" crystals that will diffract X-rays well enough that they can gather diffraction data. Graduate students desperately need their proteins to crystallize quickly so that they can solve at least one structure in time to graduate. They have limited funding and limited time to spend in the laboratory.[55] Crystallization is one step in the process that students have very little control over, and as such it can be a particularly challenging stage in the process of gathering data. Their hopes and anticipations are often met by frustration, and they often have to modulate their emotions to help them cope with the daily stress.

Crystallization experiments are often run by placing a drop of protein solution in a drop of crystallization solution on a thin glass slide. The slide is inverted over a well in a plastic multiwell dish, and the edges of the well are sealed so that the hanging drop does not evaporate. These droplets are examined regularly under the microscope to see if any crystals are growing. Not any crystal will do: viable protein crystals must be well ordered to diffract X-rays. The crystals can't be too thin or long, otherwise it will be hard to collect data while the crystals are being rotated in the X-ray beam. There are some proteins that form crystals with almost no intervention. Lysozyme is the canonical example. Found in abundant supply in animal secretions, such as tears, saliva, milk, and mucus, lysozyme protects tissues from pathogens by breaking down bacterial cell walls. Most students learn crystallization techniques by working on this protein because it is so easy to handle. According to Brent, a postdoc in Diane's lab, and a former college football player, it crystallizes "like gangbusters."

But most crystals will grow only under narrowly constrained conditions. Many proteins, particularly those that form fibers or those that are too large, too small, or too disordered, do not form crystals. Protein crystallographers are thus limited to working with proteins that can be coaxed into crystalline form.[56] One result of this limitation is that some large families of biological molecules are underrepresented in the Protein Data Bank. Practitioners must examine a very large number of different crystallization solutions, each with different concentrations of salts and other additives. It is possible that a crystallographer will never find the perfect conditions for a protein to crystallize.

First attempts at crystallization can be intensely frustrating. Brent recounts his experience contending with the time it took to get the right conditions for crystallization: "For months and months and months I got nothing. Nothing. Nothing. Nothing. Nothing. Nothing. I was getting all these other things. I was spending all this time to optimize this crap that turned out to be salt. I figured out I was using a buffer [chemical solution] that likes to crystallize. I had to change to a new buffer. At the time I didn't know. Now it makes sense. And bam! Real crystals after that."

Subjected to continuous cycles of hope and failure over such an extended duration, Brent's resolve was put to the test. Once he figured out the glitch and switched to a new buffer, the crystals started to grow. With a huge sigh, he re-enacted for me the relief he had experienced when it finally worked. Anne, a third-year graduate student in Diane's lab, recounted her experience with the indeterminate temporalities of the crystallization process. Suspense mounted as she waited for crystals to form. Sometimes it was hard for her to tear herself away from the experiment, her anticipation was so intense: "The crystals we were working on would appear overnight. So what I would do is the night before I'd set up a tray and go home and try not to think about them. Then the next morning I'd like run in and see if they were there [laughing]. Brent once pointed out to me that the hardest part of this deal is waiting. Because you really just have to wait for things to work."

Anne had to learn how to cultivate patience and perseverance in the face of the indeterminacies of crystallization. Marcel Mauss might call this "test of stoicism" an "initiation" that teaches students "composure, resistance, seriousness, [and] presence of mind."[57] She tried to control her anticipation, and keep her emotions in check. Yet, waking each morning in the hopes of finding crystals, it was clear that she was not the one running the experiment; it was running her. She was caught, subjected to the untimely whims of her molecules.

When students encounter proteins that are difficult to crystallize they will try anything and everything. Each protein seems to require a different kind of attention. Some crystallographers have to get creative. The documentary *Naturally Obsessed* features a comedic scene where a British postdoc in Larry Shapiro's lab reads the methods section of a recent publication in disbelief: "There doesn't seem to be much logic in it. These people couldn't get their protein to crystallize, so they added pickle juice to the drops. [Laughter.] From the Sweet and Snappy Vlasic brand. Does that make sense?"

Rob, an American, familiar with the brand replies, "Yeah, yeah. Vlasic. They're great." The postdoc continues, "And they found that without 1 percent pickle juice, they didn't get any crystals. So you just have to try everything!" Once they found that "magic potion" they could coax their protein to form crystals and generate reproducible results. But crystallization is also precarious. What works one day may not work the next. Anne explained: "My friends are probably sick of hearing about the fact that there are a million variables. Like if the temperature in the room slightly changes, or if there's a little bit of vibration, or if your solution has evaporated a tiny bit. You know all of a sudden your protein crystals are not going to work. You can just tear your hair out trying to figure out what the problem is."

Students must investigate where their protocol is going wrong. This requires placing their every action under minute scrutiny. Anne describes how she figured out the optimal crystallization conditions for her protein: "The variable I found for my protein—which was very important—was the timing between adding the precipitant [chemical solution] and sealing [the droplet in the well]. That was a variable in my crystallization! When I tell people that in lab meeting they say, that's not true. But it really is true, and I have the data to prove it."

For Anne, it wasn't just what she added to the crystallization solution; it was the rhythm at which she worked that determined whether she would grow crystals. The lag between adding the precipitant and sealing the well was a crucial determinant. She could have learned this only by bringing scrupulous attention to her technique. She had to ensure her performance was "fastidious." This "fetish" for "precision" gave a "ritualistic tinge" to what could be called her "evolving theory of practical action."[58]

Confronting the challenges of crystallization, practitioners often develop superstitions about crystallization conditions. They find themselves invoking magical rites and cultivating complex rituals that they then feel obliged to follow religiously. Experienced crystallographers regularly joke that protein crystallization requires "voodoo magic." Some insist on playing techno music while they mix their biochemical media. Others have found that proteins will crystallize only if they are wearing their "special sweater." Some even talk and sing to their crystals. For Brent, "there's nothing you can say: 'Well, if I do all these things it's going to work.' You can do everything imaginable, and on one day it will crystallize under this condition. And another day it won't. Why? Maybe an eyelash fell in there. Or dust from the ceiling. You know. Happens all the time. There's this one guy, he said he didn't get crystals anymore after he shaved his beard. So he thought foodstuff and particles were falling into his trays. But once he shaved he never got them anymore." This story about the crystallographer shaving his beard resonates with an apocryphal story one senior crystallographer told me about "the old German professors" who would come over to check on their students' experiments to help them figure out why their protein wasn't crystallizing. They would lean over their students' crystallization trays and ponder the situation while scratching their beards. Miraculously, after that a student's crystals would start to grow.

Protein crystals are not only hard to grow; they are also quite unstable and can disintegrate easily. It takes great care to move crystals from the crystallization tray to the X-ray diffraction machine. In *Naturally Obsessed*, Larry admits that when he is mounting crystals he always listens to the 2002 hit "Yoshimi Battles the Pink Robots" by the psychedelic U.S. space rock band the Flaming Lips. This, he assures the filmmakers, is the only way to guarantee diffraction. It's as if he uses this music to generate a "vibratory milieu" that can support the precarious energetic state of his crystals.[59] Larry's seemingly irrational reliance on this song, and the other rituals recounted above, demonstrate how remarkable forms of "magical thinking" take shape in the mundane spaces of the laboratory. It seems in these contexts that magic, ritual, and superstition are not a "threat" to science; rather, they are integral to its practice.[60] They form a part of life in the laboratory, where practitioners must contend with often-unruly substances that have captivated both their dream lives and their working lives. Indeed, the affectivity of matter is entangling and gives form to a wide range of affective labors.

EXCITABLE MOLECULES AND LIVING CRYSTALS

Edward, the postdoc in Diane's lab who described proteins as "breathing entities," gave me one of the first clues about the ways that modelers grapple firsthand with the affectivity of matter, that is, its wily, excitable, and indeterminate nature. Yet, it is not just the proteins that are lively; crystallographers treat crystals as living, breathing entities. They tell me that once successfully "seeded" in the proper media, their crystals "grow." Even when they are packed within the ordered array of the crystal lattice, individual protein molecules vibrate energetically. During my fieldwork I followed Diane as she gave guest lectures to students in a range of different courses. Her aim was to teach students in various disciplines about the critical skills they would need in order to evaluate the convoluted arrays of data and statistics that are published alongside crystal structures. During a guest lecture for students in an advanced biology course, Diane told a story that animated this excitability of molecules and crystals.

B-factors are a statistical measure of how much movement a protein molecule has inside of a particular crystal lattice. The more movement there is in a molecule, the more "disordered" the crystallographic data will be, and this makes it much more difficult to build sound models. The value of the B-factors in each region of a data set indicates where there is disorder. Molecules that have "floppy" side chains are highly disordered in that region. An interested reader must consult the B-factors listed alongside crystallographic structures and use them as a guide to determine if there's a good fit between the published model and the data.

Diane explained that cartoons of ducks were commonly used to teach concepts like symmetry and asymmetry in crystallography textbooks. But ducks wouldn't do for Diane. She is a dog person and used a photograph of her own dog, Max, to teach students how to interpret B-factors: "This is my favorite example. Those of you who know me know that I'm not a fan of the duck. I'm a fan of the dog. And so I've decided that crystallographers should use dogs as examples rather than ducks. So anyway, this is my former dog, Max. He passed away recently. He was a great dog." Diane points to a picture of Max on the beach, projected on the screen behind her. It captures vividly the fine detail of the sandy beach and the waves crashing up against the shoreline. Max, however, appears as an indistinct blur of fur in the center of the picture. Diane continues:

> I don't really know if you can make out that this is a dog, cause there's a shadow. But this is a leg; there's another leg. This is the tail. Ah, that's an ear. Here's the nose. And so, this is what would always happen when I'd go out and try and take a picture of my dog. The dog is like sitting nice, and you know smiling, and looking all happy and very calm. And then the camera comes out, and whoop! He's on his back, and then his legs are going like this. [Diane mimes her dog's movements.] So I have a series of pictures, high resolution in the background, you know really good camera, really good developing, focus, everything, fantastic: overall a really good resolution picture. But the thing I was trying to capture . . . um. . . . If this [points to Max] is like the molecule in the active site [of the protein], then it's completely disordered.

Diane flaps her arms around emphatically to demonstrate how molecules are as excitable as her dog. Diane figures the molecule as a dog, and in so doing produces an account of molecular movement that goes so far as to connote, that like her dog, Max, the molecules perform for the camera when you take their picture. Diane transposes the affective charge of Max's delight onto the protein. She continues in her lecture: "And so this is one thing to keep in mind: it's not the overall resolution of the structure but it's the B-factors that you have to pay specific attention to. Because it could be the part of the structure you care about. You know. [If it's like this (points to the photo of Max)] it's not very good information about where those atoms are located."

Crystallographers need to find ways to fix or freeze molecular movement in order to get a clear, unblurred "snapshot" of the structures packed within each unit cell of the protein crystal. The term "snapshot" is not meant to liken X-ray diffraction to image-generating camera technology; rather it connotes the production of a static rendering of a dynamic process. Cryofixation is a widely used technique to dampen molecular vibrations. A crystal is dipped in liquid nitrogen just before it is subjected to X-ray diffraction. X-ray crystallography is thus a kind of freeze-frame technique. But even in a single crystal, moving molecules may get frozen in multiple configurations. The diffraction data combine all the configurations of all the molecules packed in a crystal array as they move and change over the duration of the time that the crystal is mounted on the X-ray beam. The electron density maps produced are thus best estimates that average out the dynamic movements and subtle differences in conformation to find a single dominant structure. Crystallographers tend to select and model just one structure from the averaged data and so produce a portrait of a single, hypothetical molecule frozen in time.[61]

Affects move; they slip and slide between a modeler and her molecule.[62] Over time practitioners learn to sense how their proteins "feel." Different conditions can make proteins "happy," "unhappy," "strained," or "relaxed." Mangled molecules apparently experience "pain" (see chapter 3). Proteins need to be kept "happy" in their buffers and in crystalline form, and it is the crystallographer's obligation to keep it content under all conditions, even when it is placed on the high-intensity X-ray beam for diffraction experiments. Unfortunately it takes a long time to run these experiments. Some crystallographers devise little rituals to ensure the crystal doesn't get damaged on the beam. Brent found that his proteins diffract better if he disrupts the flow of liquid nitrogen that is used to keep the crystal frozen once it has been mounted on the beam. He uses his plastic debit card to deflect the stream of liquid nitrogen and allows his crystal to melt ever so slightly. According to him, this gives the crystal an opportunity to "relax" so it can produce a cleaner diffraction pattern.

Jamie, a second-year grad student in Diane's lab, is quite anxious about transferring her crystals from liquid nitrogen storage to the diffraction beam. And once they are on the beam she worries that her crystals "will die." In one case she had to rotate the crystal in the beam for six hours in order to collect the full data set. She watched with horror as her crystal was "getting killed" partway through the experiment. She reenacted her anxiety in our interview: "Nooo! I still need you, I have three hours left!" Apparently crystals not only grow, but they can also expire:

Well you know you start with this little drop [in the crystallization well] and . . . usually it starts clear . . . it looks like water, then out of it grows . . . it's like babies forming! Wow! Then you're like, "I've got one and it looks good. I'm going to get going!" And you put it in cryoprotectant, to protect it. And you put it in the beam and then you have the cryostream [liquid nitrogen] on it to keep it cool. And it is sad when you take forever and you put them on the beam and they don't diffract at all, or they diffract, but they don't diffract well enough. Or they are basically ruined before you get the data you need. Which is what we mean when we say die. And it is sad, a very, very sad situation.

Jamie is affectively entangled in the life and death of her crystals. She panics when her crystals are getting "fried" in the X-ray beam and mourns their death when they disintegrate. Crystallographers must learn to come to grips with the wily, lively nature of their molecules. Their obligations to expend great effort to take care of their proteins and crystals at every stage of this long, drawn-out process fills the laboratory with a range of energies and emotions. Anxiety, anticipation, failure, elation, defeat, sadness, and hope are some of the affects that contour life in this laboratory. Remarkably, these affects extend to crystallographers' entanglements with their machines.

MACHINE LOVE AND THE TECHNOLOGICAL SUBLIME

It is no surprise that machines are focal and vocal research participants in these laboratories: machines outnumber people, and they demand constant attention and interaction; they are never left to their own devices. Machine love and machine anguish are well documented in accounts of laboratory life. Machines have biographies. They have birthdays. They are celebrated when they work and reviled when they don't.[63] The machines in the laboratories documented here ingather energies and emotions with particular intensity.

Crystallographers were among the first life scientists to make use of computers, initially for alleviating the massive labors they faced with calculation, and later for reducing the physically laborious process of data collection.[64] In each of these applications, computers introduced important changes in the ways modelers did their work. Computers have, however, never completely replaced the modeler. The "human-machine lens," that complex configuration that tethers practitioners and their well-trained intuitions to a suite of computers, machines, techniques, and algorithms, remains the only way to resolve cryptic X-ray diffraction data (see figure 1.3).

Above all other machines, computers are crystallographers' constant companions. At each step of the process modelers rely on computing power: they use online databases to access the DNA and amino acid sequences of proteins of interest; and they upload the coordinates of their completed models to the Protein Data Bank. For data collection and analysis they rely on an extensive array of software programs and algorithms tailored to transform their diffraction data into workable three-dimensional electron density maps; and they use interactive computer graphics interfaces to build atomic resolution models into these maps. Some of this software is cobbled together in-house or shared through open source channels, and propriety programs often have sleekly designed graphics and weighty user manuals. Labs are equipped with dedicated computer rooms, where crystallographers may sit for hours in the dark, physically tethered to their computers through a mouse, keyboard, or other manipulative device. To visualize their maps and models in three dimensions they wear stereoglasses with electronically coordinated flickering lenses that make particular features of their models pop out in 3-D.

Computers are their closest allies and worst enemies. These machines are variously cursed, praised, coaxed, and stroked. Most of the time they are a source of incredible frustration; only occasionally do they arouse elation. Computer interfaces are designed as conduits that facilitate intimate encounters between modelers and their data. But more often than not software "misbehaves" or "acts up," sometimes deleting files, propagating errors, or generating impenetrable data. One of the significant challenges a student faces is learning how to use each new piece of software required for each step in the process. The problem is compounded as new iterations of each program are released regularly. The manuals are often impenetrable, and novices can get stuck on a simple glitch for days. When their professors are busy, hands-on help can be hard to come by, and students will have to ply one of their more experienced colleagues in the lab to help coach them through a snag. Students rely heavily on informal peer mentorship and support from other students to make day-to-day progress. On occasion a computer program will miraculously cooperate and facilitate, rather than impede, the model-building process.

Protein crystallographers must grapple with other kinds of machines as well. In *Naturally Obsessed* the audience gets a good dose of the technological sublime as they follow lab members back and forth to the Brookhaven Synchrotron, the National Synchrotron Light Source in New York state. This synchrotron is a massive, factory-scale assemblage of instruments geared toward high-energy physics experiments.[65] Forty-five minutes into the hour-long documentary the viewer is brought into the synchrotron for the third time. Doctoral student Rob Townley sets up a video camera to capture the day's events and document his progress for the filmmakers. The stakes are high for Rob. He's there to test his latest batch of protein crystals, which he has spent the past four years desperately trying to grow in the lab. Every other trip he's taken to the synchrotron has ended in failure: even his best-looking crystals failed to produce data. This beam at Brookhaven is the one locally available X-ray source that can give him the data he needs to build his model. In the words of his lab mate Kil, "the synchrotron is the be-all, end-all. It's the de-cider. It makes decisions on your life and your career. You put things in front of it, and it decides for you. You know? There are no intermediate steps. It's all or nothing. You've either got a structure, and it's good, or you've got nothing." Political theorist Susan Buck-Morss cites Kant to explain the concept of the sublime: "faced with a threatening and menacing nature—towering cliffs, a fiery volcano, a raging sea—our first impulse, connected (not unreasonably) to self-preservation, is to be afraid. Our senses tell us that, faced with nature's might, 'our ability to resist becomes an insignificant trifle.'"[66] Kil is humbled before the synchrotron as if it were a massive display of "nature's might."[67] He has reason to fear: he has been laboring for years and even his best crystals have yet to diffract.

But not all X-ray diffraction machines inspire such reverence and awe. Diane's lab is lucky enough to have an X-ray beam in-house. Though members of the lab have to travel to a nearby synchrotron to collect high-quality, publishable data, it is useful to have a working beam close at hand so that modelers can check on the viability of their crystals. When active, the machine in Diane's lab hisses and whirs, foaming with liquid nitrogen in its well-protected room. X-ray diffraction machines are finicky devices. This particular machine was frequently on the fritz. Occasionally students had to borrow beam time from another detector on campus. I attended a training session where members of the group were instructed how to use the machine housed in the biology department. It took a number of hours to walk through the idiosyncrasies of that machine, and many of the group members left feeling unsure and confused.

Amy is a fifth-year PhD student in Diane's lab. Her experience with their in-house machine offers a counterpose to the overwhelming awe inspired by the Brookhaven Synchrotron and the anxieties that the biology department's machine induced in her colleagues. Amy's dissertation project had recently come to a dead end. She was trying in vain to make sense of an impossibly mangled data set. And yet, she was buoyant in our interview. She confessed that she had fallen in love with the X-ray diffraction machine in the lab. When she first joined the lab this machine intimidated her. But she quickly realized, with some trepidation, that one day all the students who knew how make the machine work would graduate, and the responsibility for its maintenance and repair would fall on her.

> The one thing I'm really proud of myself for is that my first year I went to [the senior students], and when they were fixing things I made them show me what they were doing. And I stood there, and I was irritating and annoying. In that little sister way, I hope; I hope there was something endearing about it! [Laughter.] They taught me how to change the filament. As soon as they knew I could do it on my own they had me do it most of the time, or a significant amount of the time. When things would break and it wasn't too complicated for them, they would show me what they were doing. And I think just because there was no one else there when [they] graduated, it fell into my hands. I put myself into that position.

She took it on herself to learn how to fix the machine, and in the process became its primary caretaker. She insisted on qualifying her skills: "It's the only thing I ever use tools on, ever!" Even though she doesn't fashion herself as a handy person, "people trust me to be in charge of it." "I love the machine. It is great. I love it when it's up and nothing's broken. . . . It took me a long time to not be afraid of it. And now I feel like it's mine. And when other people know how to fix it better than me, I kind of get jealous."

She offered to show me the logbook where she carefully recorded the machine's daily events. She documented when it was working, when it was broken, and all the repairs it had sustained. The logbook was a lovingly recorded machine biography. It was a record of her involvements with this machine, entanglements that were full of affection and admiration, as much as frustration, impatience, and anxiety. So whereas in one encounter a synchrotron-scale X-ray machine becomes a god-like figure that decides one's fate, in another kind of encounter, a laboratory-scaled instrument becomes a sometimes disruptive, sometimes tame companion, a pet, and even a love object.

CONCLUSION

Protein crystallography is more than a molecular visualization technology. The indirect nature of diffractive optics makes it necessary for modelers to get fully entangled with their instruments and materials as they rend imperceptible substances into visible and palpable forms. Recall that "to render" is also "to submit oneself" and "to give oneself over" to a task. It is here in the laboratory that we find students giving themselves over to the task of crystallizing their protein. It is in this act that they are also submitting themselves to training, a process that not only hones their techniques, but also reconfigures their ethos and habitus. Students learn the affects, attitudes, postures, and comportments that help them cope with the imperceptible and wily nature of their materials and the indeterminate, if mostly prolonged, temporalities of their experiments. These practitioners teach me about the intimate association between rendering techniques and affective labor. Their molecules and machines become "matters of care." As practitioners struggle to purify their proteins, grow their crystals, and keep their computer programs in check and their X-ray diffraction machines running, they demonstrate how a well-functioning "human-machine lens" in protein crystallography relies on intimate forms of care as much as on rational judgment. And as they confront repeating cycles of hope and frustration in the face of the indeterminacies of matter they also open up sanctioned spaces for the practice of ritual, superstition, and the invocation of magical thinking. It is in the midst of their experiments that these students get entangled with the affectivity of matter: their machines and molecules are no longer deterministic objects that obey mechanical laws. They are wily, unpredictable, and full of personality.

It is tempting to treat crystallographic technologies as apparatuses that "capture" and contain living phenomena, taming them as objects that can be swept up into capital-intensive life science industries. This approach however, casts experimental objects as passive captives of the apparatus. Rather, as these practitioners have taught me again and again, their apparatuses can never fully contain their objects; the wily phenomena they try to pin down continually escape and evade attempts to extract clean, clear data. It is through ethnographic attention to forms of affective labor in the laboratory that practitioners' relations to their objects can be seen in a new light. A gestalt shift makes it possible to see that it is not so much the phenomena that are caught in the apparatus. Rather it is the scientists who are caught: they are the ones arduously entraining their bodies, imaginations, and instruments to the rhythms of phenomena they desire to know. Philosopher of science Isabelle Stengers calls this push and pull between scientists and their objects a form of "reciprocal capture." It is in this strange space of reciprocal capture that molecules, crystals, and machines seem to acquire behaviors, intensities, moods, and feelings that mirror the affective states of their modelers.[68] As they lean into their data and get swept up by the energetic movements of these excitable materials, modelers can be seen hitching a rides on and being pulled in by the phenomena they struggle to comprehend. This shift from the idiom "capture" to one that sees researchers "hitching onto" and "getting caught by" brings attention to the ways that modelers must learn to *move with and be moved by* the phenomena that they attempt to draw into view.

Skeptical readers may wonder whether this description of molecular affects is just a glorified form of anthropomorphism. Questions about the propagation of affects, and forms of anthropomorphism, animism, and animation, will be taken up in later chapters of this book. For now it is important to begin to see that modeler, molecule, and machine are all in the making in these entangled relations. Where this chapter detailed affective entanglements in crystallization techniques, the next takes us into the kinesthetic entanglements involved in the work of deciphering crystallographic data and building maps and models.

Recall the comedic scene depicted in the cartoon in the introduction to this book (see figure 1.1). Michaels conducts a *body experiment* as he contorts himself into the shape of a DNA helix. How do protein crystallographers' bodies become resources and repositories of knowledge about molecular structures? To answer this question we must first understand the myriad ways that modelers' bodies are involved in the work of interpreting crystallographic data. Making crystallographic models involves the careful construction, manipulation, and analysis of complex three-dimensional structures made up of thousands of atoms. This process cannot be automated: there is no computer algorithm that can render a model of a complex molecular structure directly from X-ray diffraction data, or directly from the protein's DNA sequence. Rather, model building demands the kinds of skills and practices associated with craft production.

Anthropologist Heather Paxson defines craft as a "hands-on" practice, one that "resists the steady creep of standardization" in the production of "well-functioning objects with utilitarian value."[1] Much like an artisan who sculpts true-to-life renderings out of wood, clay, or bronze,[2] modelers use ready-to-hand materials to craft highly valued objects that can stand in as proxies for molecular structures. In order to build three-dimensional atomic-scale models of protein, crystallographers use a range of materials in their modeling work, including physical materials and computer graphics media. Model building is a reflexive, improvisational, and exploratory practice. Modelers figure out the structure of the molecule in the very act of building models. The best modeling materials are those that are tangible and interactive. Early developers of interactive molecular graphics technologies went to great lengths to develop environments that preserved the interactivity and tangibility modelers had come to depend on when working with physical modeling materials. The introduction of computer graphic technologies for protein modeling certainly shifted the distribution of labor in this craft; however, these technologies did not automate model building. Rather, modelers remain tightly tethered to their computers in the building process. Malcolm McCullough, a practitioner of the digital arts, proposes a definition of craft that invites technological mediation. In *Abstracting Craft: The Practiced Digital Hand*, he suggests that craft hinges on the "invention of technologies that support the subtleties of the hand," and for him, this includes interactive digital media.[3] Digital environments that provide a "continuously workable medium" allowing "continuous interactions" with the objects on the screen can thus retain the craft character of physical model building techniques.[4] No matter what the medium, an artisan's hands can "pick up experience."[5] And yet, the medium does matter. Interactive molecular graphics have changed what it means for a virtual object to be tangible.

It is the dexterity of the artisan's hands that is one of the most remarkable features of craft production. Yet it is possible to see that craft practices engage an artisan's entire sensorium. Crystallographers' demand for tangible media reveals their reliance on forms of synesthesia, including multisensory modalities through which vision folds over into touch and movement.[6] Synesthesia involves the "cross-registering of sensory experience," as when one sense folds over into another.[7] Modelers working with physical or digital media combine their keen visual and spatial sensibilities with their sense of tactility ("the sensation of pressure"), drawing these together with forms of proprioception ("the perception of the position, state and movement of the body and limbs in space"), and kinesthesia (a visceral sense "originating in muscles, tendons and joints" that allows for "the sensation of movement of body and limbs"). Haptics, referring to the "sense of touch in all its forms," names the synesthetic modality that ingathers forms of tactility, proprioception, and kinesthesia.[8] Art historian Jennifer Fisher suggests that the "haptic sense functions by contiguity, contact and resonance"; it "renders the surfaces of the body porous, being perceived at once inside, on the skin's surface, and in external space," and so facilitates "the perception of weight, pressure, balance, temperature, vibration and presence."[9]

As a craft, molecular model building is an inquiry that is simultaneously an enactment of a form of knowing. Art historian Pamela Smith asserts that there is an "epistemological status" to such "craft operations," and she sees an artisan's technique as the enactment of a "bodily form of cognition."[10] McCullough concurs, finding that "skill is sentient: it involves cognitive cues and affective intent."[11] Paxson suggests that craftwork is reflexive, and that its practitioners move "between what is *sensed* (apprehended through sensory input and subjective evaluation) and what is *being sensed* (the empirical conditions and materials that are manipulated by 'tweaking')."[12] Craft practices demand multisensorial forms of trained judgment that she calls "synesthetic reason." In this sense, protein modeling is akin to artisanal production: it is "a reflexive, anticipatory practice guided by synesthetic evaluation" at every stage of the model-building process.[13]

The protein modelers I work with call this a form of "manual thinking." It is through the slow, reiterative, interactive work of building maps and models that researchers come to grips with—and so make sense of—molecular forms and functions. Resolving X-ray diffraction patterns into electron density maps and molecular models requires that modelers exercise their synesthetic reason in an open-ended, improvisational, and intuitive mode. To do this, they rely not only on their eyes and hands, but also on their memory and imagination. In this sense modelers must cultivate both a kind of haptic vision,[14] and what anthropologist Joseph Dumit and I call a kind of "haptic creativity."[15] Michaels's body experiment enacting the twist of a DNA helix is one example of a form of haptic creativity that is integral to the work of experimental reasoning with molecular structures. He enlists not only his hands, but also his arms, head, torso, and even legs in a kind of synesthetic bricolage in order to figure out the contours of a molecule.[16]

Over time, and through the slow and iterative process of trial-and-error, model building with tangible materials eventually reconfigures practitioners' *habitus*, articulating their sensorium and honing their intuitions. Drawing on Bourdieu's definition, Paxson asserts that "*habitus* names a reflexive feel for strategic action under contingent circumstance."[17] This chapter and the following one explore the fine details of how students acquire a habitus tuned to model building. What becomes clear is that their labor is not only geared to the task of fashioning scientific facts; it also serves as a way to train modelers' kinesthetic and affective dexterities. Indeed, it is only in the practice of building their first crystal structures that a novice comes to learn the intricate specificities of protein biochemistry.

The history of protein crystallography is rich with accounts of model building, and yet limitations in the historical record make it difficult to access the affective and kinesthetic dimensions of this practice. The ethnographic encounters documented here invite a reexamination of this history and raise new questions about entanglements among modelers, their models, and machines.

This amplification of otherwise tacit accounts of haptic creativity in protein crystallography serves to pose new questions about modes of embodiment and the senses in scientific practice.

Structural biologists have built and used three-dimensional crystallographic models of protein structures since the late 1950s.[18] Since then, a diverse array of materials has been used to amplify molecular phenomena to human scale. Historian of science Eric Francoeur has documented the use of molecular models in chemistry and biochemistry, detailing the materials used in their construction and how particular conventions and aesthetics have been standardized and disseminated.[19] The media forms used to model molecules have changed significantly over the years, and yet these materials are consistently selected for their tangibility and manipulability. According to Francoeur a special feature of three-dimensional molecular models is that that they "embody, rather than imply, the spatial relationship of the molecule's components":[20] "Like many other types of object handled by scientists in the field or the laboratory, they can be touched, measured, tested, dissected or assembled, and tinkered with in many different fashions. In other words, they act as a material analogy."[21]

In this sense, a molecular model acts an analogue of a protein, and the materials used to construct it act as a "material analogy." In this sense each modeling material *embodies* molecular configurations in distinct ways. Each invites particular kinds of interactions and modes of attention. At the same time each model is a distinct *embodiment* of a modeler's knowledge about a specific atomic configuration or structural relationship.

As Francoeur aptly notes, the "working out" and "sorting out" of structures with physical models demand a kind of "thinking with the hands" that has long been an integral part of the work of chemists and biochemists.[22] This practice of manual thinking is well exemplified by an oft-told story in the history of structural biology. American biochemist Linus Pauling is remembered for his exceptional skills modeling proteins in three dimensions: his "discovery" of the structure of the α-helix in Oxford in 1948 has become the stuff of legend.[23] The α-helix is one of the most common secondary structures in protein molecules, formed spontaneously when amino acids along a peptide chain make delicate hydrogen bonds with one another (see plate 1 and plate 2 for examples of α-helices). Max Perutz, who had been in a race with Pauling to figure out the same structure, wrote an obituary for Pauling after he died. In it he fondly describes how Pauling worked out the structure of the α-helix while lying in bed recovering from a bad flu: He "amus[ed] himself by building a paper chain of planar peptides" until he "found a satisfactory structure by folding them into a helix."[24] It was by touching, measuring, testing, dissecting, and assembling this paper chain that he was able to figure out the spatial organization of atoms faster than Perutz. Solving the structure hinged on his haptic creativity with ready-to-hand materials.

TRAVELING MODELS

A three-dimensional model can be thought of as a record of the intimate knowledge its modeler gained during its production. Yet models have lives that extend beyond the immediate grasp of their makers.[25] Models compel interaction and are readily picked up by the "curious hands" of many other users and engaged to different ends.[26] Models can become instructors, teaching new users how to think about a particular protein structure in novel ways. A protein model can itself become a site of inquiry and experiment. Models and modeling practices are mobile: they move from research into teaching contexts and back again as they are built, used, remodeled, and repurposed.[27]

In 2004 I conducted ethnographic research in an undergraduate class that examined protein structures and processes involved in protein folding. The course was taught by Jim, a protein biochemist, and his colleague and collaborator Geoff, a mechanical engineer who also works on protein structures.[28] During a lecture on the major folds or secondary structures that give proteins their shape, Jim demonstrated the structure of the α-helix with a well-worn, colorful ball-and-stick model (see plate 2 for an example of such a model). He claimed that the relic in his hands was brought to this campus some fifty years ago from the Laboratory of Molecular Biology in Cambridge, England. The model Jim held up in front of his class re-membered the molecular structure that Linus Pauling rendered in 1948. Indeed, this was one of many iterations of the α-helix that had been produced to help researchers grasp the fine details of this special fold.

To elaborate on this point, Jim recounted a story about his experience as a postdoctoral fellow at Cambridge. There he worked with British protein biochemists Cyrus Chothia and Arthur Lesk. At that time they were investigating Pauling's proposed structure in order to learn how α-helices might pack together to form larger tertiary structures within proteins. In this case, Pauling's original model of the α-helix was transformed. Hans-Jörg Rheinberger makes a distinction between an "epistemic thing," the phenomenon that the experimenter aims to describe, and a "technical object," something that is known well enough that it can serve as an instrument or component of an apparatus in an experimental system. Before the experiment, an epistemic thing remains elusive and unknown; it is not quite yet within the grasp of the scientist and it has not yet solidified into an object. A technical object may at one time have been an epistemic thing, but once elucidated it could then be put to the task of determining other epistemic things. In this sense, the α-helix was no longer an unknown entity that called out for inquiry; it had become a "technical object" in an experimental system that could pose new kinds of questions.[29] Rather than simply inquiring into the nature of this molecular structure, Chothia and Lesk organized their inquiry into the ways this structure related to other structures. Made over into an experimental tool, the α-helix became what Rheinberger might call a "vehicle for materializing questions" and generating new knowledge.[30]

Models of the α-helix became instructive objects that taught their users new ways of seeing. According to Jim, they spent "years and years and years just *looking* at this structure." This required rebuilding physical models of the α-helix and trying to figure out how multiple helices might pack together to form larger assemblages. He reminisced:

One thing was very clear in that group: some people were just able to sit and look at the structures. But most people could not do that. They had to get up and get a cup of coffee and do an experiment. Some people could just look at the structures. And finally they saw things that nobody else saw. Because that discipline of sitting and looking is something that is very hard. And it is something that has been lost. I worked with [a protein crystallographer] who won the Nobel Prize for a three-dimensional structure. He used to sit there and say to us, he'd say, "You Americans you can't sit still long enough! You go off and do an experiment. . . . You don't *look*."

Back in "America," Jim used a ball-and-stick model to teach his students how to see. In so doing, he offered insight into the corporeal discipline involved in working with and learning from molecular models. When Jim presented the α-helix to the class he made it clear that "sitting and looking" is anything but passive activity. You can't just gaze lazily at the model: "If you just *look* at it you don't see anything." Playing on the double meaning of the verb "to grasp," he told the class: "Now . . . this is not easy to grasp, and that's why it's so important to *grasp* these structures."[31] He picked up the model, which was approximately two feet tall and a foot and a half in diameter, and rotated it around in his hands, assuring the class, "soon you will start to see." He demonstrated that "seeing" requires active handling. With his hands and eyes he showed the class how to "walk through" the model amino acid by amino acid. Examining it atom by atom, he used his hands to feel around the grooves and ridges formed by the side chains as they spiraled up the helix. For Jim, "sitting and looking" involved his whole body, and so he showed the students what they would have to do with their bodies in order to get a handle on protein models. From Pauling's sickbed, to experiments on higher-order protein structure, to lessons on manual thinking and haptic visuality in an undergraduate classroom, this model of the α-helix has traveled well.

MODELING MEDIA

Different materials afford distinct kinds of interactions among modelers and models. In 2008 I interviewed Nobel Prize laureate Sir Aaron Klug at the Laboratory for Molecular Biology in Cambridge. He told me that modeling materials "are anything that comes to hand." At eighty-two he was still active in the laboratory and was eager to share the vast collection of models he had built and used over many decades. He spent many years working on the molecular structures of viruses, proteins, and ribonucleic acids. Some of his many working models are stored in an adjacent lab. I had the opportunity to open up the storage cabinets and hold the models in my hands (see plates 3–6). They were quite old and fragile, and some were barely hanging together. They were crafted from a remarkable range of materials including folded paper, various kinds of pushpins, bits of wood, plaster, glue, metal screws, thermosetting plastic, rubber tubing, electrical wire, metal rods, and ping pong balls. Some of the paper models were marked up with pencil or pen, and some had thick lines drawn with black markers indicating all sorts of ways that these models were actively used in reasoning through structural relationships. One cabinet in Klug's laboratory housed a set of more recent models. These intricate and colorful models were made from the adjoinable plastic parts of standardized modeling kits. Like the Tinkertoy models, so familiar from high school chemistry laboratories, these models afforded more intricate forms of manipulation and analysis at the atomic scale.[32] When I remarked on the variety of models Klug had built, he explained: "you use whatever you need. A model is to illustrate certain points. You don't just make one model, you make several models." And later in the interview he elaborated: "you make different physical models to illustrate different *aspects* of the problem and how you solved it. No one model conveys everything." Each model makes a different kind of argument about protein structures and interactions. They are what historian of science Sophia Roosth, in the context of her study of multimedia fabrications in the life sciences, has called "materialized theories."[33]

Sometimes modeling materials are hard to come by. In addition to collecting blood and tissues from freshly killed animals, Mike Fuller, the long-term member of the Cambridge protein modeling teams whose experience was documented in chapter 1, was charged with sourcing modeling materials. He recounted how he rode his bicycle around Cambridge desperately looking for "white card" (a thin cardboard) and pipe cleaners to bring back to the lab. These materials were apparently scarce at the time because of wartime rationing. A couple of years later he was called on by Kendrew to buy up all the "mechano-clips" from the toy stores around town to help build early models of myoglobin.

Crafting atomic resolution models in three dimensions requires great patience. This work also reveals that what in the end gets called a process of "discovery," actually appears in practice as a wayward form of improvisation. In his account of the "discovery" of the helical structure of DNA in the early 1950s, Watson reminisced that his and Crick's "first minutes with the models" were "not joyous."[34] As he and Crick got closer to determining the structure they sometimes spent whole afternoons "cutting accurate representations of the bases out of stiff cardboard" to produce models of nucleotide pairs that could be shuffled in and out of different pairing possibilities.[35] This work was frustrating: "Even though only about fifteen atoms were involved, they kept falling out of the awkward pincers set up to hold them the correct distance from one another."[36] Indeed, they had to keep "fiddling" with the models to get them to hold together.[37] This improvisational practice eventually enabled them to figure out how DNA's nucleic acids adenine and thymine, and guanine and cytosine, could pair to form the double helix.[38] A time-consuming process involving trial-and-error, model building with physical materials requires intensive labor, physical engagement, and exploratory interaction with often finicky materials.

Some materials even record the handiwork of their makers. In 1957, John Kendrew's laboratory in Cambridge, UK, produced the first model of a protein molecule. The protein was myoglobin, the molecule that carries oxygen in muscle tissue. This low-resolution model traced the structure of α-helices as they wound through the molecule. The model, molded out of thick tubes of dark Plasticine and supported on wooden pegs, was nicknamed "the sausage model." Figure 2.1 is a frame shot from an in-house movie of the making of

2.1. Sculpting the "sausage model" of myoglobin. A screenshot of the making of the first model of a protein molecule from a movie produced in Kendrew's laboratory at the LMB. Used with permission from the MRC Laboratory of Molecular Biology.

the sausage model. Here it is clear that the model-builder is not Kendrew but one of the many women employed in the laboratory.[39] Remarkably, the model itself offered a record of the modeler's interactions with the modeling material; that is, the pliable Plasticine medium re-membered the movements and gestures of her handiwork.

This first model of myoglobin offered a shocking and "visceral" view into the molecular realm. Those who encountered this early model remarked that it looked like "abdominal viscera" and had a "rather repulsive" appearance.[40] Many expressed surprise at the "unexpected twists the protein chain was performing."[41] Biophysicists had assumed that proteins were symmetrical chemical structures. After all, they, like other simple chemical substances, could be coaxed to form "beautiful" crystals. The broader scientific community was thus not prepared for the surprise when photographs of a more carefully crafted version of the sausage model hit the press in 1958 (figures 2.2 and 2.3).[42] Figure 2.4 shows Kendrew peering over a massively amplified version of the more refined model crafted for audiences at a public lecture. It seems, however, that the crude low-resolution model offended biophysicsts' molecular aesthetic. As Max Perutz recounted in *Scientific American*: "It was a triumph, and yet it brought a tinge of disappointment. Could the search for ultimate truth really have revealed so hideous and visceral-looking an object? Was the nugget of gold a lump of lead? Fortunately, like many other things in nature, myoglobin gains in beauty the closer you look at it. As Kendrew and his colleagues increased the resolution of the X-ray analysis in the years that followed, some of the intrinsic reasons for the molecule's strange shape began

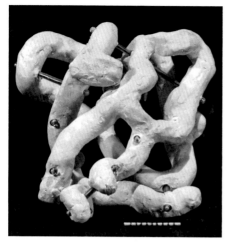
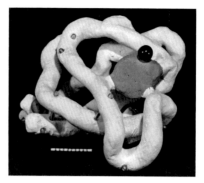

2.2 and 2.3. Photographs of John Kendrew's models that circulated in the press. Used with permission from the MRC Laboratory of Molecular Biology.

to reveal themselves. The shape was found to be not a freak but a fundamental pattern of nature."[43]

Max Perutz was working alongside Kendrew in Cambridge trying to model hemoglobin, the protein that carries oxygen in blood cells. Hemoglobin is four times larger than myoglobin, and was proving much more difficult to model. Max Perutz also tried Plasticine when attempting to build his earliest, low-resolution models in the early 1960s. But this material proved too unstable for his larger and more complex structure. Instead he cut thermosetting plastic into topographical sections, stacked these on top of one another and baked the model to set it permanently into shape (see plate 7).[44] One of the downsides of thermosetting plastic is that it produced a rather clunky, inflexible structure that made it difficult to analyze the structural and functional properties of the molecule.

Over the years, crystallographers were able to intensify their X-ray beams, modify their techniques, and apply augmented computational power to their data to achieve higher-resolution crystallographic maps and models. Eventually they were able to swap their Plasticine and thermosetting plastic for standardized molecular modeling kits with metallic machined parts. In the process, their models acquired new properties. Proteins no longer looked like grotesque bodies. Rather, they began to acquire the aesthetic allure of mechanical architecture.[45] These highly articulated models had clean lines, precise angles, and movable elements that could be clicked in and out of func-

2.4. John Kendrew contemplating an amplified model of myoglobin. Used with permission from the MRC Laboratory of Molecular Biology.

tional conformations.[46] The mechanical properties of these models could be engaged dynamically and performatively as a means to make arguments about molecular function. Eventually Perutz was able to produce atomic-resolution models of hemoglobin using a standardized ball-and-stick modeling kit.[47] In a video interview conducted toward the end of his life, he demonstrated the mechanism of oxygen binding using an atomic resolution model of the hemoglobin molecule. As he moved the model in and out of its conformations, he showed with delight the effect of oxygenation on the structure of the iron-bound heme group (see figures 2.5–2.8).[48]

INTERACTIVE MOLECULAR GRAPHICS

We decided to develop programs that would make use of a man-computer combination to do a kind of model-building that neither a man nor a computer could accomplish alone. . . . It is still too early to evaluate the usefulness of the man-computer combination in solving real problems of molecular biology. It does seem likely, how-

2.5, 2.6, 2.7, and 2.8. Stills from a video of Max Perutz demonstrating the chemical mechanism of hemoglobin binding oxygen in an atomic-resolution model. Video produced by the Vega Science Trust.

ever, that only with this combination can the investigator use his "chemical insight" in an effective way.

—Cyrus Levinthal, "Molecular Model-Building by Computer"

Physical materials often posed challenges for model builders.[49] Even mass-produced modeling kits became cumbersome for rendering large structures, subject as they were to the unfortunate effects of gravity and mechanical stress. This was a lesson learned by one group of molecular modelers based in Manchester, UK, in the early 1960s. They were frustrated when their elaborate molecular model made of balsa wood and elastic bands collapsed in the dry and dusty basement in which they were working. They went so far as to contemplate building it underwater in a swimming pool to cancel the effect of gravity.[50]

C. David Barry was among the members of this group in Manchester. Following this failure he joined MIT biologist Cyrus Levinthal at Project MAC to help develop the first interactive computer graphics workstation for visualizing, manipulating, and predicting protein structures.[51] Between 1963 and 1967, Levinthal, Barry, and others developed an interactive molecular graphics machine they jokingly nicknamed "The Kluge."[52] This interface made use of a "crystal ball" (an early mouse) and light pen to enable control of rotation and the selection of specific coordinates of the structure. This proved to be a significant improvement over the swimming pool option. Interactive graphics satisfied the desire for a practical zero-gravity chamber for molecular modeling.[53]

Robert Langridge, a protein crystallographer and key supporter of Levinthal's work at Project MAC, articulated the benefits of interactive graphics over physical models in a 1981 paper reviewing advances in computer graphic modeling. He writes: "Space filling or wire models are satisfactory up to a certain level of complexity, but purely mechanical problems cause serious difficulties since the model on the bench and the list of [atomic] coordinates in the computer are not necessarily closely related (*especially after the model is degraded by many curious hands*). Particularly difficult is the restoration of a structure after simple modifications. With computer graphics, the display and the data are directly related, storage of prior configurations is simple, and pieces do not fall off."[54]

Apparently, the very pliability of physical models was both their greatest strength and their greatest weakness. Tangibility was key for the production of models that could allow modelers to "grasp" the structure and dynamics of the molecule. It offered a means for researchers to think with their hands while they incorporated new structural knowledge. However, once available to "curious hands," these toy-like structures tempted continuous reworking and tweaking, eventually leading to conformational distortion. Researchers were motivated to overcome the challenges faced working with physical models and so struck up collaborations to develop interactive computer graphics interfaces that would allow them to build models on-screen.

It was crucial that the haptic vision and creativity that protein modelers had cultivated and come to rely on when working with physical models were not lost in the transition to virtual media. Accounts of the development of computer hardware and software reveal how practitioners approached the problem of preserving the tangibility and manipulability of models. What becomes clear from these accounts is that an intimate relationship between user and computer had to be engineered into a workstation interactive enough to keep the modeler physically engaged in model building. With interactive computer graphics techniques the crystallographer is intimately coupled to the computer screen through an array of input devices that aim to mimic some aspects of physical model building. Eric Francoeur and Jerome Segal's history of the development of interactive molecular graphics suggests that while this interactive technology offered a medium distinct from the physical models previously used to investigate structures, it also preserved some of the tangibility of these modeling materials.[55]

This tangibility was not, however, immediately obvious to the uninitiated. At a Gordon conference in 1965, Langridge presented the Kluge system to an unenthusiastic audience. As he recalled, one crystallographer "objected that a graphics display would simply not do as a substitute for physical models, since he had to have his hands on something, something physical, so that he could understand it."[56] For Langridge, "standing up at a conference and showing 16mm movies, in the early days, was really not a good substitute for sitting in front of the computer and actually using it. When you first got your hands on that crystal ball at Project MAC and moved the thing around in three dimensions it was thrilling. There was no question."[57]

Early developers of these programs sought to generate the "smooth handling" of graphic models in "real time" on the computer screen. They aimed "to produce an illusion (a hand-eye correlation) strong enough that the operation required to manipulate the model via the computer" could become "instinctive."[58] In this way the molecular graphics "map and model" could be "manipulated almost by hand."[59] For Langridge "smooth rotation" of three-dimensional objects was key for making use of the display seem "natural" to persons used to handling "real" models.[60] He noted that there was at that time "no precise definition of the terms real-time and interactive." The definition he offered was this: "The difference between interactive and noninteractive uses of computer graphics depends on how long you are willing to wait to see a result."[61] "Satisfactory" interactions demanded advancements in the speed of computer processors, and since these improvements were slow in coming, the user's patience was required.[62]

In the hardware systems that emerged later, a whole new array of input devices were developed and used to enhance the human-computer interface in the simulation of a "real-time" interactive modeling experience. In an article called "The Human Interface," M. E. Pique wrote: "Ideas spreading from Xerox PARC and Atari, through the Apple MacIntosh and the Commodore Amiga, will reach molecular graphics during 1986: pop-up windows, pull-down menus, more than one thing going on at a time. During the next 5 years, users and builders will make molecular systems more like video games, with mice and trackballs, some joysticks that are specialized by function, and the working system easier to use and more fun."[63] Switches, knobs, joysticks, tablets—and a range of apparatuses to generate the experience of "3-D vision" through stereoscopic technologies—connected the user to the maps and models they could manipulate on screen.

By the early 1980s, a number of technologies were available to produce 3-D effects. Langridge and his coworkers describe one stereovision system, where "left and right perspective views are presented alternately," such that "when they are viewed through a synchronized shutter, each of the observer's eyes sees only its associated image, and the result is perceived as a stereoscopic image with a strong sense of depth of field."[64] Creating the three-dimensional effects and the "illusion" of depth through stereoscopic techniques, however, assumes that the user has binocular vision. So attentive to the interaction between users' bodies and the graphics hardware, and convinced that "a good ten percent" of the population has difficulty seeing in stereo, a group of researchers at the LMB in Cambridge devised an elaborate optical system to simulate three-dimensional perception for "one-eyed guys."[65] Reworking the physiology of vision for one- or two-eyed researchers, such innovations of stereoscopic techniques attest to their inventors' recognition of users' varied modes of embodiment and abilities.

Remarkably, interactive computer graphics developers achieved more than "an illusion" of connection between modeler and model: the interactive graphics workstation prosthetically extended a physically engaged modeler into a tangible world of graphic molecules. Ensuring that protein researchers experienced the physicality they had come to expect from their molecular modeling work, interactive molecular graphics developers offered a successful alternative to modeling with physical materials. In the process they also produced a new kind of tangibility for virtual objects.

BETWEEN THE PHYSICAL AND THE VIRTUAL

Take another look at the diagram of the "human-computer lens" in chapter 1 (figure 1.3). Note the area of the diagram that labels the "computed three-dimensional electron density map." Producing electron density maps is a crucial step in the work of making X-ray diffraction patterns legible. The crystallographer must use the data from diffraction experiments to calculate an electron density map. Though much of the labor of calculation is relegated to the computer, this remains a craft practice that requires both haptic vision and haptic creativity. This electron density map is a map of probabilities, read almost like a topographical map. Regions of high electron density indicate the likelihood that an electron will be found in this area. The assumption is that these electrons belong to specific atoms from the individual amino acids that

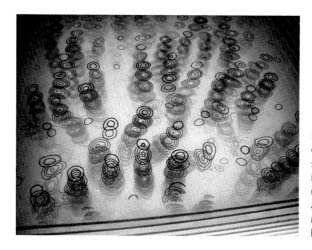

2.9. Hand-drawn electron density maps stacked between sheets of Plexiglas and illuminated from below. A model housed in the archives at the MRC Laboratory of Molecular Biology. Photograph by the author.

make up the polypeptide chain. The "human-computer lens" diagram shows how an electron density map corresponds to the ball-and-stick model of an amino acid hovering above. Deciphering this map from diffraction data and then figuring out which amino acids fit into each region of the map are both challenging steps that entangle modelers kinesthetically and affectively.

In the early days, crystallographers built large-scale three-dimensional electron density maps out of physical materials (figure 2.9). The data they gathered had a tomographic logic, that is, it produced slices of electron density data through incremental depths of the molecule.[66] Modelers would trace each slice of electron density data on transparency paper, and stack these between Plexiglas sheets, building up a physical model of the electron density map layer by layer.[67] When illuminated from below, this translucent stack gave ghostly visual cues as to the location of electrons throughout the molecule. John Kendrew describes his efforts to model myoglobin from such stacks of electron density as a process of "dissection," such that "from the map it was possible to 'dissect out' a single protein molecule."[68]

This method proved "almost unbelievably cumbersome," according to the authors of a review published in 1975, which assessed developments in the field.[69] This is particularly true given crystallographers must construct and compare many different electron density maps in the course of their work on any single model. The review outlined how, in 1968 at Yale University, Frederic M. Richards had come up with an innovation that radically transformed the work of mapping and modeling. Richard's "optical comparator," "Richard's box," or "Fred's Folly" as it came to be known, "revolutionized the interpretation of protein electron density maps."[70] The device enabled them to perform an optical illusion to facilitate model building from physical maps. They used a half-silvered mirror to make it appear as if the model under construction was "embedded" within the three-dimensional electron density map.[71] The model could then be manipulated until its projected image fit within the electron density. The coordinates of the atoms could then be measured and calculated from the model itself. And while this "arduous work" was both "highly tedious and inherently inaccurate," it was a step up from the method Kendrew first employed.[72]

Novel systems of interactive computer graphics overcame the practical limitations of the solid Plexiglas electron density maps and wire models. For example, they enabled the "fitting" of a digital model directly into the electron density map, rather than having to "dissect out" a structure from a solid object.[73] These interfaces, nicknamed "electronic Richard's boxes," made it possible to visualize and manipulate variously sized volumes of electron density on the computer screen. Users could wear stereoglasses to facilitate three-dimensional perception and they would be able to "superimpose" stereo images of atomic models over the maps, rotating and translating them "until an optimum fit of the model to the map [was] achieved."[74] In this way, "fitting model to map" could be "far more convenient and faster than the mechanical operations in the Richard's box."[75] An added benefit, and indeed what the developers saw as the most important feature of this interactive system, was that the spatial coordinates of atoms in the constructed model could be recorded automatically in the computer. This replaced the time-consuming and error-prone work of trying to measure the atomic distances from scale models. Indeed, the grid logic of computer graphics systems could accurately locate each atom in the structure.

By 1977 Demetrius Tsernoglou and his collaborators at Wayne State University and the University of North Carolina could report in the journal *Science* the first successful rendering of a protein entirely through interactive molecular graphics.[76] These technologies, however, did not make physical models obsolete. Diane was a PhD student in Susan Fielding's protein crystallography lab in the 1990s.[77] By the time Diane arrived in her laboratory, interactive molecular graphics programs were readily available for constructing three-dimensional electron density maps on-screen. And yet, Susan insisted that her graduate students first learn how to build physical models of electron density using Plexiglas sheets. Diane explained Susan's rationale for this pedagogical

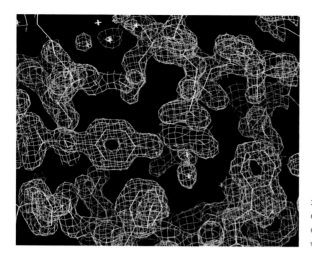

2.10. A virtual electron density map of Fernando's crystallographic data. Used with permission.

exercise: physical models were the best way to get a feel for the map and molecule as a whole because digital graphics could present only small pieces of the map at a time. In this way, novices just beginning to work in virtual media would always have a physical reference to ground them. When students went on to use interactive molecular graphics tools, they would already have cultivated a sense of the multidimensionality of the electron density topographies they were navigating on-screen (see figure 2.10).

While today almost all crystal structures are built on-screen, physical models do retain their pedagogical value. In her office, Diane showed me one small molecular model she uses to make a particular argument about a chemical interaction. Other physical models are scattered throughout the lab and meeting rooms. Fernando, a fifth-year PhD student in Diane's lab, was frustrated by the scarcity of physical models in labs and classrooms. He felt that physical modeling techniques should be maintained more actively alongside molecular graphics to help crystallographers-in-training gain experience manipulating three-dimensional objects. According to his observations, molecular modeling kits were less prevalent in chemistry classrooms and tutorials in his department than when he was in school, and he saw this as the product of an intellectual culture that devalues physical models, treating them as mere Tinkertoys rather than as serious tools. While they were indeed scarce, I did find evidence that physical models were still actively used in lecture courses and teaching laboratories.[78] However, these were often deployed in conjunction with interactive molecular graphics and two-dimensional diagrams.

Diane makes the embodied nature of computer modeling work explicit when describing her experience building crystallographic models on-screen. She invokes the same language and gestures one might use to describe model building with physical materials. In the midst of an interview in her office, she explained: "And physically you are sitting at your computer, often with the stereoglasses on. And you are *physically dragging* pieces of protein structure, like amino acids, and sticking them in. You drag it in and you stick it there. And then with your dials or your mouse, you are adjusting it, moving the pieces to get it to fit. So you are *physically building* with the stereoglasses and the mouse. You are physically building in a model into this electron density."

As Diane described building the model, she stretched her arms out in front of her and reenacted the activity of model building. She used her hands to mime her work at the computer. Her hands clasped and pulsing around invisible objects, she carved out the space of the computer screen, the amino acids and the shape of the electron density map that she rotated in her hands. In the open, gestural space in front of her body she built a model "on-screen," and so expressed how tangible the graphic model is for her.

When crystallographers are caught up in the task of building protein models on-screen they never leave the model hovering in virtual space: it is kept in motion through rapid and restless gestures of the mouse and the quick-paced, sometimes clumsy, tapping out of keyboard commands that pull up new windows and views. In one window, data will be streaming up the screen; in another, the crystallographer holds a skeleton-like interactive rendering of a model. Modelers keep it alive in space and depth, rotating it on screen and zooming in and out, keeping it visible at multiple angles, constantly shifting their visual and haptic relationship to it. This dynamic practice appears to offer a means for modelers to keep the three-dimensionality of the model visible and tangible.

But more than modelers' hands and eyes are in play. Though more subtle than their hand movements, their entire bodies are kinesthetically and affectively entangled in the task of manipulating the model on-screen: with subtle movements initiated at the head and neck, modelers move as they rotate the model, leaning in, pulling away, and even peering around behind virtual obstructions in order to see and feel their way through the intricate structure. Moreover, as they parse the thicket of this dense visual field for another witness, either for a curious, novice onlooker or another expert viewer, they pull the model off the screen through elaborate gestural choreographies that animate the structure's salient features. Expert modelers' performative gestures instruct novices (who haven't already experienced the embodied interactivity of a molecular graphics interface) how to see and feel molecular forms.[79]

In this sense, interactive computer graphics interfaces reconstitute what it means for a virtual object to be tangible. Virtual media theorist Katherine Hayles argues against the prevailing assumption that users are drawn out of their bodies, or disembodied, in interactions with virtual media.[80] Indeed, what I have seen suggests that rather than "dematerializing" the molecule into some body-less virtual reality space, over time the interactive molecular graphics workstation enables a particularly effective kind of handling for molecular models. Virtual models acquire a materiality and tangibility through their manipulation on-screen. While it is true that novices have a hard time experiencing the tangibility of computer graphic models, over the extended duration of laboratory training and with the experience of constant interaction with these objects, their perceptions shift. This reconfiguration of their sensorium is crucial in the process of becoming a protein crystallographer.

LOST IN THE MAP

Most of the time you are in a fog. You know. It's very confusing. It's sort of on the frontier. It's . . . it's difficult. Right? It's complicated. You are stumbling around in the dark.

Rob Townley in *Naturally Obsessed*

Where some of his lab mates lose out on the glory, Rob Townley, the awkward, disheveled, forty-year-old grad student featured in the documentary *Naturally Obsessed*, makes the big win.[81] A former member of the U.S. Navy, an avid mountain climber, and a self-described "loose cannon," Rob is the unlikely hero of the story. He comes up from behind, having struggled for years to produce protein crystals capable of diffracting X-rays. Finally, after bouts of desperation and frustration, he succeeds in engineering a protein and growing crystals that diffract beautifully. Once he finally has his hands on crystallographic data, however, he finds himself faced with another daunting challenge: he must transform this data into a three-dimensional model that defines the coordinates of every atom in the molecule.

Toward the end of the documentary, the filmmakers bring us into a darkened room where we find Rob tethered to a computer, wearing stereoglasses and rotating a three-dimensional map of his protein on the screen. As described in the epigraph above Rob ponders the liminal state he inhabits as he wanders lost through this seemingly endless maze of flickering images on-screen. He is in the early stages of the model-building process. The electron density map he has generated from his data only faintly traces a tentative arrangement of atoms in the molecule. The filmmakers edit in a voice-over, and we hear Larry Shapiro describing what Rob is looking at on the computer screen: "There are 9,073 atoms [in this molecule]. And that's why there's so much data in this, because we are actually finding the three-dimensional position of, in this case over nine thousand atoms. This file, which has ten thousand lines of data in it, is a description of that molecule that he is manipulating."

The model-building process is itself a rite of passage in the process of becoming a protein crystallographer. The common lore in the lab is that even if well versed in the theory of crystallography, modelers remain novices until they have fully built their own structures. As British crystallographer Dorothy Hodgkin recounted in her 1964 Nobel lecture, while the techniques of structure determination can be "formally" represented as a cycle of mappings followed by "rounds of calculation" and modeling, "the outline hardly gives an accurate impression of the stages of confused half-knowledge through which we passed."[82]

Crystallographic models are built slowly through a recursive and iterative interplay between increasingly refined electron density maps and molecular models. Moving back and forth between different kinds of maps that correct for various errors, crystallographers actively cycle between techniques of mapping and modeling. As they insert amino acids into the electron density map, they use the growing model as the basis to generate a "calculated" electron density map. They then use this to compare and contrast with the "observed" electron density. In other words, they can construct hypothetical electron density maps of the models they are building as a means to test the model against the observed data. Thus they move through rounds of mathematical refinements, recalculating the density peaks, refitting the model, and continuously comparing calculated electron densities with observed electron densities. Layered into this process is the corroboration of their model with the known amino acid sequence of the protein. Gradually a clearer and clearer image of the map and model emerges.

Much of the difficulty in this work lies in the fact that the model is never self-evident from the map. Faced with an electron density map, the crystallographer has very few clues as to which parts of the protein fit into which parts of the electron density. It is up to them to recognize which amino acids fit into particular configurations of electron density. According to Diane, the modeler must use "known knowledge" to "interpret what otherwise would be completely un-interpretable." Fitting a model into the map involves a wayward and intuitive process, and a crystallographer must draw on formalized knowledge of allowable molecular geometries, including the distances and bond angles among atoms within the polypeptide chain, and intramolecular forces that hold the whole molecule together. For Diane, model building requires the modeler to get comfortable with the experience of meandering through the electron density map, never really knowing for sure "where you are." Doing work that a computer alone cannot achieve, crystallographers must first get lost in the map and feel their way around familiar and unfamiliar forms in order to connect the model up atom by atom. This is a central aspect of the work of the human-computer lens (see chapter 1).

I interviewed many of the students in Diane's group during my years in and around her lab, and during this time I watched novice modelers build their first structures, as well as more experienced crystallographers negotiate challenging data sets. As they progressed from the early phases of their projects, students eventually discovered just what it was that they had to bring to their crystallographic data. Dehlia, a fourth-year PhD student in Diane's lab whose first structures were already published in *Nature*, explained that it wasn't until she started building that she realized "the extent" to which she had to participate in the model-building process. She described her early experiences in front of the computer as a disorienting walk through unfamiliar territory: "It's so densely packed that really, you're walking through a jungle of amino acids" (see figure 2.11).

Edward, who has had lots of expertise building crystallographic models, explained how hard it is to learn how to see in such a jungle: "The amino acids don't always jump out" at you, and you can't always "pull them out" into view. Samantha, who was in the early stages of building her first model, was in the thick of this confusion: "How do I know where that backbone would go? How would I figure out where this amino acid would go?" Dehlia explained that at some point "you have to start making executive decisions."

Amy, whose love of the lab's X-ray diffraction machine was documented in chapter 1, had been having incredible difficulty solving the structure of a new protein she was working on. Amy confirmed Dehlia's observations and emphasized that "a lot of guesswork" goes into building a structure: "And guesswork isn't the best word to use; maybe 'subjective' would be the best word to use to describe it. And that's not something you can understand until you

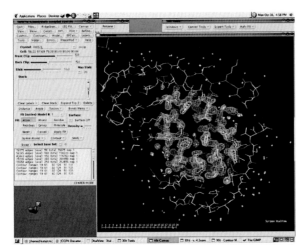

2.11. A screenshot of the interactive computer graphics interface Fernando uses for building his model into the electron density map. Used with permission.

actually have a structure that you have done yourself, or are in the middle of doing. The first structure I did was an easy one. I was really surprised that it was up to me to put in the [amino acid] residues. It was up to me to put the [polypeptide] backbone in. I was just really surprised. . . . That it was something . . . that I could make a mistake and no one would know. It's kind of scary, and it makes you really wary about other structures sometimes."

Not only do they need to be able to discern where and when the computations are off, they need to make "executive decisions" about where amino acids go in the structure, which confirmations of atoms are energetically feasible, and which defy allowable bond angles or inter-atomic distances. To do their work well students must cultivate sound judgments that demonstrate a respect for known atomic interactions and viable molecular configurations.

This situation presents a steep learning curve for crystallographers-in-training, and students often make mistakes when attempting to build their first structures. Their expertise is still in-the-making. This raises significant questions: At what point can they start trusting themselves? What are the grounds for trusting one another's models? Interviews with lab directors and graduate students suggest that in addition to their dexterities and intuitions, novice modelers need to build confidence in their skills.

GROPING TOWARD INSIGHT

Best known for his articulation of the role of tacit knowledge in scientific practice, philosopher of science Michael Polanyi offers insight into crystallographic model building. Polanyi was a physical chemist who used X-ray crystallographic techniques in his experiments. Indeed, he developed elements of his thinking about tacit knowledge with reference to crystallographic practice. Drawing on gestalt psychology, Polanyi found "inarticulate manifestations of intelligence" beneath the surface of scientific practice. For him, this intelligence "falls short of precise formalization," and experimental progress is made incrementally, by trial and error, in such a way that researchers "grope" their way toward insights.[83] Diane's experience modeling maps onto this description well: for her, the structure can remain obscure for a long time, until a shift in perception opens it up to view. Once you have started building your model, she explains:

> Then you'll look at it and go, "Okay, there's a big side chain here." And three residues down there's something long. And this looks like an arginine [an amino acid] and down there [points] that looks like something big. And you'll go through your sequence, and go, "Okay, where are the arginines? What's four residues away? Oh, lysine [an amino acid]. That's no good." And you will work your way through. And you'll sort of build some of it, and then go, "Okay now I'm lost and I don't know where I'm going next." . . .
>
> And there are certain folds that people know. Like TIM barrels [see figure 2.12]. One time I could see some helices in an early map, and I was putting a couple in, and I put a couple more in. And then, I think I got up for a minute and came back and just sort of saw from a distance what I had done. And I looked and there was a whole bunch of helices around in a row. And I said, "That's a TIM barrel!" And there's got to be strands in the middle. And then I pulled in a TIM barrel and went, okay, it's a little off, it needs some adjusting, but yeah, that's what it is.
>
> And so, sometimes it takes a long time to recognize the fold, because sometimes it's not a very standard fold. And other times it can come out relatively quickly, you'll all of a sudden see the connections by how things are, or you'll find a region where you can see the density of beta-strands. And you know that you can pull in the model and try to get it to fit.

For Diane, model building is like a "detective story" where the crystallographer has to search for clues about their structures: "that's why you never know you are done until you are *done*. Because at the end stage you go, 'Okay if that's correct we should be able to connect [amino acids] five and six, and it's all there!'" Diane's account suggests that there is a gestalt shift in seeing that occurs through the immersive work of modeling, where the form of the folds jump out at her, emerging whole from a piecemeal process.

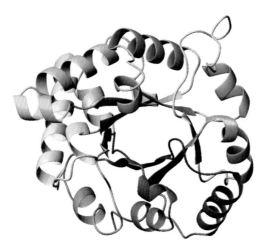

2.12. A ribbon diagram of a TIM barrel structure found in the enzyme triose phosphate isomerase. Note the barrel structure formed by the circular arrangement of alpha-helices. PDB ID: 8TIM. Structure deposited by P. J. Artymiuk, W. R. Taylor, and D. C. Phillips.

The interactive molecular graphics systems that Diane uses engage her in an "intuitive," "trial and error" mode of experimentation that is familiar to her from working with mechanical models. As their early developers advertised, these systems exploit an "interactive mode," one that is "able to take advantage of the powers and versatility of the 'human computer' for pattern recognition and inductive thinking."[84] It is this ongoing tactile interactivity with molecular graphics, this form of "manual thinking" involved in Diane's modeling of the TIM barrel, that makes explicit the ways that haptic creativity is integral to the formation of an expert crystallographer's habitus. As we will see in the following chapter, it is this tactile, kinesthetic, and proprioceptive visuality that makes crystallographers' expertise so essential to the work of molecular modeling.

CONCLUSION

Where chapter 1 described the ways that scientists-in-training get affectively entangled with their materials and instruments, here the craft practices and material cultures of molecular modeling demonstrate the many ways that modelers get kinesthetically entangled with modeling media. Even today, as crystallographers make use of twenty-first-century computer power, proprietary software, and flashy graphics, modelers remain tethered to their computers and intimately involved in the modeling process. The following chapter outlines how the kinesthetic knowledge they build up through their modeling efforts allows them to sculpt an embodied twin of their model alongside those that flicker on their computer screens.

In our first interview, Diane told me about the challenges involved in learning how to model proteins in three dimensions (see figure 3.1). She explained that it is hard to learn how to "think intelligently about structure." She recounted the steep learning curve her students face trying to master X-ray crystallographic techniques and build molecular models. What can be even more challenging is interpreting the functions of proteins, that is their biochemical activities, from the models. Acquiring the skills to "see what the structure is saying" is "hard to do, and it takes time," but, she assured me, eventually one does "get better at it." She described what often happens when her graduate students show her computer graphic renderings of their models in the early stages of the building process. "Look I connected it!" they would proudly declare, presenting their models on-screen. Yet, when she would examine their models in detail, looking closely at the bond angles between the amino acids and the direction of the polypeptide chain that winds through the protein, her response would often be anguished: "What did you do to that side chain? No! No! Let me move it back!"

As she told the story, she contorted her entire body into the shape of the misfolded model. With one arm bent over above her head, another wrapping around the front of her body, her neck crooked to the side, and her body twisting, she expressed the strain that would be felt by the misshapen protein if it had to take this form. "And I'll just get this pained expression," she told me. "I get stressed just looking at it. . . . It's like I feel the pain that the molecule is in, because it just can't go like that!" Compelled to fix the model, she mimed for me a frantic adjustment of the side chain by using one arm to pull the other back into alignment with her body, tucking her arms in toward her chest and curving her torso over toward the core of her body to demonstrate the correct fold. With a sigh of relief she eased back into a comfortable position in her

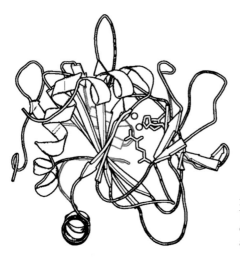

3.1. A ribbon diagram of a protein molecule. The atomic resolution chemical structure detailed in the right upper region indicates the active site of the protein. Used with permission.

chair, and the comically anguished look on her face relaxed back into a warm smile. Apparently, the students in Diane's story had not yet acquired a *feeling for* the proper molecular conformation. In a mode evocative of what Evelyn Fox Keller has described as Nobel Prize laureate Barbara McClintock's "feeling for the organism,"[1] Diane's gestures and expressions suggest that her kinesthetic and affective entanglements with proteins play a key role in her ability to "think intelligently" about their structure.

Oddly, this feeling doesn't seem to reside in Diane alone; the protein in her story also appears to have feelings. And Diane is clearly concerned about the affective state of this protein; in this case, the strain it would be under if it had to conform to the structure described by the student's model. In contrast to the "happy" and "relaxed" states that students worked hard to induce in the proteins they prepared for crystallization (see chapter 1), this protein is "in pain." It would be easy to see this demonstration as a straightforward form of anthropomorphism, wherein Diane imposes a human concept of pain on the molecule. Yet, it is possible to approach this situation differently. One could say that this molecule has been humanized as much as Diane has been *molecularized*. This chapter explores how it is possible for Diane's sensorium to be so finely articulated to molecular forms and movements that she has become responsive to their energetic and affective states.

How does she acquire this feeling for proteins? And what kind of feeling is this? While there are elements of empathy in Diane's response (she feels the protein's strain as a pain in her own body), it is perhaps better described as a "shared feeling" produced out of a sympathetic relationship, where sympathy according to the *Oxford English Dictionary* is "the quality or state of being affected by the condition of another." She is clearly affected by the strain a protein would experience if it were to be forced to fold into an awkward configuration. Sympathy involves "an affinity between certain things, by virtue of which they are similarly or correspondingly affected by the same influence."[2] She reenacts the conditions that give rise to strain in the molecule by uncomfortably contorting her own body. Why would she do this? What does her body experiment achieve? In *The Order of Things*, Foucault introduces sympathy as one of the four "similitudes," or forms of resemblance, that shaped the logics of representation in the classical period. He reminds his readers that sympathy connotes both a feeling and a way of rendering likeness, or verisimilitude. For Foucault, sympathy as a feeling could "be brought into being by a simple contact," "excit[ing] the things of the world to movement," and "draw[ing] even the most distant of them together" in similitude and shared sensation. Sympathies "traverse the vastest spaces in an instant," "drawing things toward one another in an exterior and visible movement," and in so doing, they "give rise to a hidden interior movement—a displacement of qualities that take over from one another in a series of relays."[3] Diane's sympathetic response traverses a wide chasm between her body and that of the molecule. Pain appears to move as if in a relay or circuit of affects between the model, molecule, and her body.[4] Crucially, the direction of this movement of sympathies in Foucault's description remains undefined. In Diane's performance, it becomes unclear who is animating what and what is animating whom. She moves her body to demonstrate the strain of the molecule, and in the process comes to feel an interior, visceral response, in the form of a sympathetic sensation. According to Foucault, sympathy "has the dangerous power of *assimilating*, of rendering things identical to one another, of mingling them, of causing their individuality to disappear—and thus of rendering them foreign to what they were before."[5] "Sympathy transforms": indeed, it is akin to a kind of "becoming with others," an "involution" of sorts.[6] Affecting and affected by one another, modeler and model oddly come to resonate with and resemble one another. It is in the very movements and minglings of the bodies entangled in model building that Diane can be molecularized at the same time as her molecule becomes humanized.[7]

THE KINESTHETIC IMAGINATION

Diane's feeling for the molecule demonstrates how protein modeling reworks conventional understandings of the relationships between model and modeler, mind and body, and the relationship between seeing, feeling, and knowing. Philosopher of science Ian Hacking makes a distinction between "models you hold in your hand"—material models made with "pulleys, springs, string and sealing wax"—and "models you hold in your head," conceptual models and mental images that function through analogy and imagination.[8] Yet, if we take seriously Diane's experience of the pain of the misshapen molecule, such a distinction between "models-in-the-hand" and "models-in-the-head" does not always hold. While Hacking draws these two kinds of models apart, Diane's sympathetic response suggests a deeper entwining of material and conceptual models. Indeed, she carries more than a "mental image" of what a molecule should look like in her head:[9] seeing, feeling, and moving with the chemical constraints of the molecule, she has honed an affectively charged sensibility attuned to molecular forms. Philosopher Maurice Merleau-Ponty argues that sensation and movement are intimately tied to visual understandings of form.[10] Diane's well-articulated *kinesthetic imagination* demonstrates well how this coupled nature of seeing and feeling plays out in the work of crafting structural knowledge.[11]

Diane's expression of the strain of a misfolded model suggests that several modes of body-work are involved in articulating a modeler's kinesthetic imagination. First, it is clear that she can sense structural irregularities in models that her students cannot detect. How does a crystallographer acquire such skills? This chapter explores how the hands-on, trial-and-error process of building models with various materials articulates a modeler's sensorium. The haptic creativity that is integral to molecular modeling allows her to *incorporate* structural knowledge in her kinesthetic imagination. Moreover, once embedded, her body-as-proxy can become a dynamic experimental model. Rearranging her contorted limbs, Diane conducts a *body experiment* to relieve tension in the model so she can get a feel for the correct structure. Once she has found the right form, she can use her body to reason through possible chemical affinities and movements to figure out "what the structure is saying."

Kinesthetic experiments like this come alive in her performative gestures as she shares the fine details of her structural knowledge with her students and colleagues. Incorporated models are performed, sometimes subtly in the rhythm of a casual conversation, and sometimes with considerable flourish, both within and outside of the laboratory. They come alive in conversations among experts and their students, and in conference presentations and classroom lectures. Three modes of body-work are operative here: the body-work of incorporation, reasoning, and performance. These modalities raise questions about forms of knowing in science, and the relationship between seeing, feeling, knowing, and making. The often-tacit dimensions of these corporeal practices suggest too that novice modelers face significant challenges in acquiring this know-how.

In this chapter, Diane is an exemplar and a guide who helps pose a new set of questions about the role of researchers' bodies in scientific practice. Some may argue that this "feeling for" the molecule in general, and Diane's embodied intuitions in particular, are extraordinary. Indeed, if one knew that Diane had studied drama in addition to chemistry as an undergraduate at college in the 1980s, she might be construed as a particularly "expressive" scientist. And yet, interviews with both male and female graduate students and a range of expert practitioners show that this "feeling for the molecule" is in no way exceptional; rather, it is part of a collective phenomenon observable among experienced protein crystallographers. Chapter 8 of this book explores some of the forces that constrain how practitioners come to move their bodies. That discussion raises crucial issues about the relations of gender and power and explores some of the different challenges that male and female modelers encounter when using their bodies as proxies for their models. Forms of body-work are indeed constrained by norms that dictate what counts as proper conduct. What is particularly striking is that women may actually feel these pressures more acutely and be more reserved in their molecular performances, whereas male modelers may have more room and more permission to express their kinesthetic knowledge. Before these issues can be fleshed out, it is necessary to return to the question at hand: how is a modeler's kinesthetic imagination articulated? What do modelers learn by building models?

GIVING BODY TO MODELS

Historian of biology Nick Hopwood's account of early embryological modeling techniques offers insight into the intimate relationship between modelers and their models.[12] He documents embryologist Wilhelm His's (1831–1904) techniques for sculpting scale models of embryos in wax. In defense of a mechanical theory of embryological development, His developed a method for precisely reconstructing the form of embryos from the details derived from microscopic examination of thin slices of tissue. These thin sections gave him detailed visual insight into the contours and hidden depths of embryonic tissues. He then used projective drafting and freehand wax sculpture techniques to flesh out the sectioned images into exquisite three-dimensional forms. His's craft demanded artisanal skill and enrolled his entire sensorium.

As Hopwood argues, it was "the experience of modeling" that was "the most compelling evidence of the importance of mechanical principles in development."[13] In other words, His gained his insight into the mechanical processes of embryogenesis by working with the forces of his body and those of his modeling materials. He built models so that he could access a "more direct bodily apprehension of form," and he insisted that model building could give him a "three dimensional mental image" of the phenomenon."[14] For him, "the pictures in the memory that have once made their way through the hand stick much more firmly in the head."[15] In this process, His honed his kinesthetic imagination and cultivated an affectively charged sensibility of embryonic form. Hopwood sees His's commitment to modeling as "a passionate argument for doubly embodied knowledge":[16] His had to "use his fingers" in order "first to make his problem," and it was by this method that he was able to "'to give body' to his views."[17]

His's assertion that model building sculpts insight at a corporeal level resonates well with molecular modelers' haptic vision described in chapter 2. Recall how Jim walked through the α-helix in his protein folding class to teach the students how to "grasp" these structures. Protein models thus share some fascinating similarities with His's wax sculptures: these models are not merely representations of proteins or embryos; they are also not merely representations of scientific knowledge; they are also enactments that engender new forms of knowing.[18] A full discussion of this enactive approach to model making, and the ways that it troubles conventional theories of representation, will have to wait until chapter 4. For now, it is necessary to examine how crystallographers articulate their intuitions and learn to exercise their judgment in rendering protein models from crystallographic data.

A FEELING FOR THE MOLECULE

Feminist and queer theorist Sarah Ahmed insists that affects and emotions are not innate phenomena arising from within the physiology of individual bodies; they are, rather, distributed and relational phenomena that circulate and propagate among bodies and objects. It is in the movement of affects, as they slip, slide, and sometimes stick to bodies, that the very boundaries between objects and bodies are constituted. Ahmed builds on Freud to suggest that "it is through the intensification of pain sensations that bodies and worlds materialize and take shape"; in other words, it is by coming into contact with things that cause pain that bodies acquire their surfaces, forms, and volumes. Consider Diane's feeling for the molecule. She contorts her body to come into contact with the source of the protein's strain. For Ahmed, "the recognition of a sensation as being painful" is a process that "reconstitutes" one's bodily space and requires a "reorientation" of one's body "to that which gets attributed as the cause of pain."[19] Feeling the strain in her molecularized body-as-model, Diane readjusts her limbs to simultaneously correct the model and relieve the protein's pain. What is crucial here is that "the recognition or interpretation of sensations" such as pain build on "past bodily experience."[20] To relieve the pain in both her body-as-model and in the molecule that the model is supposed to render, Diane draws on an already articulated repository of embodied knowledge about proper molecular form.

It takes time, but students eventually learn how to "give body" to molecular models. Novice modelers must entrain their bodies, imaginations, and intuitions to the contours of the substances they are modeling. In order to do this, they must first assimilate all the available formalized knowledge about the chemical and physical properties and behaviors of protein molecules. They must learn what "happy," "relaxed," and well-folded proteins look like, and get a feel for how these molecules hang together in their watery cellular worlds. They need to cultivate the kinds of skills that allow Diane to recognize when a model defies allowable parameters. They have to be able to see and feel where the model is off; and to do this they need to fully grasp the wide range of chemical affinities and repulsions that hold a protein together.[21]

There are rules they need to follow. Their models must obey the laws of chemistry and physics: attractive and repulsive forces in a molecule set the parameters for allowable distances and feasible bond angles between atoms. Modelers should know how particular chemical groups in one part of the polypeptide backbone interact with other groups. They must account for the ways hydrogen atoms from one amino acid can bind to the hydrogen atoms of another amino acid, or how bonds between sulfur atoms form to stabilize folds in the polypeptide chain (for more on the primary, secondary, and tertiary structures of proteins, see the appendix to this book). Crystallographic data do not always give a modeler a solid basis for determining where these bonds are or, for example, the specific orientation of a particular amino acid. As Diane's visceral reaction to the mangled model makes clear, practitioners must learn to respect protein forms; that is, they must stay true to defined molecular possibilities.

One technique that modelers use to simultaneously build up their intuitions and learn how to evaluate a crystal structure is to "walk" through it atom by atom. Justin is a fifth-year PhD student in a laboratory that models protein dynamics through computer simulations. He collaborates with protein crystallographers and has done some work with members of Diane's group. While he doesn't collect X-ray data and solve crystallographic structures, he does use published structures avidly, accessing the coordinates of known structures from the Protein Data Bank. His research is focused on an area known as "protein design" or protein engineering, and he relies heavily on computer simulation experiments to test the molecular dynamics of his engineered molecules. At the same time he has gone to great lengths to bring the results of these simulation experiments into the wet lab where he has developed assays sensitive enough to test his *in silico* hypotheses in vitro.

In order to learn how to design new proteins, Justin has relied extensively on crystallographic structures to teach him how proteins found "in nature" hold themselves together. He walks through the atomic configurations of a crystal structure atom by atom. During an interview he explains how he relates to this data. He takes the crystal structure "for granted" in that it is a model of something that "has been seen in nature": "No matter what you might think of [the structure], at least once nature said, 'Okay, I'll use this one.' So you build up your background for what's okay." Justin must trust the crystallographer to stay "true to nature."[22] This means he must trust the experimental conditions and the intuitions of the modeler. This is a significant assumption as the structure rendered in a given data set may have less to do with what "nature uses," as much as how a particular protein folded itself up in a given crystallization solution, or the form of a molecule that is favored by a certain crystal structure as it grows. Recall Zeynep's experience documented in chapter 1. She was concerned that many crystal structures are based on peptides that have been engineered and that these may not reflect the characteristics of the original protein. It took a lot of time and effort for her to build up trust in her engineered peptides. And yet, for Justin's purposes, even though the molecule may not have folded into a biologically relevant structure he can still learn from it: he can "build up" his "background" for the atomic interactions that shape "feasible" structures.

He explains his process of learning a structure and evaluating its merits by walking through it amino acid by amino acid:

> You learn by working with this on your own over time. One of the exercises I remember working with early on, which really teaches you all of this, is the exercise [of walking] through a protein. The simplest version of the exercise is to visit, one by one, each of the histidines in the protein and look for what kind of hydrogen bonds it might make with nearby side chains [histidines are amino acids; see figure 3.2].
>
> There's a reason for this. In most crystal structures, even though [crystal-

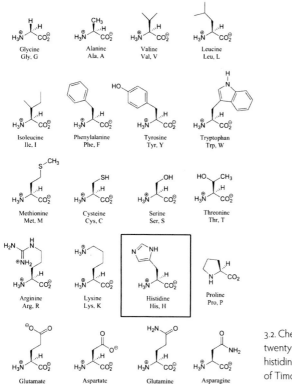

3.2. Chemical structures of each of the twenty amino acids. The structure of histidine is highlighted. Figure courtesy of Timothy Soderberg, Chemwiki Site (chemwiki.ucdavis.edu).

lographers] sort of know where the histidine is, they don't know which of four actual orientations it might be in. Because there's a 180-degree flip that might not be resolved from crystallography. And then there's another aspect to this: the nitrogens and carbons might be indistinguishable. Well, it's really the same thing because if you can't see nitrogens and carbons, you can't see the flip, and then you don't know where the hydrogens are. And the hydrogen could be on one nitrogen or the other, so that makes really four or five possibilities for where the histidines are.

So the exercise gets you looking at one particular functional group and what's in the neighborhood around it, and understanding, "Oh, that's about three angstroms away, that could be a hydrogen bond. And that's five angstroms away, no, that's not a hydrogen bonding group. That's just a carbon. Oh, yes it is, because that's a carboxylase."

Justin needs to know the specific orientations of each amino acid and precisely where the hydrogens bond. This is crucial information for someone in the field of molecular dynamics. If Justin wants to simulate molecular forces to be able to predict potential intermolecular and intramolecular interactions in and among engineered proteins and their substrates, he needs to be able to correct any errors and resolve ambiguous regions of a model. These are details that may not have been critical to the protein crystallographer, or perhaps there were regions in the molecule that the crystallographer was unable to resolve from their data set. He explains how he figures this out by hand and by eye:

> So you can see how far away things are. You understand very quickly which are the functional groups. . . . You recognize what groups can receive hydrogen bonds, what can donate them, and that's a lot of what we're looking at. . . . It's this 3-D jigsaw puzzle. . . . You get a sense for, "Well, that looks strange because the rings from two side chains are interacting at some strange angle." Or, "If it's a hydrogen bond, the distance is good but the angle is bad." There are programs that could tell you, "Distance good, angle bad," but you can also see it by eye once you've played with it more and more.

Justin could use a computer program to determine if the protein model is happy, that is, at "low energy," or if it is under strain. But his judgment and intuitions are also honed just by "playing" with the model. This ongoing kinesthetic engagement with atomic-scale crystallographic data allows him to assess whether his corrections to a crystallographic model are feasible. Eventually he can begin to detect contortions "by eye." It takes a long time to develop the skill set required to feel the pain of a misshapen protein model. But Justin's approach of walking through each structure atom by atom gives him a way to entrain his sensibilities to "proper" molecular configurations.

MOLECULAR EMBODIMENTS

> To get used to [things] is to be transplanted into them, or conversely to incorporate them into the bulk of our own body. Habit expresses our power of *dilating* our being-in-the-world, or changing our existence by approaching fresh instruments.
>
> —Maurice Merleau-Ponty, *Phenomenology of Perception*

> We may say that when we learn [a] probe, or a tool, and thus make ourselves aware of these things as we are of our body, we *interiorize* these things and *make ourselves dwell in them.*
>
> —Michael Polanyi, "The Logic of Tacit Inference"

Justin's kinesthetic insights into molecular configuration give us some clues for how to think about Diane's ability to feel the strain of a misshapen protein model. He shows how a modeler's body can become sensitized to molecular configuration; how, in short, modelers get *molecularized*. A phenomenological approach to model building helps to draw out the fine details of this process. Exploring the prosthetic nature of tool use, Maurice Merleau-Ponty and Michael Polanyi offer insights into the intimate association of bodies and tools in learning.[23] According to them, instruments can shift our perception of the contours of our own bodies. In Merleau-Ponty's sense of the word, a modeler can learn how to "dilate" the perceived contours of her or his body. Learning to use new instruments involves folding them into the modeler's own "corporeal schema," and also by extending this schema outward to meet the tool as an extension of the modeler's self.[24] These insights suggest that bodies are open to the world, porous to new possibilities, and adaptable to new kinds of tools. For example, Diane can apparently dilate her corporeal schema to such an extent that the computer graphics interface becomes a prosthetic extension of her body. In this process she can simultaneously learn how to "interiorize" the products of her body-work.

In a key moment during an interview with Diane, she offered this insight into her experience incorporating molecular forms. She told me:

> The person who builds a structure . . . understand[s] the structure in a way that I don't think anyone else ever will. And I try now as an advisor, I try to get inside the structure and really try to understand it at that level. And I have for a few of them, but it is really time consuming, I mean, to sort of have the structure in your head in *three dimensions*, which is how I felt about some of the other structures that I actually did build myself. And I would be at a meeting and people would be discussing a mechanism, and I would kind of close my eyes and try to think about it and go, "No. Too far away."

A number of striking insights emerge from Diane's description. For her, the person who builds the model knows the structure in ways that others will never fully grasp. She insists that no one will ever understand the molecule as intimately as those who built the model. She acquires remarkably local and personal knowledge of a molecule by grappling with it through interactive graphics. This helps her get "inside" of the model. Diane makes clear the frustration she feels about the limitations of the two-dimensional figures she must construct to communicate to others what she identifies as the most salient features of the structure. These visualizations are merely abstractions of modelers' intimate knowledge of molecular form. Getting to know a structure requires more than perusing the list of a molecule's atomic coordinates in the

PDB; a viewer must handle and manipulate each model as a tangible, if virtual, three-dimensional object in order to acquire at least some of the knowledge that the crystallographer who built the model possesses. And this is why, when new structures are presented in the literature or at meetings, they are "too far away." To bring them closer, the modeler must go to the PDB and download the coordinates of the model into an interactive molecular graphics program so that she or he can examine the structure personally and get a feeling for its folds.

As Diane described in other conversations, it is only by actively handling the model through interactive molecular graphics programs that she can project herself "inside" of it and figure out "where she is" within the structure. She achieves this intimacy with the model by dilating her corporeal schema to meet its form. Indeed, it seems as if she is able to morph the perception of her body enough that her own limbs become effective proxies for chemical structures. Feminist philosopher Judith Butler calls attention to the pliability and plasticity of the corporeal schema through the concept of the "morphological imaginary."[25] Butler suggests that we have the capacity to actively transform how we experience our bodily configurations. The physical contours of our bodies do not necessarily constrain our capacity for imagining our morphologies otherwise. Modelers' bodies need not look anything like molecules to form a sympathetic relation with them. As they expand and extend their kinesthetic imaginations, modelers get *molecularized*. Diane's experience exemplifies this well:

> And you know, it's really this vision that you have of the active site, and sort of this sense of how tightly packed it is and how much flexibility there might be and where those regions of flexibility are. To have this sort of sense that you have. And you can think about it then *moving* in a way because you sort of know something about what the density was, so that you know that part is definitely mobile right in there, but that this part would not be mobile. And this information is kind of like stored in your brain in some way, and it's not something that is easy to communicate, because, you know you can't explain something in three dimensions to someone.

Although she indicates that she "stores it" in her "brain" and can rotate the molecule around in her "head," her whole body is engaged in descriptions of the model's flexibility, intramolecular forces, tensions, and movements. Once incorporated, she has both a "vision" of the active site and a "sense" or feeling for the forces within the molecule that exceed what could be described as a "mental image." It is through the multidimensionality of her body that she is able to appreciate the form and movements of the protein model. That is, she feels the spatiality and movements of the molecule by virtue of the spatiality and movements of her own body. While she is frustrated by how hard it is to relay three-dimensional, structural knowledge, Diane's body provides an articulate medium for vivid expression of the fine details of molecular structure: inflected and informed by the molecular models that inhabit her body, she demonstrates with clarity the twisting helices and the movements of the peptide backbone meandering through the molecule. Throughout our conversations, during class lectures, and in informal discussions with members of her lab, her gestures and affects animate the forms, textures, and tensions within the protein.

Diane's molecular embodiments are in no way exceptional. In-depth interviews with her male and female graduate students, and with other crystallographers, show that those who have made it through the rite of passage of model building, those who have "solved" their own structures, can carry specific knowledge of the configurations and chemical mechanisms of their proteins in their bodies. For example, an interview with Diane's postdoc Brent makes it clear that molecular embodiments are not restricted to a stereotyped "expressive woman scientist."[26] Though he admits to teasing his colleague for using a physical model to make arguments about molecular structure, his own body has become a proxy for the model he has built on-screen. When I asked him to describe one of the proteins he had modeled before, he proceeded with an elaborate demonstration of its chemical mechanism. He leaned across the table between us and drew his hands together, carving a small pulsing sphere out of the space in front of him. In order to describe the specific intramolecular forces between a small cluster of amino acids in the active site of his protein he tenderly drew the middle finger of one hand across an invisible force field on the palm of the other, indicating the exact site where charged amino acids interact with each other. Throughout his demonstration he held a buoyant tension in his hands that extended along his arms and into his whole body. He had cultivated a profound feeling for his protein in the course of building the model, and his gestures and affects reflect the intimacy of his molecular knowledge.

A novice's habitus has not yet been inflected with such precise sensibilities. This was clear in the context of my observations in an introductory laboratory course for undergraduate students taking up a major in biological engineering. Several modules of the lab focused on protein structure. Students worked on a protein called β-galactosidase (β-gal), an enzyme involved in the metabolism of sugars, and one commonly used in molecular biology as a marker to report gene expression. The students used it in their biochemical assays and protein engineering experiments. One of the lab modules explicitly addressed the structure of β-gal. Each pair of students had a laptop at their bench. During one module, they were instructed to use interactive molecular graphics software to inspect the structure of the molecule. Many had never worked with protein structures before. They played with the models distractedly but quickly became bored and moved back to the bench to continue their experiments. I asked several students if they found the structures useful for their experiments. They shrugged with indifference, telling me that the simple cartoon diagrams in their lab manuals were all they needed to understand their experiments.

While they are not indifferent to protein structures, graduate students and postdocs new to protein crystallography have also not yet cultivated a feel for structure. These included those who had not yet solved their own structures, those who were still struggling to get their proteins to crystallize, or those stalled at the stage of trying to, as they say, "massage" poor quality diffraction data into meaningful electron density maps. When I asked them to describe proteins they had learned about but had not modeled they rarely used their bodies, and if they did, their gestures were vague and imprecise, as if their hands loosely circumscribed the general form of an object at a distance. They were familiar with the model, but it did not yet "belong" to them.

When crystallographers build models with interactive computer graphics they are not only producing visual facts. The interactive graphics workstation is also a pedagogical site for training new protein crystallographers. The students who presented their misshapen models to Diane, and elicited from her a cry of pain, were crystallographers-in-training: they were still in the process of acquiring a feel for the possible geometries, forces, and movements within proteins. Of the advanced students I interviewed, all recognized Diane's skills but said they were still "nowhere near her level." This was a skill they understood as her ability to look at the data and model and intuit which residues were right and which were wrong. Diane's skill, as she says, to "see what the structure is saying" doesn't rely on her ability to memorize the correct chemical configuration. Keen molecular vision is, for her, a haptic practice of observation and manipulation, where seeing is also a way of feeling what the structure is expressing in its form.

Thus the richest, most detailed model of the molecule resides in the modeler. A number of specific protein models inhabit Diane's body—those models she worked on herself. Her body has become a proxy for the models that otherwise exist only on-screen. These *molecular embodiments* are the product of her intense involvement in the modeling process over long periods of time. As she leans into her data and folds her body around her model, she is able to give what is otherwise a virtual structure a physical body, a place for it to dwell. The practice of building protein models has thus *articulated* Diane's body with specific molecular knowledge. Rachel Prentice, an anthropologist of science, technology, and medicine, calls this a process of "mutual articulation." Prentice's concept, which was developed in the context of the design and use of surgical simulations technologies, can be extended to the interactive practice of sculpting molecular models on-screen. In this way, the recursivity of model building can be seen to articulate the modeler's sensorium. A model is built into one's body just as it is sculpted piecemeal on-screen. In this sense, molecular embodiments are "infoldings" of the model and the modeler.[27]

Once built, the structure of a protein model remains to be interpreted. "Thinking intelligently about structure" requires hypothesizing how a protein carries out its functions in the cell. This phase of research also depends on the trained intuitions and embodied knowledge of experienced crystallographers. As Diane demonstrates, the crystallographer carries the model within her body, but she also can "get inside" of the model in order to "figure out" how it "works." Her goal is to have a three-dimensional atomic-resolution understanding of "how nature has tailored" proteins to do chemical and biological "work" within the cell. She uses her kinesthetic knowledge of the specific molecular geometry and chemistry of the protein in order to reason through possible biological mechanisms. Once molecularized through sympathetic emulations, modelers' bodies become experimental media for reasoning through molecular forms. The phenomenon of body experiments will be taken up in later chapters. For now, it is necessary to continue to explore the intimacies that shape modelers' relationships with their models.

FALLING IN LOVE

Imagine sailing for years through uncharted water, and then suddenly you see land rising on the horizon. And this model emerging was like this. So one morning in September in 1959, our results came out of the computer at the Cambridge Mathematical Laboratory. Thousands of numbers, which we plotted on sheets of paper. And then we drew contours round them, and there emerged a landscape of peaks and valleys. So, I built this model. And then, suddenly saw this thing, you know, which I'd been working on for twenty-two years. And it was a fantastically exciting moment. I always say it was like reaching the top of a mountain after a very hard climb and falling in love at the same time.

—Max Perutz, "Face to Face with Max Perutz"[28]

Protein crystallographers invest themselves in their models. A craft product of labor and love, crystallographic structures are artisanal objects. For Diane, and others, the sudden emergence of the model after the arduous "labor" of construction warrants a "birth announcement." In an interview, Diane described it this way:

> I don't know, some other people say that they want birth announcements when the structure [is coming out] . . . because it is kind of like being in labor. . . . And often a building process will take nine months. And it is, it's sort of as it's coming out . . . you're all of a sudden, "Oh! Look at where that conserved patch is. . . . Yes! Oh! Oh! That makes so much sense! That other group was wrong about what those residues do." And so it's sort of this unveiling. And then you finally give birth to your molecule. And what I've started doing is putting our structures on refrigerator magnets, and so then for Christmas you can share with your family and friends. . . . Right. Everyone sends out their pictures of their kids, and you send out pictures of your kids. It is kind of like that in a way.

An early meaning of the term "to render" was "to give birth to," and here it seems that Diane has literally rendered her model. Her maternal relationship with the model is also evocative of the various forms of affective labor and care that are involved in the work of crystallizing proteins. Recall how Diane's graduate student Jamie approached the crystals as if they were "babies forming" and "growing" in their crystallization solution. This care and attention is extended all the way through to the final "birthing" of the model, when its structure is finally made visible as crystallographic structure.

Diane's transformation of a completed model into a refrigerator magnet also suggests that once built, models acquire totemic status. As a totem, the model is not only a proxy for the molecule; it also becomes an emblem that can stand for its modeler. A model is *of* its modeler as much as a modeler's identity is bound up with the models she or he renders. Totemic models come in a myriad of forms. Some use 3-D printers to produce precise physical renderings of their proteins that they can proudly flaunt on their desks. There are even companies that specialize in producing handsome, personalized trophies that display 3-D etchings of protein structures inside transparent resin, which itself resembles a crystal. Diane puts in a special order for one of these each time a student completes a structure.[29]

Modelers also acquire a possessive relationship to the products of their labor: the models they build belong to them in specific ways. They refer to their experimental objects as "my protein" and "my molecule." Brent explained that it's not until you can produce crystals that diffract well, and start working with the data, that the molecule becomes "yours": "It's hard to learn how to build models on stuff that isn't yours. Like everyone learns on lysozyme because it crystallizes like gangbusters, and you can get diffraction. But if it's more personal to you, you are more apt to go at it longer. You are like, 'Oh. This is my data. I got to figure this out!' You know, on my own, or with some help. 'This is mine. Mine!' You can hug it, kiss it and call it yours. . . . I guess you just become attached to it because you spend all your time on it [laughing]. And then once you have that final product it's like, 'Wow! Yeah. That's mine!'"

He explained that he always keeps a number of projects running and maintains a kind of emotional distance until a protein shows promise by forming "beautiful" and reproducible crystals that diffract well. For him, it is only once a project is well on its way that he feels as if the protein is "his." In the process of relaying his story, he emphasized his intense sense of ownership of the model by drawing his arms powerfully into his chest and emphatically repeating the word "mine." This evocative gesture also served to remind me how the model belongs to him. The model is not the product of disengaged rationalization, and it does not hover in his head as a mental image. The model belongs to his body because it is a product of his intensive physical and emotional labor, and because that is where his knowledge of it resides.

Brent's kinesthetic and affective entanglements with his protein and the model he has built raise questions about the nature of intellectual property in this field. Larry Shapiro, the laboratory director in the documentary *Naturally Obsessed*, describes well the precarious position that graduate students, like Brent, inhabit as they commit themselves to the task of solving protein structures. The film documents a scene where Larry and Rob have just started working with crystallographic data and are in the midst of building the model on-screen. It has taken Rob many years to get to this point. They are very close, but Larry is cautious: "We are not going to rush into publishing before we understand what we want to publish, but we are going to move as fast as we can just to make sure we get there first. I mean, imagine, someone could have a similar thing like this. In that case, this is no longer of interest really.

Can you imagine, really? Your five years of day and night work, and it's just, 'Uhh . . . missed that, sorry.'"

Members of Larry's lab work on highly prized proteins involved in diabetes and obesity research. His concern is warranted: there are many labs competing to solve the structures of the key enzymes known to be involved in the metabolic pathways that are implicated in these diseases. This leaves students in a precarious position. Each morning before they commence their lab work, students will check the Protein Data Bank to make sure no one has "scooped them" and uploaded "their" structure.

Once the coordinates of crystal structures are uploaded into the PDB, a modeler has to let her or his model go. The PDB makes data available to others, including protein engineers like Justin, to pick up and use the models to their own ends. Researchers in fields such as molecular genetics, biological engineering, protein simulation, and rational drug design make extensive use of the PDB. Protein Data Bank entries list the atomic coordinates of the protein model, including relevant statistics and experimental data. Interactive molecular graphics interfaces built into the PDB make it possible for a researcher to access and play with the intricate chemical configurations over which another crystallographer had long labored. As mobile, interactive objects, crystal structures can be manipulated, tweaked, tested, and evaluated. As their structures begin to propagate, and enter wider circulation, crystallographers voice some anxieties.

For one thing, these PDB files do not carry the "thickness" of the modelers' chemical intuitions or kinesthetic imaginations. Recall Edward's concern that practitioners who download the coordinates to his crystal structure might treat it as a "static structure," rather than, as he described it, a "breathing entity" (see introduction). His embodied knowledge is thicker and livelier than the data that can be transmitted through the PDB: in a sense it is he who keeps his model alive, both in his kinesthetic imagination, and through lively performances of its form. Thus, in this process of letting their models go into circulation, much of their artisanal labor is made invisible as their craft-productions are, in a sense, picked up off the shelf and swept up by drug developers and biomedical researchers. Once sent out into circulation, protein structures become fetishized commodities; that is, they appear to have value in and of themselves, and the labors (physical, affective, and otherwise) involved in their production are concealed.[30]

Protein crystallographers' modeling efforts demonstrate many facets of the verb "to render." This and previous chapters have shown how protein crystallographers give themselves over to the labor of sculpting models. The kinesthetic and affective dimensions of this work illuminate the ways that modeler and model are intimately entangled and co-constituted in this practice. Modelers gain their expertise and credentials in the very process of giving their bodies over to the effort of model building. The following chapters build on these insights to explore the ways that protein modeling reconfigures conceptions of scientific representation and objectivity.

PART TWO | **ONTICS AND EPISTEMICS**

Diane wants to be able to see what proteins "look like" so she can understand what they are up to in the cell. Yet, she and her colleagues are confronted by the ever-present limits of molecular vision. Diffractive optics leave crystallographers working in the shadows cast by interference patterns, where they must rely on computer power coupled with kinesthetic knowledge to fashion their atomic-scale models. And as they engage ready-to-hand tangible materials to amplify protein structure in Styrofoam, Plasticine, metal, plastic, wood, wire, paper, and colorful computer graphics, one might begin to wonder, do these models tell us anything about what proteins actually look like?

Max Perutz once recounted a story that makes palpable the crisis of representation that plagues protein crystallographers.[1] In 1962 Queen Elizabeth II paid a royal visit to formally open the Laboratory of Molecular Biology in Cambridge. As winners of the Nobel Prize in Chemistry that same year, Perutz and John Kendrew had the opportunity to present their already-famous molecular models to the queen. They had on hand a set of larger than life physical models of myoglobin and hemoglobin. The scene is documented in a press photo showing Queen Elizabeth seated and looking up inquisitively as Kendrew demonstrates his wire model of myoglobin (figure 4.1). In his account, Perutz recalls the moment when they "proudly" presented their models to the queen and "her party." Just at that moment, one of the queen's "ladies-in-waiting" gasped and exclaimed, "Oh, I had no idea we have all those little colored balls inside us!"

In Perutz's account, this unwitting "lady" mistook the models for lifelike representations, as if they were precise amplifications of the very substances in her body. Perutz's story scripts gendered roles for an impressionable (female) novice and the expert (male) scientist in a comedic scene that pokes fun at the gullibility of lay perception. In Perutz's rendering, the "lady" is the butt of a

4.1. Queen Elizabeth, seated, looks on as John Kendrew demonstrates his model of myoglobin. Used with permission from the MRC Laboratory of Molecular Biology.

joke she didn't realize she was making. It is possible, however, to interpret this scene otherwise. Perhaps it was she who was cracking the joke, deliberately mocking expectations about what women and nonscientists can know. If so, this joke would have made palpable how the skewed relations of gender and power in that room mapped on to cultural assumptions about the boundaries between experts and novices.

Regardless of how we interpret this scene, Perutz's story raises crucial questions about the representational status of crystallographic models. Given that molecules are not directly visible, how does a protein model relate to its referent? If modelers want to know what proteins "look like," what kinds of "likeness" do their models effect for their makers and users? Moreover, what constraints do modeling materials impose on what and how a model represents? How do practitioners negotiate the distance between a model and its object?

Philosopher of science Ian Hacking proposes that it was Democritus, the ancient Greek philosopher, who first posed the problem of representation. His theory of the atomic structure of matter, which postulated an "inner constitution of things," was "an extraordinary hunch"; but it was one that could not be verified by direct visual inspection.[2] Atomists could contemplate and explore atoms moving around in the void in their imaginations, but they could represent atoms only indirectly. Hacking stages Democritus performing his theory: "This stone, I imagine Democritus saying, is not as it looks to the eye. It is like this—and here he draws dots in the sand or on the tablet, itself thought of as a void. These dots are in continuous and uniform motion, he says, and begins to tell a tale of particles that his descendants turn into odd shapes, springs, forces, fields, all too small or big to be seen or felt or heard except in the aggregate. But the aggregate, continues Democritus, is none other than this stone, this arm, this earth, this universe."[3]

Democritus's dots in the sand were all that could stand in for otherwise imperceptible particles that make up the aggregate phenomena we encounter in the world. But, according to Hacking, "representations are first of all likenesses."[4] How could Democritus's dots be anything "like" atoms? Physicist and feminist philosopher Karen Barad builds on Hacking's insight to speculate that "the problem of realism in philosophy is a product of the atomistic worldview": "With Democritus's atomic theory emerges the possibility of a gap between representations and represented—'appearance' makes its first appearance. Is the table a solid mass made of wood or an aggregate of discrete entities moving in the void? Atomism poses the question of which representation is real."[5]

Contemporary practitioners have inherited Democritus's atomic theory and continue to struggle to represent that which remains unseen. Much effort has been invested in honing visualization techniques to reduce the gap between molecular representations and their referents. And while protein crystallographers and others now have tools that can trace probable contours around individual atoms, they cannot produce a mirror-image reflection of a molecule. Molecular vision is too indirect to close the gap between the model and its object. And this is the gap that protein crystallographers must negotiate as they attempt to train their students and novice viewers how to see what they see in their models.

Earlier chapters explored the concept of rendering by examining the affective labor involved in deciphering diffractive optics (chapter 1), the haptic creativity of modeling with tangible materials (chapter 2), and the articulation of modelers' kinesthetic imaginations (chapter 3). This chapter examines rendering in relation to the broader problem of representation in the sciences. Here I foreground the epistemic and ontological issues at the core of a practice that aims to make otherwise imperceptible phenomena visible, tangible, and workable.

Perutz's encounter with the queen's party demonstrates well the kinds of epistemic anxieties practitioners face; that is, the troubles they confront documenting protein structure and communicating their results. If protein crystallographers aim for verisimilitude, for a kind of "truth to nature" that stays true to protein form, how do they grapple with the limits of their vision and the constraints of their modeling media? This chapter examines the peculiar ways that protein models relate to their referents by juxtaposing these models to older three-dimensional modeling traditions in the life sciences. Here we find modelers engaging a range of media forms to find ways to secure various kinds of "likeness," or similarity relations, between their models and their objects.

The concept of rendering provokes yet another line of inquiry. While it is clear that the representational status of these models is fraught, it is also important to ask: what is it that these models help modelers achieve? In other words, what else, beyond (mis)representing proteins, do these models facilitate? A performative, material-semiotic approach to representation can offer the critical tools needed to grapple with these issues.[6] Where conventional theories of representation can only sediment epistemic anxieties—such as those performed by Perutz in his recounting of the queen's visit—the concept of rendering offers an expansive view. This chapter locates protein models in the history of three-dimensional models in the life sciences, paying close attention to the ways that forms of verisimilitude and aesthetics shape modelers' claims to truth. To understand the relationship between protein models and their unseen referents, I explore how models and the molecules they index are "phenomena" in Karen Barad's sense of the term; that is, both are materialized only in specific experimental configurations. Molecular model building in this sense can be understood as a world-making practice that manifests some forms of life, if not others. Practitioners' ways of seeing, feeling, and knowing are reconfigured in the very moment they reinscribe the material world in the image of their model. The idiom of rendering thus makes the ontological stakes of protein crystallography palpable. Indeed, crystallographic models render the substances of life molecular; that is, they make matter come to matter as molecular. This and later chapters take up the issue of how protein models *rend* the stuff of life, in order to ask what materialities are in-the-making in practitioners' hands.

EXTREME MIMESIS

Molecular models share some similarities with other three-dimensional models built and used throughout the history of the life sciences. Three-dimensional models are essential visualization tools for teaching, learning, and research. Familiar examples of these include models of animals and plants preserved in natural history museums, and those used for teaching and research in anatomy, embryology, and evolutionary biology.[7] The enduring materiality of these models, and their stabilization through the institutional structure of natural history museums, served to secure public access to "nature's panoply."[8] In order to illustrate taxonomic classification schemes, theories of evolution, and morphological specificity, three-dimensional models of animals and plants aimed for a realistic, descriptive mode of representation. It was by replicating or simulating the likeness of a given phenomenon that these models functioned as scientific tools.

In her study of the shifting institutional organization of natural history museums, feminist science studies scholar Susan Leigh Star quotes a turn-of-the-twentieth-century taxidermist attesting that his task was "not to depict the mere outline of an animal on paper or canvas and represent its covering of hair, feathers or scales." This would imply a reduction or abstraction of the phenomenon. Rather, it was his work "to impart to a shapeless skin the exact size, the form, the attitude, the look of life."[9] Taxidermy is a good example of what philosopher of science James Griesemer calls "remnant modeling," where the skins of the very animals modeled are stuffed and posed to mimic lively bodies. He suggests that such models are able to "serve certain sorts of theoretical functions *more* easily than abstract formal [models] by virtue of their material link to the phenomena under scientific investigation."[10] The measure of an effective model was thus its verisimilitude, its likeness to the phenomenon under investigation. In this sense the explanatory force of a model depended on its capacity to successfully mimic natural processes, and it is in this "similarity relation" between world and model that models could generate experimental evidence through "equivalence."[11] This similarity relation was thought to provide experimental proof of the theory in question.[12]

Historian of science Lorraine Daston includes taxidermy in a larger modeling tradition she calls "extreme mimesis."[13] This approach to modeling is best exemplified for Daston by the Ware Collection of Glass Flowers, an extensive set of exquisitely crafted glass botanical models produced by glassblowers Leopold Blaschka (1822–1895) and his son Rudolf Blaschka (1857–1939). These intricate, life-like models of plants were commissioned by Harvard University to serve as pedagogical tools. As glass models they could preserve the lively forms of plant life more effectively than specimens pressed and dried in a herbarium. As replicas of botanical forms, these models are nearly flawless. They document the colors, textures, and gestures of plants with remarkable precision, including decaying leaves, flowers, and fruits as well those in robust form. Admiring the glassblowers' craftsmanship and the visual effect of the models, Daston remarks that, "though the actual deception of appearance taken for reality lasts only for a moment, the pleasure of potential deception lingers long."[14] In this sense, a model's potency was a measure of its ability to deceive the viewer into believing that the model was the thing itself.

Daston suggests that there was indeed a time in the history of biology when "the verisimilitude that is called illusionism in art" could become "scientific accuracy."[15]

Crystallographic models register the configuration of protein molecules at atomic resolution. Are these models then part of a tradition of extreme mimesis? The answer is both yes and no. Where the mimetic modeling traditions in the life sciences have been obsessed with reproducing the textures, tones, colors, and qualities of an object, protein modelers can't do this. They can't know what molecules "look like": diffraction data generate only indirect molecular vision. So what kinds of likeness, other than verisimilitude, do protein models effect?

To appreciate the difference, let's consider how chemists engage molecular models. According to historian of chemistry Eric Francoeur "no chemist" worth her or his mettle "would propose that models, even in their more elaborate forms, are about what molecules 'really' look like."[16] As he explains, in the hands of expert crystallographers, molecular models operate through "homology" rather than "homomorphy";[17] that is, they indicate forms and relations between elements—a molecule's configuration.[18] Once they have the coordinates for their structure, modelers are open to a wide range of representational conventions to give body to their data. They readily move back and forth between distinct renderings. These include, among others: ball-and-stick, space-filling, molecular surface, and wire-frame renderings; ribbon diagrams; and those renderings that make molecules look like edible "licorice" candies. In a single publication or during one visit to the Protein Data Bank, a modeler may use a number of distinct notational configurations to get a handle on the model. The decision to use one form rather than another is not just a matter of aesthetic choice. Each modality is designed to open up new material and semiotic spaces for experimental manipulation. Each convention is a technique for making a different kind of argument about the same body of data, and each medium produces a distinct material analogy that articulates chemical configurations and atomic relationships in different ways. No expert modeler would suggest that any of these renderings resemble "actual" molecules.

In the modeling culture of extreme mimesis, the measure of a good model was its verisimilitude, and so modelers could take pleasure in the deceptions they fabricated. Where mimetic models of plants could produce what Daston calls the "pleasure of potential deception," protein crystallographers, like Perutz in his story about the queen and her party, are anxious about the deceptive power of their models. No chemist would want people to think we have little colored balls inside our bodies. As we shall see, crystallographers go to great efforts to ensure that novices do not mistake the model for the thing itself, as if the model were a life-like amplification of an actual molecule.

AESTHETIC SEDUCTIONS AND EPISTEMIC ANXIETIES

During fieldwork for this study, I was a research fellow on a collaborative National Science Foundation (NSF) study on computation, visualization, and changing professional identity across science, engineering, and design.[19] Our meetings brought together life scientists, architects, nuclear weapons designers, materials scientists, and engineers in the fields of biological, aeronautical, astronomical, and marine research. We were interested in hearing from them how their computation and simulations technologies posed new kinds of challenges in their respective fields. Did practitioners in these distinct fields face similar problems? In May 2005, Diane attended one of the meetings we organized for this project. Diane's experience resonated with the challenges that others encountered. Diane was anxious about the "pretty pictures" that people in her field could generate so easily, with all their new "fancy" software. She explained: "The way that we usually present our X-ray data is by making ribbon drawings of the protein structure [see figure 3.1] that just trace the backbone of the structure. It used to take a long time to make those pictures."

In an earlier interview she told me stories about the labors of image making when she was just starting out in the field in the late 1980s and early 1990s. In those days it was hard to make an image look good. She described what it took to get even a preliminary image for a talk or a publication: "You would put in your coordinates, . . . and then you get a kind of printout of your image, and then you would look at it and say, 'Oh, now that's rotated, you can't see that residue anymore.' Then you would have to go back and [go] into your code and guess that, maybe if you rotate back three degrees you could see that better. And, so then you would open up the program, you'd change it three degrees. You'd close it. You'd run it. It would generate it, and [it would] come up. . . . And no you had the wrong three degrees, it should have been plus and not minus." That, she told me, "was *really* frustrating": "it took forever!" "In the really old days," she recounted, "you would have a printout somewhere else and you have to go and physically pick it up and walk like five minutes, and you'd come back. And there would only be so many printouts per day. And you only have one shot." The glacial pace of this work affected the quality of the images people produced, so much so that people would "get close" to what they wanted, and "go, 'okay that's done.'" Innovations in software today enable crystallographers sitting at computers to "spit out" fancy graphics at high speed. Now, she told me, "you can really be a perfectionist in terms of the image you show."

At the NSF meeting she explained how this recently acquired speed and ease in making images is generating a new range of problems for adjudicating truth claims in her field. To demonstrate what was at stake, she discussed using the new images at scientific meetings:

> [In the past] if you were presenting the initial structure, say at a meeting, where it wasn't published yet, . . . [say if] it was kind of the original model and you weren't quite sure of everything yet, the picture you would show would represent that it wasn't really done. It wouldn't be a fancy picture yet because it would take such a long time to make it. Now you can make it in two seconds, you know . . . the program spits out pretty pictures and when you show that the people go "Oh! It's all done!" And you can stand up there and say, "These are sort of the distances but don't believe them. Big error bars! Not finished yet! Just a rough idea!" And they'll just hold on to it and go "This is done because look how pretty it is!"[20]

Beautiful images are powerful; they have a force that goes beyond their descriptive potential.[21] Compare the delight that intricate, atomic-scale models instilled in their users, to the repulsion viewers expressed when images of the Plasticine "sausage model" of myoglobin first circulated (see figures 2.1, 2.2, and 2.3). Beautiful images produce a kind of satisfaction that both sedates and seduces. Inscribed with significance, pretty pictures can override practitioners' critical faculties. Diane's rather drastic response bespeaks the severity of this problem for practitioners in the field: "So we now on purpose make ugly figures to show it's not really done yet. Because they don't listen to you: they see it with their eyes. [Laughter in the background.] You have to show them something ugly if you don't want them to set on this and have it be the truth forever."

Diane goes to extraordinary lengths to circumvent the stunning effect beautiful images have on the adjudication of facts in her field. She is concerned with the power of these alluring visualizations to secure assent. As she phrases it, pretty pictures speak louder than words: louder even than her urgent appeals to qualify the limits of the data. In this sense, her interventions aim to recalibrate how members of the protein structure community assess visual claims to truth. Chapter 5 explores how crystallographers adjudicate the truth status of their models more fully. For now, it is necessary to turn to the question of the limits of how these models relate to their referents.

Modelers employ a range of notational conventions and modeling media when they build molecular models. While each produces a distinct style, look, and feel, the choice of which form to use is not based on aesthetic preference; each form is recognized to generate particular effects. Anxieties escalate when stylistic conventions are seen to concretize the look and feel of molecular phenomena. In an interview with Diane, I asked about the challenges she faces as an educator teaching students how to think critically about how and what crystallographic models represent. Her response was telling:

> I've found that people will take a picture as a fact in a way that they really shouldn't. Because it is just one image, *and the protein can move and it can change*. And people come and say, "But I thought this distance [between amino acids] was 3 angstroms. Period."[22] And like, that's the end of the world! "It's three point zero, zero." You know, a structure can be incredibly valuable, but it is a *model*! A model of something that *moves*! And that really you need to think about it, and think about where the data came from, the quality of the data and all these other things. And it sometimes can be too powerful, and kind of stops people from thinking about something that they still should be thinking about.

Diane wants practitioners in this field to acknowledge that a molecular model is *just a model*; users must remember that it is a static rendering of a dynamic phenomenon. In this assertion, she appears to recognize the performative power of representations. That is, she acknowledges that representations don't just depict a world "out there"; they also have the recursive power to condition and sediment how people come to see that world.

Protein models are undeniably representations of molecules. Yet, standard accounts frame scientific representation as a practice of describing objects as if they are ready-made, existing "out there" just waiting to be discovered. In this view, to represent the world well scientists must work hard to reduce the epistemic uncertainties that plague their experiments: they must design better tools and employ more effective language to enable them to deliver a clearer picture of those objects that appear to hover just out of reach. Such an approach, however, assumes the world has a fixed ontology that preexists its encounter with the scientist. To engage model making as a rendering practice is to insist that protein models do more than just re-present molecular phenomena. Indeed, these models *rend* the world in particular ways: they pull, tear, and torque the world in some ways (if not others). In the process they shape how and what we come know about the stuff of life.

Karen Barad draws on queer and feminist theories to argue for "performative alternatives to representationalism."[23] For her, performativity "shifts the focus from questions of correspondence between descriptions and reality (for example, do they mirror nature or culture?) to matters of practices/doings/actions."[24] She develops the concept of "intra-action" to call attention to the impossibility of disentangling experimenters from the objects, apparatuses, and practices they engage to draw phenomena into view.[25] The experimental scene is for her a thick entanglement of bodies, machines, discourses, and ways of seeing, through which some phenomena are materialized, to the exclusion of others. The entire experimental configuration embodies a mode of inquiry, a line of questioning, and a kind of curiosity. A crucial point is that depending on the experimental configuration, some kinds of phenomena *come to matter*, while others are rendered invisible.[26] Where conventional approaches to representation assume that scientific visualization technologies capture and register the "signatures" of nature, for Barad, Haraway, and other science studies scholars, scientific visualization is not a passive act of observation; it is an intra-active practice of world making.[27] And where Barad is ready to eschew representation, I engage the concept of rendering here as an intra-active, performative approach to representation that keeps in view modelers' desire to produce proxies that can stand in for otherwise unseen phenomena. In this sense, the concept of "intra-action" makes palpable the ways that protein crystallographers' models are not "discoveries" of an already-existing molecular world; rather, they are renderings that pull and tear the world into forms manageable and imaginable within the contemporary biosciences.

How then can this approach transform how we understand the very things that protein crystallographers struggle to model? What is "a molecule" in an intra-active, performative theory of representation? Hans-Jörg Rheinberger's distinction between "epistemic things" and "technical objects," which helped in chapter 2 to illuminate the various ways that models can be enlisted in experimental inquiry, does not hold so well in this context.[28] In the case of protein crystallography, one might presume that the epistemic thing is the unknown molecular configuration of some biological substance. However, Rheinberger's distinction between epistemic things and technical objects is not so clear in the case of protein crystallography. This is because before a biological substance is crystallized for X-ray diffraction experiments, practitioners have already anticipated the inner constitution of this substance as molecular. This might seem like a trite assertion; however, it is necessary to acknowledge how the atomic model of matter is itself materially and semiotically embedded in the crystallographic apparatus.[29] In other words, the stuff that protein crystallographers handle in their laboratories is not some epistemic thing or unknown substance; it is already an aggregate of technical objects, theories, and ways of seeing.

How does this distinction matter to protein crystallographers? First, modelers must recognize that before they apply themselves to the task of visualizing a substance at the atomic-scale, they have already secured the "object-hood" of this substance. To divide a substance up into its molecular elements is already to perform an abstraction that organizes a substance into the parts that matter; *molecules are what matter* to these practitioners and *what come to matter* in their hands.[30] Once anticipated as molecular, the substances of life acquire specific physical, chemical, and biological properties, and crystallographic models sediment and materialize molecular phenomena as such. In this sense, molecules don't just exist "out there"; they are the robust, material effects of a constellation of experimental apparatuses, discourses, and theoretical frameworks intra-acting with material substance. Thus, as modelers grapple with the power they have to (mis)represent the molecular realm through their models, they must also recognize that the object-hood of a substance is in-the-making in their hands.

Diane's concern about crystallographic models being "too powerful" suggests that she recognizes how protein models rend the world into molecules. Her attempt to deflect the power of these models by producing ugly images resonates with Alfred North Whitehead's efforts to point out the "fallacy of misplaced concreteness" alive in philosophy and the sciences. Isabelle Stengers's "constructivist reading" of Whitehead is illuminating in this regard: "Whitehead implies that if we are not prisoners of our abstractions, then we may well become prisoners of the false problems they are bound to create if we extend, outside their specialized domain, the trust they deserve only inside this domain. . . . Whitehead maintained that the challenge for philosophy was to resist this fallacy, that is, to resist the concrete character that our modern epoch has attributed to its most powerful abstractions."[31] If we understand abstractions to include things like atomic-scale molecular models, it is possible to see what kinds of problems arise if you "trust" these models to tell you something about the world beyond the "specialized domain" in which they were crafted. Whitehead would insist that these models should not be allowed to stray beyond this limited use for which they were designed. The person who picks up a model and registers it as a realistic description of a molecule has reproduced the fallacy of misplaced concreteness. This slippage effects a closure of the gap between the model and its object, sediments ways of thinking, and limits how the stuff of life can be imaged and imagined. Models can accrue a density and a rigidity that do not do justice to the ephemerality and energetics of molecular phenomena. As Diane continually reminds her students, these are *just models*, and modelers must keep their interpretations open: both the inner constitution of living substances and the models practitioners render must be understood as dynamic *phenomena* that are subject to change with shifting theories and experimental apparatuses.

RENDERING AS ENACTMENT

> [T]he term "model" is probably best understood as a verb, with the authors as subject, and the experiments and the conceptual schematic as a single, unparseable, composite object. Only at the end of the process do we have a separable entity— a model as a noun . . .
>
> —Evelyn Fox Keller, "Models of and Models For"

What if anxieties about the representational status of these models were held in abeyance for a moment? What if, rather than worrying about the power of these models to deceive their users, attention was turned to how these models are productive? Rather than looking at what these models fail to do (that is, replicate "real" molecules), we might pay attention to what it is that they accomplish. We can ask: what other than a (mis)representation is generated when crystallographers build protein models? Evelyn Fox Keller's insight in the epigraph above is a great place to begin.

As representations, protein models can be taken as what Keller calls "models as nouns," that is, as "separable entities" at the "end of the process" of X-ray diffraction experiments and model-building efforts.[32] Yet Keller introduces a crucial distinction that shifts attention to the significance of model building as a practice. Rather than focusing on models as representations at the end stage of experiments, she treats models as "verbs"; that is, as actions performed by modelers. In this view, models are not just things that stand in for knowledge or phenomena; they are also enactments that generate new ways of knowing and things known.[33] This view has implications for how we understand crystallographic models. Indeed, it is in the activity of model building that a modeler both sediments the object-hood of the molecule and acquires a freshly articulated feeling for the molecular realm. Recall Wilhelm His's approach to modeling embryological development (chapter 3). It was by working directly with tangible modeling materials that he was able to formulate kinesthetically and affectively informed theories about developmental processes. An embryological model was for him not only a representation; it was also a form of knowing.

Examining "models as verbs" requires a shift from a discourse of representation—which too easily slips into the realm of "model as noun"—to one of rendering, an idiom that can gather up modelers, their objects, and modeling media in the very activity of model making. In ways similar to the double meaning of Foucault's concept of "sympathy," which generates both a "likeness" and a "feeling," the concept of rendering tethers the representational to the kinesthetic and affective dimensions of model making. A protein model, then, is not just an object at the end stage of model building; modelers articulate their knowledge of molecules through the model-building process, and in so doing they also inflect their models with their own sensibilities and intuitions. A model is a rendering in the sense that it embodies, performs, and sediments a modeler's form of knowing. In the making, models are inflected with the affects of their modelers, and these inflections engender further effects as models are put into circulation. If models are simultaneously material, semiotic, kinesthetic, and affective, then approaching them as renderings provokes inquiry not only into what and how a model represents, but also into the ways a model rends perceptions, and in the process reconfigures how we conceive of the order of things.

FROM DECEPTIVE MODELS TO MODELS AS LURES

The epistemic anxieties that propagate among protein crystallographers relate to serious issues that they confront daily. Engaged uncritically, protein models can easily concretize the look and feel of the subvisible molecular world—not just for lay viewers. And yet a model's power is also generative. Its power to deceive is matched by its power to produce new kinds of inquiry. The example of Perutz and the queen's party is instructive here. Engaged as mirror-image representations, molecular models can indeed induce anxieties. Yet treated as renderings, they not only generate new forms of knowing as they are being built; they also can provoke new ways of thinking in their makers, users, and viewers. Isabelle Stengers introduces the concept of the "lure" to distinguish differences in the ways that experts and novices engage abstractions in mathematics.[34] She describes the different ways that experts and novices approach mathematical problems:

> Just think of the difference between the mute perplexity and disarray of anybody who faces a mathematical proposition or equation as a meaningless sequence of signs, as opposed to someone who looks at this same sequence and immediately knows how to deal with it, or is passionately aware that a new possibility for doing mathematics may be present. In order to think abstractions in Whitehead's [constructivist] sense, we need to forget about nouns like "a table" or "a human being," and to think rather about a mathematical circle. Such a circle is not abstracted from concrete circular forms; its mode of abstraction is related to its functioning as a lure for mathematical thought—it lures mathematicians into adventures which produce new aspects of what it means to be a circle [in] a mathematical mode of existence.[35]

Stengers's example of a mathematical circle shows how such abstractions operate as "lures" most especially for trained mathematicians; that is, for those who are prepared to be pulled into an "adventure" into the world of circles as mathematical objects. For Stengers, abstractions, like models, are propositions that propel the imagination toward the experience of "sheer disclosure." Effective models can produce what she calls an "empirically felt elucidation of our experience"; they "act as 'lures,' luring attention toward 'something that matters.'" As a lure, a model sets the expert modeler in motion, "vectorizing concrete experience."[36] Indeed, Diane's experience feeling the pain of a misfolded protein, or Edward's insistence that his molecule breathes, helps make clear what Stengers is suggesting. Consider how protein models can "vectorize" the affects and kinesthetic imagination of an expert crystallographer. When renderings are engaged as lures, it is less important what they look like; rather, what matters is how they draw modelers into new adventures in molecular realms.

CONCLUSION

Rendering can be understood as an intra-active, performative practice. This concept helps us see how both protein models and their molecular referents are phenomena that are manifest sometimes precariously, and sometimes robustly, in the complex "ontological choreography" of the laboratory.[37] Pushing past the epistemic anxieties that plague modelers' efforts to represent the molecular world well, I suggest that models' representational failings do not necessarily limit their usefulness to modelers. A good model is one that can open up rather than limit inquiry, one that can lure practitioners into new ways of seeing, saying, feeling, and knowing. When modelers approach their models as more than merely flawed representations of "real" molecules, they are explicit about model building as a site for training perceptions, sensibilities, and dexterities, and for generating lures for new adventures in experimental inquiry. Perhaps this approach to the luring, performative nature of renderings helps practitioners avow their contributions to model making. The next chapter explores more fully how practitioners adjudicate the truth status of their models while situating the limits of their knowledge and their claims to truth.

"THE GREAT PENTARETRACTION"

In December 2006, Geoffrey Chang, a young and stunningly successful principal investigator heading a protein crystallography lab in the Department of Molecular Biology at the Scripps Research Institute, published a letter in *Science* retracting five high-profile research papers in which he and his lab had presented novel structures of a series of membrane-bound proteins.[1] Three of the papers had been published in *Science* (one in 2001, two in 2005), one in *Nature* (in 2006), and another in the *Proceedings of the National Academy of Sciences* (in 2004). These retractions came in response to a paper published in September 2006 in *Nature* by a team of Swiss researchers. That paper provided evidence that the structures that Chang's laboratory had produced were wrong.[2] When Chang investigated, he was "horrified to discover" that a "homemade data-analysis program" had "inverted" two columns of data, skewing the data and massively contorting the protein structures his lab had built. This piece of software, which had been "inherited" from another laboratory, had been used on all five of the structures he had published.[3]

In January 2007, Chris Miller, a biochemist and protein crystallographer based at Brandeis University, published a letter in *Science* in response to what he called Chang's "Great Pentaretraction."[4] In this letter, Miller turns Chang's "devastation" into a cautionary tale presenting a tough lesson in pedagogy and training for students in the arts of molecular visualization:[5]

[W]hile an embarrassment to the authors, [it] nevertheless provides the rest of the field with some small measure of comfort beyond mere schadenfreude. The mistake so clearly illustrates two lessons that we aging baby boomer professors ram down the throats of our proteomically aroused graduate students: (i) that those lovely colored ribbons festooning the covers and pages of journals are just models, not data, and (ii) that you invite disaster if you don't know what your software is actually doing down there in the computational trenches. Students have a hard time subsuming these dicta into their souls for two reasons: the tyranny of authority (the vanity journals occupying the vanguard) and the inherent beauty of the macromolecular models that emerge, as if by magic, from the user-friendly crystallographic software accumulated over decades through the generous labor of the field's talented reciprocal space-cadets.[6]

Recall Diane's anxiety about the seduction of beautiful models documented in chapter 4. Miller shares her concern. For him the "beauty of the macromolecular structures" seems to overpower students' critical faculties. The "proteomically aroused" graduate students Miller writes about here risk activating the fallacy of misplaced concreteness when they forget that the models they are working with are "just models." Miller points to two key challenges crystallographers currently face in training their excitable, easily enchanted students: (1) how to teach them to be wary of the awesome and dangerous beauty of their models; and (2) how to get them to think critically about how their labor-saving computer software helps them to build these structures.

In this brief manifesto, Miller tethers the problems of visualization to problems in pedagogy and training and locates this knot as a central matter of concern for the future of his field. Miller has his finger on the pulse of what is at stake in structural biology today: How, in the face of increasingly alluring graphics and automated programs, can crystallographers cultivate and propagate the craft skills that are crucial to producing robust visual facts? What would it take to train students to ensure that their affective and kinesthetic dexterities are attuned both to proper technique and a respect for protein form?

Getting the structure of these proteins "right" is a high-stakes endeavor. Three of the protein models that Chang's group botched belong to the MsbA group of proteins, an "ancient family" of membrane proteins whose biological function—the active (energy-dependent) transport of molecules across cell membranes—has captured the interest of biochemists and pharmaceutical researchers.[7] These proteins are referred to as "membrane pumps," because they "pump" unwanted molecules out of the cell. Molecular pumps are of "great clinical interest" because they enable certain cells, such as bacterial and cancer cells, to continually rid themselves of unwanted substances.[8] These pumps help pathogenic cells dispose of toxins, including the anticancer drugs and antibiotics that researchers have invested so much energy developing.[9]

Chang's retractions show that visual seduction is a serious problem for expert practitioners, not just gullible students. Visual evidence is so "powerful" that it can often override even expert critical faculties. For example, when Chang published the first MsbA structure in 2001, it caused a stir among biochemists familiar with the molecule. Chang's structures just did not make sense in light of available biochemical evidence. In an e-mail exchange several months after the retraction one crystallographer told me that "no one in the field believed his structures. I am not in the field, but I overheard people talk. His structures went against ALL the biochemistry—that tells you that something is wrong."

As reported in a news article that appeared with Chang's retractions in *Science*, one biochemist admitted: "When the first structure came out, we and others said, 'We really don't quite believe this is right.' It was inconsistent with a lot of things."[10] This same biochemist told a journalist at the *Scientist* that in 2001, when he and his colleagues first looked at one of the structures, they "knew it was wrong, but couldn't prove why."[11] Others reportedly experienced difficulties "persuading journals" to accept their publications especially if their results did not agree with Chang's structure.[12] One biochemist who had served on funding panels suggested that "Chang's work was influential" and shaped decisions made by funders: "Those applications providing preliminary results that were not in agreement with the retracted papers were given a rough time."[13] It appears as though the allure of visual evidence blinded members of Chang's group, the reviewers in high-profile journals, and the very funding bodies that set directions for research. Biochemical studies "published in unglamorous workaday journals" were "dismissed as just old-fashioned biochemistry," and as such could not stand as persuasive evidence in the face of visually compelling data.[14] Chang's models demonstrate the "persuasive" power of a picture to silence even expert critics.[15] Once published and affixed in the literature, his structures—whether people believed them or not—stood as the officially sanctioned models of these proteins for six years. The enduring allure of these models had serious consequences for practitioners in several fields.

Chris Miller's pedagogical campaign is directed to experts, novice modelers, and visually stunned peer-reviewers. With Diane and other practitioners working in a range of disciplines, Miller shares a concern about the seductive capacity of computer-mediated visualization technologies. He and Diane both insist that a protein model is *just a model*. It is at its barest an approximate set of atomic coordinates, points in space that have been overlaid by a range of diagrammatic conventions that instruct the viewer how to look and how to interact with the data in different ways. The issues that Miller raises here go beyond anxieties about the ways that models concretize the look and feel of the molecular world. Here he homes in on the problems of quality control, the accuracy of the data and the model, and the looming possibility of technical error and interpretive failure. So, while the last chapter focused on the crisis of representation in protein crystallography, and how modelers contend with the gap between their renderings and the phenomena they hope to describe, this chapter examines the peculiar culture of objectivity that shapes how modelers adjudicate the veracity of their models.

Miller reminds us of a crucial problem: it is possible to get the model wrong. Recall Diane's anguished response to her students' mangled models. She makes clear how embodied knowledge plays a role in evaluating the truth status of a model. How is this "feeling for the molecule," this affectively charged kinesthetic knowledge, bound up in a larger constellation of disciplinary values, norms, and mores that shape how practitioners adjudicate their facts? On what basis do practitioners evaluate whether a model is good or bad? What is the measure of a good model and a good modeler? How are claims to truth asserted and policed? This chapter examines how efforts to resolve the problem of the limits of knowledge in this field revolve around pedagogical efforts to instill "epistemic virtues" and decry "epistemic vices."[16] Objective knowledge in protein crystallography is in this sense conditioned by *moral, affective, and kinesthetic entanglements* that entrain students to proper conduct in the laboratory.

In addition to the "great pentaretraction," this chapter documents protein crystallographers' reactions to another event, the retraction of twelve published structures in response to accusations that the data were fabricated. These two retraction events raise serious allegations about the hazards of poor technique, improper conduct, and the moral failings that thwart attempts to secure "objective" truth. They point to the profound epistemological uncertainties that plague protein crystallographers, and both events spur renewed vigilance and caution about published structures. This preoccupation with *epistemic uncertainties* can, however, leave intact a view that the world is made of determinate objects, hovering about somewhere "out there," just waiting to be described by better tools and more virtuous modelers. Yet, perhaps not all the error inheres in the modelers and their technologies.

Approaching crystallography as a rendering practice helps to keep the ontological stakes in view. This chapter moves to document an episode when two models of the same molecular assemblage were published side by side in the journal *Science*. It shows how crystallographers must simultaneously grapple not only with epistemic uncertainty, but also with the *ontological indeterminacies* that make it hard to render molecules well.[17] Remarkably, these three events demonstrate that in practice objectivity bears no resemblance to the mythical idealizations of detachment and neutrality that pervade popular conceptions; rather the practitioners described here assert themselves as *modest modelers*, avowing a *situated objectivity* founded on local, partial truths, and committed to rendering the excitability of an indeterminate molecular world.[18]

OBJECTIVITY'S ENTANGLEMENTS

Objectivity has a history, and what has come to count as objective knowledge has changed remarkably with shifting scientific practices and cultures.[19] Standard accounts of objectivity tend to set up a dichotomy between facts and values, where values "mix with scientific knowledge, but only as a contaminant."[20] This analysis begins rather from the premise that mores, norms, and values are, as Lorraine Daston insists, "integral to science: to its sources of inspiration, its choice of subject matter and procedures, its sifting of evidence, and its standards of explanation."[21] Close examination of responses to the retraction of Chang's papers gives us a more nuanced understanding of the complex relationship between ethics and objectivity in science and makes palpable some of the "social facts" that hold protein crystallographers together as a scientific community today.[22] In other words, this episode makes explicit the otherwise tacit mores and norms that condition how crystallographers adjudicate their facts and make claims to truth.

Daston insists that efforts to ascertain objective knowledge cannot be separated from the mores and norms that govern claims to truth. Indeed, what gets to count as objective knowledge at any moment in history is shaped by a *moral economy* that conditions what counts as good and bad practice, what counts as a good and bad practitioner.[23] For Daston, a moral economy is "a web of affect-saturated values that stand and function in well-defined relationship to one another."[24] Moral economies take shape around particular disciplinary problems, techniques, and aspirations. She notes, for example, that augmented capacities for "precision measurement" in a variety of disciplines shaped a "moral economy of quantification," one that was "stern in its call for self-discipline."[25] Practitioners were encouraged to cultivate "certain personal idiosyncrasies, namely those of skill and, especially, the character traits of diligence, fastidiousness, thoroughness, and caution."[26] Where some attitudes were celebrated as "epistemic virtues," others were denounced as vices.[27] The characteristics that Daston describes suggest that a moral economy does not just secure the norms of a community, but also conditions the *ethos* and *habitus* of its practitioners.[28] In this sense, practitioners' affects, attitudes, sensibilities, and comportments are oriented to what counts as good and bad science, and what is to be valued, and what is to be disparaged. In this sense mores and norms condition membership, recognition, and authority in science, and these values take shape through a complex constellation of relationships, rather than through a set of explicit codes or strictures. In this view values, sensibilities, and postures are always in-the-making in a given scientific practice and culture.

Let's take a closer look at the concept of moral economy. Consider how the term "economy" operates here. For Daston, economy does not necessarily refer to "money, markets, labor, production, and distribution of material resources."[29] Moral economies are economical for her in the sense that they are "regulated" in the form of a "balanced system of emotional forces, with equilibrium points and constraints."[30] Historian of science Robert Kohler on the other hand explicitly develops the economic dimensions of this concept to explore how the moral economies of early twentieth-century genetics laboratories were dedicated to the mapping of *Drosophila* chromosomes. He outlines how a moral economy explicitly governed exchange relations, especially the exchange of laboratory materials, postdoctoral researchers, living organisms, and data.[31] At the same time, for Kohler, moral economies include the "social rules and customs that regulate such crucial aspects of community life as access to workplaces and tools of production, authority over research agendas, and allocation of credit for achievement."[32] He foregrounds the ways that moral economies shape how recruits are socialized, how problems are identified as "doable," and how resources are "mobilized."[33]

Kohler also offers a remarkable description of the ways that *Drosophila* researchers were captivated by the remarkable fecundity of their breeding flies. The "breeder reactors" they set up in their experiments had the unforeseen effect of capturing the researchers inside their experiments, to such an extent that the rhythms of researchers' lives became entrained to the breeding cycles of the flies. His account offers a potent reminder of the ways that laboratories entangle humans, nonhumans, and machines in ways that are perhaps not so regulated or reducible to economic relations. As part I of this book explored, affects move and values coalesce in surprising ways as modelers work closely with their molecules, models, and machines. In an effort to move away from functionalist analyses, which tend to treat mores, norms, and affects as "regulators" of social functions, my account pays attention to the ambivalences and indeterminacies that contour subjectivities and objectivities in science. The dense thicket of moral and affective entanglements I observe in the protein crystallography lab can perhaps be better theorized on the topology of an *ecology* rather than *economy*.[34]

Ethnographic attention to the moral and affective ecologies of the laboratory draws out crystallographers' often-ambivalent relations not only with other practitioners, but also with nonhuman objects, instruments, and metaphors as they try to stay attuned to the task of crafting robust facts. Consider, for example, the affective entanglements that thrive in protein crystallography laboratories, including the sheer effort, exhaustion, frustration, excitement, devastation, and exuberance that shape modelers' relations with their lively crystals and machines (chapter 1). These entanglements take shape in surprising ways. Crystallographers must go beyond satisfying the technical requirements of an experiment: they are bound in relations of obligation to care for all elements of an experimental configuration. Indeed, a core epistemic virtue that practitioners must cultivate is a kinesthetic and affective attunement to protein form. In this context at least, it seems as if "objectivity" is this very obligation modelers feel toward careful description of their rather evasive objects. In other words, objectivity names the peculiar ecology of moral and affective entanglements that tether modelers to the substances they aim to model.

EPISTEMIC VIRTUES

Viewed as an infraction in a moral ecology, it is possible to see the nature of Chang's offense: he and his students were not properly attuned to their molecules, models, or machines. What then are the epistemic virtues celebrated by protein crystallographers? What is the ethos and habitus modelers hope to cultivate to secure access to objective facts? Which dispositions, sentiments, attitudes, and postures are modelers-in-training encouraged—both tacitly and explicitly—to cultivate in the lab? In a discipline that grapples without a visible referent, what counts as "right depiction," and what is the measure of a robust model? And crucially, who has the authority to decide?

Protein models are renderings; they are inflected by the sensibilities and intuitions of their modelers. Rather than policing these subjective inflections, practitioners celebrate their contributions to model-building efforts. Theirs is not a "mechanical objectivity" that insists on "self-effacement" and "emotional detachment"; rather protein crystallographers exercise a practice of modeling that relies extensively on "trained judgment."[35] Modelers recognize their contributions to model building as an "epistemic virtue."[36] This means that they celebrate their active participation in model building, recognizing that they are integral elements in the "human-computer lens." Crystallographers' expert judgments are crucial at every stage of the process, from designing and purifying proteins, to testing crystallization protocols, to maintaining X-ray diffraction machinery and assessing the accuracy of their computer algorithms as they build their interactive computer graphic models by hand and by eye.

Given its reliance on a carefully honed intuition and trained judgment, it is no surprise that concerns about what constitutes a sound molecular model are closely coupled with a growing anxiety about what constitutes "proper training." In interviews with novices and experts, I learned that a properly trained crystallographer appears to be one who can discriminate between good and bad practice, and good and bad data and models. Throughout their training protein modelers are instructed how to stand (both literally and figuratively) in relation to their instruments, materials, and models in order to ensure the renderings they generate are robust. This requires staying vigilant against errors introduced through automated computer algorithms, and cultivating sound judgment to ensure that the "human" part of the human-computer lens is entrained to protein form.

Best practices in protein crystallography are conditioned by the virtues of caution and care. Expert judgment must be cultivated and applied *modestly*; knowledge must be *situated*; facts must be qualified, located, and recognized as *partial*; and renderings must be recognized to have performative effects. This culture of objectivity and its associated ecology of moral and affective entanglements resonate remarkably with Donna Haraway's calls for "situated knowledges" and "modest witnessing" in science.[37]

AUTOMATION AS EPISTEMIC VICE

A generational transition is underway today as increased reliance on computer technologies sets up a new division of labor in protein crystallography laboratories. Miller and his "aging baby boomer" colleagues grew up in an era when they had to write their own codes: until the late 1990s protein crystallographers developed their own algorithms to process and visualize crystallographic data. Diane is also part of this generation: her students call her "an old-school crystallographer," and she proudly dons this moniker. As a graduate student she wrote her own code using FORTRAN. She knows how her programs work; she can anticipate the kinds of bugs that are hard to avoid when writing code; and so she understands how easy it is to introduce errors in a crystallographic data set.[38]

In other areas of scientific visualization, automation is seen as the best way to remove bias and produce objective results.[39] This assumption does not hold in the field of protein crystallography, however, where resistance to complete automation renders modelers as artisans with craft skills. Recall, for example, the lesson learned in the Chang case where the models were mangled by a glitch in a computer program: left to their own devices machines and algorithms can easily propagate errors and botch the model-building process. In protein crystallography the only way to ensure quality data and prevent error is through constant attention and intervention in computing processes. These constant interventions are based on a fundamental mistrust of computers. Trust is a crucial issue for crystallographers, who must work in the shadows of their diffraction patterns to resolve models of objects that are otherwise imperceptible. Recall Zeynep's challenge learning to trust her engineered peptides (chapter 1), and Jess's mistrust of the "executive decisions" novice modelers make in the building process (chapter 2). A novice must learn when trust is warranted, and what should not be trusted. Trust and distrust take form in modelers' ongoing negotiations with their materials, models, and machines. This is part of the training of a novice's ethos and habitus.

Diane does not trust software to handle the data correctly: she insists that crystallographers must be vigilant and constantly keep computed processes in check. Today, many of the model-building programs have been standardized, and entire suites of software have been made available to crystallographers to process and visualize their data. "Master crystallographers" (whom Chris Miller refers to as "reciprocal space-cadets") have "generously" given their labor and their skills over to developing software and interfaces that facilitate model building. However, once these skills and this knowledge get embedded in computer algorithms they become "black boxed."[40] As such, they can no longer play an explicit pedagogical role in training future crystallographers. Working only at the "surface" of a user-friendly interface, students are no longer trained to "dig into the guts" of the codes.[41] Many have no clue how their algorithms manipulate the diffraction data they gather from their X-ray experiments.

Diane resists the automation of what she sees as a crystallographer's irreducible contribution to model building. She recognizes that the human part of the "human-computer lens" is critical to the production of sound structures.

Recall my conversation with Diane's postdoc Edward featured in the introduction to this book. He was particularly frustrated with the "garbage" that his automated computer program was spewing out. Like him, many expressed reservations about automation, particularly in the form of automated modeling tools. Where some algorithms are useful for predicting the structures of small molecules, they are not trustworthy for modeling large, complex molecules like proteins.

Diane wants her students to cultivate sound intuitions and judgments, and so she banned the use of automated model-building programs in her laboratory. One story was repeated to me on several occasions in conversation with her and in interviews with her students. At some point earlier on in the lab's history, she discovered that one of her students had used predictive modeling programs to help build part of a structure. She was so angered that she made the student repeat the modeling work by hand. Her mistrust of automation was vindicated when the student discovered that the computer program had failed to predict the correct structure. Automation amounts to "cheating" for Diane. This is in part because it sidesteps the laborious training process necessary for crystallographers to cultivate their skills and to develop and exercise expert judgment. Though they do crave shortcuts that could get their work moving faster, Diane's students generally support her approach. Several were concerned that predictive software may one day make their labor and knowledge obsolete. In some ways their resistance to automation can be interpreted as a means of preserving professional jurisdiction over their craft.

THE ANATHEMA OF FABRICATION

Diane takes seriously her role as an educator. She has put extensive effort into training members of her laboratory, as well as graduate and undergraduate students across her campus, how to stay attentive to the sources of error in crystallographic structures and to cultivate and exercise the skills they need to adjudicate those structures published in scientific papers. I sat in on her graduate course in macromolecular crystallography and attended her guest lectures in other courses from 2003 to 2005. In each of these venues, she aimed to give students a sense of the kinds of skills they would need in order to make sense of the crystal structures and data they encounter in the pages of prominent scientific journals. She wanted to give them some important clues to navigating the tricky terrain of multifarious data forms, calculations, and statistics. And she also wanted to make it clear why they needed these critical skills. To do this she liked to give them a glimpse into the "dark, dirty secrets of crystallography." She wanted them to be able to see "where structures can go wrong": "I do a recitation [tutorial] on how you evaluate a structure. And so I tell them all the ways that they can cheat. You know, what are good statistics, and how can you make your statistics look better when they are not good. And so, I go through this whole lecture on how to really evaluate a paper. And what are the important caveats in the structures? And, you know, how do you know what's real? How do you know what to believe? You know, part of me is on sort of a little bit of a crusade, because I think there is a lot of wrong information out there."

As Chang's retractions demonstrate, there is indeed "wrong information out there." Diane's efforts to educate her students are a call for vigilance against a range of "bad" practices. Chang's mangled models appear to be the product of negligence and a lack of rigorous evaluation. But Diane isn't concerned only with sloppy work. I didn't fully understand Diane's concern about "cheating" until December 2009, when the University of Alabama at Birmingham requested that the Protein Data Bank retract twelve entries that had been filed by one of their former employees. The university's website published this statement: "After a thorough examination of the available data, which included a re-analysis of each structure alleged to have been fabricated, the committee found a preponderance of evidence that structures [12 PBD entries] were more likely than not falsified and/or fabricated and recommended that they be removed from the public record."[42]

This statement concluded the university's intensive two-year investigation. Krishna Murthy, the crystallographer whose work was in question, had trained as a postdoc in Nobel Prize laureate Thomas Steitz's laboratory at Yale and in a prominent lab at Columbia University before founding his own lab at the University of Alabama. He was by all measures a well-trained protein crystallographer. And yet, Murthy's findings first came under scrutiny when questions were raised about a structure his group published in *Nature*.[43] The protein was C3b, a key molecule involved in human immune response. This molecule was of significant interest for pharmaceutical development, and competition among other labs in the field was intense. Murthy's structures were published simultaneously alongside structures of the identical protein generated by two other groups.[44] One of the groups detected some odd anomalies between their data and Murthy's findings and set up a team of experts to take a closer look at Murthy's data. They published "a brief communication" in *Nature* in 2007 noting major inconsistencies "with the known physical properties of macromolecular structures and their diffraction data."[45] They found that Krishna

Murthy's data were "physically implausible"; they were "consistent with protein molecules in a vacuum but not with those surrounded by disordered solvent, as is always seen for macromolecular crystals."[46] The physical implausibility of the data suggested that it was fabricated rather than generated from X-ray diffraction experiments with crystalline proteins. Murthy's group published a response defending their structures.[47] However, the university committee appeared to have gathered further evidence of fabrication, and the case moved forward.

Whether or not Murthy's structures were fabricated, what is clear is that the mores and norms that condition what gets to count as a sound model, and who gets to stand as a good crystallographer, are made palpable precisely in the ways that a community handles perceived infractions. One response to the announcement about these retractions came from Stephen Curry, a protein crystallographer at Imperial College London, UK. His blog on the website nature.com featured a review of the documentary *Naturally Obsessed*. He and his lab group had viewed the documentary around the same time news of the retractions broke:

> Our screening was quite timely, following hard on the heels of a day that had seen the best and worst of crystallography, confirming it as a tortuous vocation that offers huge rewards but only at the price of severely stressing its adherents. On Thursday the Nobel Prize for chemistry was presented to [Venkatraman] Ramakrishnan, [Thomas] Steitz and [Ada] Yonath for determining the crystal structure of the ribosome, and another potentially prize-winning structure appeared on the cover of Nature. But that day also brought news that no fewer than twelve crystal structures published . . . were to be withdrawn from the protein data bank and the associated papers retracted because the university had a "preponderance of evidence" that the structures had been fabricated. This was a truly shocking revelation, one that the community is struggling to absorb.[48]

As another commentator in *Nature* put it, "The fraud is the largest ever in protein crystallography."[49] And the consequences were significant. These papers had accumulated more than 450 citations, and "the disputed structures had important implications for discovering drugs against dengue virus and for understanding the human immune system." Indeed, the results of these publications sent the "hunt for drugs" against the denge virus "down a blind alley."[50]

One outcome of this event was increased mistrust of models and modelers, with the decision to implement new policies around structure validation and peer review. The World Wide Protein Data Bank (wwPDB), a consortium of international protein repositories, "convened expert, community-driven Validation Task Forces" "to advise on the most suitable criteria to use for validating structure entries (model, data and fit of model to data) when they are deposited." The outcome of these task forces was the development of "dedicated validation pipelines that use community-agreed software to assess the quality and reliability of the atomic models, the underlying experimental data, and the fit of the models to the data."[51] As of August 2013, groups must validate their models using approved protocols prior to uploading the coordinates of new crystal structures.

Members of the protein crystallography community were clearly alarmed by news that published structures had been fabricated. Their shock makes palpable the otherwise unspoken norms that contour what counts as objectivity in this field. While it is true that protein crystallographers make extensive contributions to their models in the form of constant and engaged interventions in the data, there appears to be a clear line between proper modeling and fabrication that must not be breached. Those who fabricate their data or falsify their statistics are defying one of the field's core epistemic virtues: proper attunement to protein form.

ONTICS

The epistemic virtues and vices described above focus attention on the uncertainties that plague structural biologists as they attempt to model molecules at atomic resolution. The challenge these practitioners face is staying true to protein form. Their efforts are hampered not only because there is no immediately visible referent after which they can fashion their models, but also because they work in a context where "cheating," bad training, poor judgment, and error-ridden algorithms make it hard to describe molecules well. All the examples described thus far suggest that error inheres in the domain of the modelers: practitioners' failing tools and imperfect techniques, their lackadaisical attitudes and moral culpability, are what prevent them from properly describing the atomic configuration of molecules. Are there perhaps other sources of error in crystallographic models?

Efforts to police crystallographers' practices are grounded in a set of assumptions that hold that molecules are "determinate objects with determinate properties."[52] If only crystallographers could perfect their techniques and keep their students and colleagues attuned to the same epistemic virtues, their models would stay true to the molecules found in nature. And yet this assumes that molecules are preexisting objects "out there" waiting to be described. The world, in this view, has a fixed ontology: what holds us back from grasping the world's sheer positivity is the fallibility of human knowing. As chapter 4 explored, modelers have a sophisticated approach to the ontological status of the stuff that they aim to model. When protein crystallography is approached as a rendering practice it is possible to see that both molecules and the models that describe them are "phenomena"; that is, they are inseparable from the experimental apparatuses engaged to bring them into view.[53] Science studies scholar Astrid Schrader makes this inseparability clear. She examines the distinction between the epistemic and ontological issues that scientists confront in the context of research into toxic marine microbes.[54] For Schrader, "epistemological uncertainties" are those that "refer to 'gaps' in or incompleteness of human knowledges."[55] For her, "the very idea of an epistemological uncertainty presupposes an a priori separation of the epistemological question of 'how we know' from the ontological status of 'what we know,' where only the former, that is, our knowledge is allowed to vary."[56]

Schrader draws attention to not just the "epistemic uncertainties," but also the "ontological indeterminacies" that plague scientific practice. Recall that one feature of the moral and affective entanglements that condition the culture of objectivity in this field is a respect for and attunement to protein form. What, then, are the ontological indeterminacies that these practitioners struggle with? Consider the challenges protein crystallographers face trying to coax their proteins to crystallize. Practitioners report proteins to be obdurate, reluctant, and recalcitrant substances. Recall as well Edward's surprising assertion that molecules "breathe" (introduction) and Diane's comparison of molecular movements to her dog's excitability in front of the camera (chapter 1). If proteins move and change, if they get excited and wiggle about, then what counts as proper molecular form? As excitable materials, these substances appear to resist attempts to secure their object-hood.

MOLECULAR ANTICS

Another episode in the recent history of protein crystallography offers some insight here. In 2003, *Science* published crystal structures of the same molecule generated by two separate groups (see figure 5.1).[57] The two papers were published side by side, and a commentary comparing the structures was published in the "Perspectives" section of the journal.[58] What was remarkable was the difference between these two crystal structures and the models and mechanisms that each of the groups proposed. This was "one of the first ex-

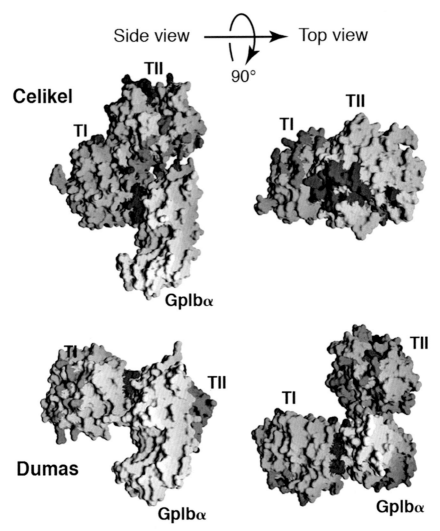

Side view → Top view

90°

Celikel

TII

TI

GpIbα

TII

TI

Dumas

TI

TII

GpIbα

TII

TI

GpIbα

5.1. Models of thrombin crystal structures compared. From Sadler, "A Ménage à Trois in Two Configurations," 178. Reprinted with permission from AAAS.

amples" of such a remarkable difference between two independently determined protein structures.[59]

The molecule that the two groups were so intent on modeling was thrombin. Thrombin is a protease, which means that it can break down other proteins by cleaving their bonds. It is an essential molecule in wound healing because it induces platelet aggregation and participates in the "coagulation cascade" that enables blood to clot at the site of injury:[60] "Without thrombin we would bleed to death, but with too much thrombin we could die of thrombosis. To avoid these untoward consequences, thrombin is regulated by a bewildering array of other molecules."[61] GpIb-IX-V is a receptor found on the surface of platelet cells and has been associated with the binding of thrombin to platelets. This receptor is a complex molecule consisting of four different polypeptide chains. The primary chain of interest, and the one that is responsible for binding thrombin, is GpIbα.

How thrombin binds to this part of the molecule matters immensely to practitioners concerned with developing therapeutic treatments for diseases related to blood coagulation. Indeed, both groups attempted to crystallize thrombin bound to a similar fragment of GpIbα. At the end of their experiments, however, the crystal structures that they determined differed remarkably. In his article "A Ménage à Trois in Two Configurations," published with the two papers in *Science*, J. Evan Sadler writes: "It is fascinating that two groups can study the same interaction with very similar reagents, only to obtain different crystal forms that display distinct modes of thrombin-GpIbα binding. Whether technical factors explain the astonishing disparity between these structures is uncertain. In any case, the differences led the two groups to offer strikingly different interpretations."[62]

Indeed, Reha Celikel and his coauthors "see in their structure a mechanism for sequential binding of two thrombins to one GpIbα." They "propose that thrombin could promote intramembranous clustering of GpIbα signaling complexes" and offer biochemical data to support this model.[63] John Dumas and his coauthors, on the other hand, "look at their structure and visualize thrombin as an adhesive ligand for GpIbα molecules on two membranes, zippering platelets together and promoting thrombus growth."[64] Distinct crystal structures of this molecular assemblage gave rise to very different interpretations of the functional properties and mechanisms of these molecules. But Sadler assures readers that "the differing interpretations are not mutually exclusive, and their breadth foreshadows the directions that future experiments are likely to explore." He insists that these studies "will be critical to determine which of the observed thrombin-GpIbα interfaces are biologically important."[65]

Of the many publications that have since cited these two papers and commented on their differences, none suggest that one structure is more sound than the other, and none infers technical error to account for the differences.[66] While the groups did follow distinct experimental trajectories, reviewers did not use these technical differences to disqualify one or the other structure or render them incommensurate.[67] I sent the original papers and commentaries to Diane to see what she thought about the crystallographic data. She hadn't come across this particular controversy yet and decided to assign the structures to graduate students in her macromolecular crystallography course to see if they could detect any errors. Apparently, the data and statistics supporting both structures were of "good quality," and the class "didn't find any issues with the crystallography."

A mystery remains. In a 2010 review of the literature, Zaverio Ruggeri, one of the principal coauthors of the 2003 Celikel et al. paper, noted that "it is striking that a bimolecular interaction studied by numerous investigators for more than 30 years and for which there are two known crystal structures, remains poorly understood with respect to mechanisms involved and functional significance."[68] This and several other papers have engaged the thrombin controversy to reassess the role of molecular dynamics in protein form and function. One group makes this explicit: "Both structures appear to have been determined at a satisfactory technical standard. For these reasons, the discrepancies suggest a fascinating plasticity in binding between the two proteins that may be biologically significant."[69] The existence of two different structures suggests that these are perhaps intermediate forms in a much more dynamic set of molecular interactions.

Mark Gerstein's laboratory at Yale has been investigating molecular movements and the dynamics of protein-protein interactions for over a decade and has compiled a searchable database and tools for analyzing "molecular motions."[70] In a 2004 paper, Gerstein's laboratory explored the plasticity of proteins in its survey of conformational changes involved in protein-protein interactions. They include the two thrombin crystal structures as evidence that "proteins can exist in an ensemble of conformational states." They hypothesize that proteins "simultaneously exist in populations of diverse conformations" and that "flexibility within regions of a protein" would enable it to "adopt new conformations and, in turn, bind structurally distinct ligands."[71] Where earlier accounts relied on a "static picture of protein structures" and on data that resolved just a "single conformation of a protein," Gerstein and his collaborators "extend the analysis to a dynamic perspective" to shed light on how proteins oscillate between "alternate conformations."[72]

Another group of researchers is actively expanding its efforts to track the wily phenomena of "intrinsically unstructured proteins."[73] Their contribution is shifting long-held assumptions about protein form and function:

Since it is generally assumed that the function of a protein is closely linked to its three-dimensional structure, prediction or experimental determination of the library of protein structures is a matter of high priority. However, a large proportion of gene sequences appear[s] to code not for folded, globular proteins, but for long stretches of amino acids that are likely to be either unfolded in solution or adopt non-globular structures of unknown conformation. . . . Clearly the assumption that a folded three-dimensional structure is necessary for function needs to be re-examined. Although the functions of many proteins are directly related to their three-dimensional structures, numerous proteins that lack intrinsic globular structure under physiological conditions have now been recognized. Such proteins are frequently involved in some of the most important regulatory functions in the cell, and the lack of intrinsic structure in many cases is relieved when the protein binds to its target molecule. The intrinsic lack of structure can confer functional advantages on a protein, including the ability to bind to several different targets.[74]

They propose that unfolded and intrinsically unstructured proteins may have key cellular functions. In so doing they highlight the limitations and biases of crystallographic techniques and databases. Crystallization and X-ray diffraction are most effective for globular proteins, which means that many other kinds of molecules fall outside of a protein crystallographer's view of the cell. These researchers encourage protein crystallographers to question the fetishization of structure, assumptions about the relationship between a molecule's structure and function, and to transform their interpretive schema to ensure that their data forms accommodate the indeterminacies of protein forms and movements.

Further calls for a shift in attention to the ontological indeterminacies of proteins come from Jane and Dave Richardson, who run a protein crystallography laboratory at Duke University. Together the Richardsons have built up extensive expertise in the areas of interactive molecular graphics and intramolecular movements.[75] Their current research examines subtle movements in the "backbone" of polypeptide chains. They call these "backrub" movements and play up the metaphor in their publications, including one titled "The Backrub Motion: How Protein Backbone Shrugs When a Sidechain Dances."[76] They attest that in solution molecules are incredibly mobile: "A major driving force for residue-scale mobility," that is, at the level of individual amino acid residues, "is the constant bombardment by solvent and other molecules, felt especially by surface sidechains that *dance between favorable conforma-*

tions (rotamers) under that bombardment and transfer some of those forces to their local *backbone*."[77] They call for a more dynamic attention to crystallographic data: "Individual crystal structures are not usually thought of as dynamic, both because crystallization selects only a subset of the conformations populated in solution, and also because, at most accessible resolutions, alternate confirmation are manifested only by lowered electron density and are thus seldom modeled in the coordinates. At very high resolution, however, multiple conformations become directly visible."[78] The lesson the Richardsons and their collaborators want to bring home is that crystallographers need to pay attention to protein dynamics even as they generate static renderings.

Insights from research in the areas of molecular dynamics have significant implications for thinking about the beings and doings of proteins. Indeed, once proteins are figured as agile, an active new view of the *molecular practices of cells* emerges, with implications for rethinking evolution. According to Gerstein's group: "This ability of proteins to adopt multiple structures allows functional diversity" and "greatly facilitates the potential for rapidly evolving new functions and structures" without relying on the slow pace of random mutation in genetic sequences.[79] From this view, molecules are not only dynamic, they are also creative: as they fold, unfold, and refold they improvise and invent new forms and functions. This claim, which gives a kind of creative agency to protein molecules, appears nearly blasphemous when held up against neo-Darwinian models that constrain the evolution of novel functions to random mutations in DNA.[80]

It is perhaps precisely when practitioners begin to explore the indeterminacies of molecular phenomena that they most explicitly confront the constraints of a scientific positivism that insists that the world consists of ready-made objects that are patiently awaiting discovery. Perhaps these are the moments when practitioners must avow not only their active contributions to modeling and the limitations of their vision, but also their participation in crafting the ontologies we all come to live by. This is when it is possible to see most acutely how, in this field, modelers' ethos and habitus are informed by moral and affective entanglements that oblige them to pay heed to the wily creativity of molecular life.

Situated knowledges require that the object of knowledge be pictured as an actor and agent, not a screen or a ground or a resource, never finally as slave to the master that closes off the dialectic in his unique agency and authorship of "objective" knowledge.

—Donna Haraway, "Situated Knowledges"

[The modest witness] is about telling the truth, giving reliable testimony, guaranteeing important things, providing good enough grounding—while eschewing the addictive narcotic of transcendental foundations—to enable compelling belief and collective action.

—Donna Haraway, *Modest_Witness*

The culture of objectivity in this field is contoured by a more-than-human ecology of moral and affective entanglements that exert a strong pull on practitioners. Producing "reliable" models that can "compel belief" and "collective action" is hard to do. Modelers must cultivate artisanal skill, judiciously attune their kinesthetic and affective dexterities, and practice subtle vision through a diffractive optics. They must care about their molecules, models, and machines, and be cautious about the limits of vision and knowledge. They must grapple with the power they have to materialize an unseen and ineffable molecular realm, and insistently qualify what they know and what they don't know in order to avoid the predicament of misplaced concreteness. These practitioners are articulate about the epistemic uncertainties they face as they struggle with ad hoc, error-prone techniques to tame unwieldy substances, data forms, and models. And just as they cultivate a respect for the uncertainties of knowledge, they also demonstrate deference toward the indeterminacies of the molecules they draw into view. Their "object of knowledge," in this sense, remains resolutely "an actor and agent," one that can be unruly, obdurate, or even moody.

Theirs is a practice that resonates remarkably with Donna Haraway's call for "situated knowledges." This form of "situated objectivity" offers a refreshing corrective to that body-less, detached, disinterested objectivity that remains a founding mythology for the sciences. Neutralized mechanical objectivities grounded in detachments, deferrals, and disavowals hold no truck among master crystallographers, as they are disciplined to be accountable for the techniques they use to make visible the otherwise imperceptible molecular world. Cautious to locate their contributions, the limits of their instruments, data, and knowledge, novices are inducted into a practice of objectivity that is founded on partial truth. They must acknowledge their role in staging the "ontological choreographies" through which they render life molecular.[81]

Where they mistrust automation, there is no better certification of the truth-value of a model than that which has been built by an expert. As objects made, as renderings, the models belong, first and foremost, to their modelers. They not only demonstrate the crucial role of embodied knowledge in the production of scientific facts, these practitioners also offer an exemplary model for reworking assumptions about the relationship between ethics and objectivity in science. If Chris Miller and Diane have their way, the next generation of protein crystallographers will be *modest modelers*. They will, in other words, embody the kind of "modesty" that Haraway advocates in the epigraph above. What this means in practice is that the "right" structure is not a mimetic reflection of "the truth" of nature: these practitioners know that there is no single model that could accommodate or communicate the lively nature of these "breathing molecules." Rather, they aim for best-possible estimates of molecular structure—robust renderings that can hold up against close scrutiny of the data. A model's relationship to truth is thus kept in motion; it is an improvisational dance the modeler makes between meaning and matter.[82] This is a dance played out with a "nature" that is recognized as an agent, not merely an inert object or resource for consumption.

PART THREE | **FORMS OF LIFE**

W ho here has taken a biology course before?" Dan, a professor of biological engineering, looked up at the eighty or so students who had crowded into a too-small lecture hall on the first day of spring semester classes at this private, East Coast university. They had arrived for a freshman seminar aimed at recruiting a new cohort of students into the school's brand new biological engineering major. Save one or two, all the students put up their hands. "Good," he responded. "But this will be a little different from what you learned in your other courses." Dan was the coordinator for this half-semester course that featured lectures by biological engineers drawn from departments across the institution. He turned to introduce the director of the program, Stan, who offered the students a taste of what this new major would offer.

"Biology has changed," Stan told the class. "When I was your age biology was *just starting* to be on the verge of being quantitative and designable." According to him, the molecular and genomics revolutions transformed biology by making biological "parts" and "components" available to manipulation at the molecular scale. "Biology today is at the point where getting the parts and manipulating them is relatively easy. Now, the hard part is: How do they *work*? Now that you know what the components are, how do they *work*? Well," he announced to the class, "they work as machines."

Stan turned to the projection screen that displayed a black-and-white, time-lapse movie of a macrophage cell migrating across a slide. He and the students watched its magnified, animal-like body undulate as it pulled itself across the screen.

If you look at a picture of a cell here migrating across a surface, you want to know how to make that cell migrate faster, to colonize a biological material, or slower to prevent a tumor from metastasizing. You have to look *inside the ma-*

chine for how the molecular components work together as a machine to transmit forces to the environment; to pull on the environment, pull the rest of the cell along. There's the actin cytoskeleton, and all sorts of proteins that link the actin cytoskeleton to receptors across the cell membrane. These all work as an exquisite, *many, many, many, many* molecule machine.

This moving image was juxtaposed to other projected images, including colorful cartoons of the "molecular machinery" of the cell, and engineering-styled electrical circuit diagrams that traced the intracellular "regulatory circuits" that "govern" the cell's large "machine assemblages." "Now that we have the components," Stan explained, "biology needs to be studied the way engineers look at things." These freshmen, interpellated as would-be biological engineers ("*you* have to look inside the machine"), were instructed to see this cell as engineers engage their objects. Over the duration of the course the classroom became a training ground for new students to learn to see through the obscuring density of the seething cellular masses that constitute living bodies. The instructors aimed to instill in them the desire to get at the underlying parts, components, and devices that "do work" in the cell to "drive" cellular life.

"These are very appealing metaphors, and this is engineering language," Stan explained. Indeed, molecular machine analogies are alluring to many. They have become pervasive in the conventional forms of writing that appear in scientific texts, as well as ways of speaking in pedagogical contexts. These analogies are also propagated through popularized accounts of the contemporary life sciences. In these contexts, proteins are ubiquitously figured as "molecular machines," "the machinery of life," and even as "nature's robots."[1] Biological molecules and their assemblages are rendered as the mechanical levers, hinges, switches, motors, gears, pumps, locks, clamps, and springs that "transduce" forces, energy, and information inside living cells.[2] These components are seen to assemble and reassemble into complex interlocking devices that act to build and maintain the cell as a higher order machine. In the hands of some biological engineers and protein modelers, cells have become the factory floors of nanoscale industrial plants.[3]

This chapter considers the practical horizons of the work of rendering life molecular. In addition to protein crystallographers, protein folding researchers, and those who study molecular dynamics, biological engineers also care a great deal about the molecular configuration of proteins and actively work with protein models to reengineer life forms. This chapter draws attention to the conjoined material, semiotic, kinesthetic, and affective dimensions of rendering practices to ask what forms of life are in the making in the twenty-first-century life sciences, more broadly. Where previous chapters were concerned with protein crystallography, this and the following chapters examine how other practitioners inflect protein models with particular textures, tones, and significances.

When a protein is rendered as a machine something more than just a likeness to machines is produced. Machinic renderings produce a range of effects and affects. Renderings tell stories: they activate modelers' imaginations and shape their perceptions; and at the same time, the models and meanings they mobilize can act recursively to sediment particular ways of seeing and storying life. This means that it matters how molecules are rendered. When practitioners figure molecules through machine tropes and stories, they are not just artfully describing the molecular phenomena that appear at the end point of their experiments. Their renderings transform how the stuff of life is made visible, tangible, imaginable, and workable.

Rather than delivering a polemic against the reductive logics of machine analogies and the ways they instrumentalize life, I begin from the premise that the manufacture of these machinic renderings is an exquisite achievement. This chapter offers an ethnographic account of protein modelers' creativity with both words and things as they learn to *put machines to work* in living organisms. I argue that rendering molecules as machines is a craft practice, one that makes it possible for practitioners to visualize and intervene in molecular worlds in particularly effective ways. This chapter explores the many layers of this achievement by taking a close look at machinic renderings at distinct moments in the history of protein science and examining the expertise required to make these renderings do work in research and teaching contexts.

Machine tropes have played a central role in shaping how molecules have come to matter as experimental objects in the history of the life sciences. Machines are, however, not the only figures that animate proteins. Chapters 7 and 8 show how molecular life has been and continues to be figured in registers that oscillate rapidly between machinic and animistic forms. Take the example of Edward's breathing molecule or the seething cell displayed in the introductory class for biological engineering students. To the uninitiated, including those who showed up for the first lecture, the gooey substances that churned inside the writhing cell looked nothing like machines.[4] Yet for those trained in the practical arts of machinic modeling, the seething cell isn't merely *like* a machine; it becomes one in their hands.

Analogies enlist imagination, experience, and intuition to illustrate complex processes or otherwise imperceptible phenomena.[5] Peter Taylor and Ann Blum suggest that modeling by analogy and metaphor enable "associations from one field to animate a scientist's thinking about another field," and so make "research 'do-able.'"[6] Some analogies can connect the familiar contexts of embodied experience to distant or imperceptible phenomena. Analogies can be thought of as producing their own form of "realism."[7] They are, in this sense, "material-semiotic," achieving a materiality that can often be more tangible than the things to which they refer.[8] Modelers' analogies are integral to their models; analogies enable modelers to tell stories about molecular phenomena. This chapter thus expands the concept of rendering to pay attention to the ways that machinic analogies shape both the aesthetics and the haptics of molecular modeling. This "haptic aesthetics" of machines provokes a range of meanings and significances, and induces particular kinds of tactility and affective response in those working with molecular models.[9]

In addition to the achievements protein modelers garner by modeling molecules as machines, this chapter also tracks contexts where machine analogies cease to be recognized as the crafty work of their makers. These are moments when researchers disavow their own ingenuity and contributions to the visualization process. In what contexts do modelers' creativity with analogies get obscured? When do analogies become so conventional that they get "frozen into literal expression" and become "dead metaphors"?[10] When do machine analogies collapse in on their referents and literalize molecules as machines? Here I take heed of Emily Martin's astute insight that perhaps the analogies that animate life science practice are not so much dead metaphors as "sleeping metaphors." This chapter follows through on her invitation to "wake up" these slumbering tropes in order to see what kind of work they are doing for these practitioners, and what machinic metaphors contribute to their renderings of molecular life.[11]

There are particular contexts in which practitioners explicitly put the metaphor of molecular machines to work and denounce more wily figurations. This chapter documents situations when practitioners are encouraged to clamp down on machinic renderings. Machine analogies, it seems, are especially compelling for engineers. Once deployed as recruitment devices, these renderings attract practitioners with engineering expertise, and students who seek training in engineering, into the study of protein biology. One effect is that these machinic renderings are reconfiguring "thought collectives" in the life sciences today.[12] Through a close analysis of the semiotic, kinesthetic, and affective allure of machine analogies, this and the following chapters aim to open up, rather than foreclose, the conditions of possibility for what can be seen, said, imagined, and felt as fact about the stuff of life at this moment in the history of the life sciences.

In the late nineteenth century, Thomas Henry Huxley drew on a long history of mechanistic reasoning when he argued that life must be analyzed according to its chemical and physical properties. In the 1870s he developed a mechanical theory of the cell that he peddled as the "protoplasmic theory of life."[13] In 1880, the *Encyclopedia Britannica* published Huxley's definitive entry on biology in which he displayed, for a wide audience, his new way of thinking about the stuff of life. He explained: "A mass of living protoplasm is simply a molecular machine of great complexity, the total results of the working of which, or its vital phenomena, depend, on the one hand upon its construction, and, on the other, upon the energy supplied to it; and to speak of vitality as anything but the name of a series of operations, is as if one should talk of the 'horology' of a clock."[14]

Huxley was influenced by his Cartesian predecessors who conjured pliers, springs, pumps, bellows, cords, retorts, and hydraulic systems inside the bodies of living organisms to help them interpret the mechanical functions of organs and tissues.[15] Huxley's contribution to this genealogy was a pronounced shift in scale, one that rendered machines inside the subvisible recesses of the cell.

Huxley's parsing of the protoplasm as a constructed machine whose aggregated parts must be supplied with energy would probably not make today's audiences swerve. Huxley's contemporaries, however, took great exception to this analogy and its implications for vital phenomena. Lionel Beale, a vitalist who argued against the mechanization of life, was the president of the Royal Microscopical Society and one of Huxley's most prominent opponents.[16] In his 1881 annual address to the society he voiced this strong objection to Huxley: "It is not most wonderful that Professor Huxley can persuade himself that a single reader of intelligence will fail to see the absurdity of the comparison he institutes between the *invisible, undemonstrable, undiscovered* 'machinery' of his suppositious 'molecular machine' and the *actual visible works of the actual clock, which any one can see and handle, and stop and cause to go on again.*"[17] Beale's primary complaint was that, given the limits of microscopic vision at the time, "molecular machines" could be no more than an elaborate fantasy.

For molecular machines to exist they had to have, like a clock, "actual visible" workings into which one could intervene. According to Beale, "magnify living matter as we may, nothing can be demonstrated but an extremely delicate, transparent, apparently semi-fluid substance."[18] Beale deplored subjective interpretations and rejected the use of what he saw as figurative language. Refusing a machinic analogy, he placed his faith firmly in a "mechanical objectivity" that entrusted his vision to the limited power of his microscope. Beale's objectivity hinged on a practice of detachment and neutrality, and he vociferously disavowed the figurative nature of his own rendering practices. Better representation would have to wait until he could augment the magnification of his microscope; only then would the true nature of living substance be confirmed.

Beale's response is a vociferous disavowal of the role analogies play in scientific vision. His is a refusal to acknowledge that scientific representation relies as heavily on semiotic invention as on optical devices like microscopes.[19] Thinking with the concept of rendering, it is possible to see the irony of Beale's denunciation. He apparently did not recognize the genius of what might be called Huxley's "working conceptual hallucination."[20] In contrast to Beale's intractable vision of the delicate, transparent substance of cells, Huxley's molecular machine could become an alluring object of analysis for the exact scientist. By conjuring the cell as a complex molecular machine, Huxley conceived of a biological object whose properties could, theoretically at least, be quantified. The metaphor of machinery offered Huxley a bridge he could traverse in his imagination between the visible, tangible, and manipulable world in which he lived and the invisible, intractable world of biological molecules.[21] The prominence of the molecular machine analogy today suggests that this continues to be an alluring metaphor that draws would-be engineers into the sciences of life.

Huxley's theory of molecular machines was not vindicated in the late nineteenth century. In recent years, however, protein modelers have been augmenting the resolution of their optical systems to visualize molecules at atomic scale, and increasingly they are modeling large assemblages of proteins. Oddly enough, the closer they look, the more machines they seem to be "discovering" "at work" in cells. Some might read this as a triumphalist tale in which molecular visualization technologies have finally vindicated Huxley's daring and provocative *premonition* of the underlying nature of molecular life. Indeed, it seems as if the mechanical "works" of the cell have finally been made as "actual" and tangible as Beale's clock.

In her groundbreaking essay "Situated Knowledges," Donna Haraway argues that there can be no unmediated access to the molecular realm: all visualization systems are "active" and "partial" ways of "organizing worlds."[22] To insist that all molecular visualizations are *renderings* is to acknowledge that those who practice this art have never relied exclusively on visual evidence in order to construct their models. That is, visualizing molecules has always involved invention: modelers must continually conjure and experiment with evocative figural vocabularies; and these analogies grant them at best a tenuous, tentative link to what is visible, imaginable, and speakable at any particular moment in the history of their science.[23] Protein modeling is in this respect simultaneously a practice of molecular storytelling.[24]

From this perspective it becomes possible to see that the visualization technologies researchers deploy do not merely "unveil" fully functional molecular machines within the body of the cell. Machinic analogies are also not merely aesthetic flourishes of language or attractive figures of speech. Rather they can be seen as powerful devices for rendering new views of life, and potent "lures" that can "vectorize" practitioners' imaginations and lines of experimental inquiry (see chapter 4). What, then, accounts for the allure of machines for these practitioners in the sciences of life?

MECHANISMS

Mechanism has held a prominent place in the history of the life sciences.[25] Mechanism is a mode of reasoning that approaches living phenomena as an assemblage of machines that perform chemical and physical functions. According to historian of the life sciences Georges Canguilhem:

> A mechanism is a configuration of solids in motion such that the motion does not abolish the configuration. The mechanism is thus an assemblage of deformable parts with periodic restoration of the relations between them. The assemblage consists in a system of connections with a determined degree of freedom: for example, a pendulum and a cam valve each have one degree of freedom; a threaded screw has two. The material realization of these degrees of freedom consists in guides—that is, limitations on the movements of solids in contact. In any machine, movement is thus a function of the assemblage, and mechanism is a function of configuration.[26]

Mechanisms have three-dimensional, temporally dynamic configurations. This definition has repeatedly been extended to living phenomena. A biological or physiological mechanism is a configuration of bodies whose movements can effect changes in the flow of movement, energy, or heat. A molecular mechanism, for example, can conduct work in a body by transducing movement or energy through an assemblage of deformable parts. Mechanical theories like this postulate an internal teleology of things and processes; that is, they recognize functional relationships between the part and the whole and infer relations between form and purpose. Protein modelers adhere to a mechanistic teleology as they narrate transformations in the chemical and structural properties of proteins interacting with other molecules. In their hands, proteins acquire mechanistic functions.

In the fall of 2005, I sat in on a full semester of lectures in a third-year undergraduate course that introduced students to the study of biomolecular kinetics and cellular dynamics. The course met in the same building that houses Diane's laboratory. Stan, the director of the biological engineering program, and his colleague Brian taught the course. Trained as a physicist and chemist, Brian runs a biological engineering lab at a computer science and artificial intelligence research center on campus. Justin, whose efforts to walk through crystal structures were documented in chapter 3, is one of his PhD students. The course was geared toward training would-be biological engineers to conceptualize and model molecular dynamics and interactions. Protein engineering and drug design require students to get a feel for the ways that complex molecules interact with each other in solution and in the viscous environment of cells. Students must learn how to parse molecular dynamics "mechanistically," and to attune their intuitions to the temporality and spatiality of complex biological machinery.

The instructors wanted their students to be able to analyze molecular dynamics with the dexterity that mechanical engineers engage human-scale worlds. In one of his lectures, Brian defined "mechanism" as the parsing of a living entity, such as a cell, into discrete, interconnected units. For him, a mechanism is an abstraction that severs a larger entity into parts and orders them by their interactions and functionality. He is invested in garnering as much mechanistic knowledge about his system as possible. He told the class: "You have to get a mechanistic understanding of everything. Because that's where the true power comes from. If you have a mechanistic understanding you really know how it works and you can change how it works. If you have kind of a philosophical understanding you can describe it after the fact. You can wrap some pretty words around it, but that understanding isn't sufficient to empower you to make the system do something different; that is, what you want it to do. So that's our mantra. The question is how deep into the mechanism do you need to know?"

As a biological engineer he is on the lookout for effective means for intervention: his aim is to be able to manipulate biological mechanisms to make them do specific kinds of work. The "true power" that Brian invokes is the ability to engineer molecular mechanisms that perform predictable functions in living systems. The biological engineering "mantra" that served as a refrain throughout his and Stan's lectures was formulated in the imperative: "measure, model, manipulate, and make." They aimed to train a new generation of scientists to build quantitative models of cellular and molecular processes that would enable the reengineering of living systems.[27]

MANIPULATING MOLECULAR ARCHITECTURES

Historian Christoph Meinel's account of the earliest construction of physical models in chemistry links the aesthetics and logics of molecular models to those of architecture and engineering.[28] He traces three-dimensional modeling practices in chemistry back to a period in the nineteenth century before chemists had fully theorized the spatiality of atomic structures.[29] According to him, the material world was already imagined as manipulable and available to engineering projects. In 1865, some of the first molecular models were presented to a distinguished audience at the Royal College of Chemistry in London by the head of the college, August Wilhelm Hofmann. Hofmann demonstrated what he called the "combining powers" of atoms by joining painted croquet balls to each other by means of metallic tubes and pins. These "glyptic formulae," as they were called, were assembled as vertical and planar branching structures and mounted on stands. The aim was to "rear in this manner a kind of mechanical structure in imitation of the atomic edifices."[30] Though they were not intended to depict the actual structure of atoms, they closely resembled the ball and stick models still in use today. According to Meinel, "this carefully composed performance" was "meant to convey the idea of the chemist as someone who knows how to manipulate matter according to his will, and who will eventually be able to build a new world out of chemical building materials that could be assembled and disassembled *ad libitum*."[31]

Meinel links the aesthetics of these early chemical building blocks to other kinds of model-building kits already in wide circulation in the nineteenth century. These included construction kits that enabled "children to create a variety of polygonal forms by connecting peas or colored balls of wax by means of toothpicks."[32] These were powerful instruction devices to inculcate children into the "conquest of space" made possible through the engineering and architectural "culture of construction" that dominated aesthetics in the nineteenth century.[33] Once made over into mechanical objects that could be assembled, disassembled, and rebuilt, chemists could reimagine themselves as engineers synthesizing mechanical structures. Thus, more than a decade before Huxley had proposed the notion of molecular machines, chemists had developed a "symbolic and gestic space" in which molecules could be imagined and manipulated as mechanical devices.[34] Thus users of the standardized molecular modeling kits, which began to circulate in the 1930s, were already prepared to engage molecules with a form of haptic creativity inspired by architecture and engineering projects.[35]

Decades after the metaphor of code reduced molecular biology to an information science, architectural metaphors are on the rise again today. Protein modelers and biological engineers are reconstituting the materiality of cells and molecules by modeling them as machines with functional and manipulable architectures. Stephen Harrison, a prominent protein crystallographer whose Harvard-based laboratory builds atomic-resolution models of protein molecules, has suggested in recent years that there are "hints" that "specific kinds of control logic are embodied in specific kinds of molecular architecture. . . . Thus, structural biology must seek to understand information transfer in terms of its underlying molecular agents by analyzing the molecular hardware that executes the information-transfer software. . . . The architectural principles of the cell's control systems and the dynamics of their operation are no less proper studies of structural biology than are the organizational and dynamical properties of the molecular machines that execute the regulated commands."[36]

Harrison remains invested in a view of the cell as a hub of information transfer modeled on a militarized computing command, control, and communication system. Yet, for Harrison, life is denser than code. Molecules have become mechanical objects that exert force and conduct "work" in order to "drive" cellular life. In his iteration, analyzing the "execution" of a cell's "regulated commands" requires an approach that goes beyond cracking codes scripted by genetic messages. Such analyses must be buttressed by structural studies of the "architecture" and "hardware" of living cells. Harrison keeps his eye on the organization and dynamics of the "molecular agents" that underlie the transfer of information and the transduction of forces and energy in the cell. These "molecular machines" are the cell's hardware, and thorough investigation of their physical and chemical properties are required in order to give full form to this model of cellular life.

As Harrison and his contemporaries invent new analogies, new machines

6.1. "A Cogwheel for Signal Transduction across Membranes." The cover of *Cell* 126, no. 5 (September 8, 2006). Copyright Elsevier.

are accumulating inside cells.[37] Like car parts, proteins are being constituted by "biological parts," "components," and "devices." These machines are built from a range of familiar devices including computer hardware, electronic circuits, and the springs, locks, clamps, pumps, and motors of modern-day mechanical devices. David Goodsell, a structural biologist and artist based at the Scripps Research Institute, is the author of a book of molecular illustrations called *The Machinery of Life*. He brings his molecular aesthetic to the PDB, illustrating their featured "molecule of the month," and producing art for their posters. His "Molecular Machinery: A Tour of the Protein Data Bank," published in 2002, documents biological molecules as an inventory of cellular components and parts (see plate 8). Machinic metaphors also gear into older imaginaries, including the cogs and wheels of early industrial capitalism. This metaphor has materialized on the cover of a 2006 issue of the journal *Cell*, which features a research paper describing a "cogwheel for signal transduction across membranes" (see figure 6.1).[38]

Mechanisms and machines continue to generate powerful tropes and logics for the study of otherwise intractable living processes. Machines are inventions that can be reworked and revised. They are like Beale's clock's "actual visible works" that "any one can see and handle, and stop and cause to go on again." They are tangible objects that afford intervention. And it appears that the "symbolic and gestic spaces" these machines incite come complete with their own *haptic aesthetics*.

Machinic renderings also thrive in modelers' kinesthetic imaginations and in the symbolic and gestic spaces of their body experiments.[39] Larry and his student Rob demonstrate well the ease with which protein modelers lean right into the language and logics of mechanical devices to cultivate what might be called a *haptic aesthetics of machines*. I engage this concept to explore how modelers use their bodies to *fold semiosis into sensation*, enacting not only a style of reasoning, but also a "sensorially nuanced aesthetics."[40] A video clip featured on the *Naturally Obsessed* website, which aims to provide background information on the science of protein crystallography, shows Larry sitting in his office in front of his computer as he explains to the cameraperson how molecular machines work in our bodies. He reaches out his left hand and gazes at it while it pulses rhythmically, animating the breathing, three-dimensional structure of a protein molecule. Looking right at this molecule, he explains: "You know, these proteins are the machines of cells. And we can look at them and understand in essence their mechanical function. So, that's basic science. But now this kind of science has become really pervasive in drug design. That's what most drugs are, specific inhibitors of the mechanisms of one of these machines."

As he continues he explains how protein structure data contribute to developments in rational drug design. He uses the example of protein inhibitors, a class of drugs that are designed to disrupt protein function by fitting into the "active sites," small crevices in a molecule that contain reactive chemical groups. Larry explains that these drugs work by "essentially just jamming a monkey wrench into the workings of that machine." He uses his body as a proxy for the molecular machine, and to punctuate his point, his right hand repeatedly slides into the crevice of his left palm. Its gears locked, this molecular machine ceases to pulse in his hand.

The film climaxes when Rob perseveres against all odds to solve the structure of his protein. Rob's resounding success is captured on film. Toward the end of the documentary viewers find him working alongside Larry as they put the final touches on a *Science* paper they are about to submit.[41] In the paper, the protein is figured as a "metabolic switch" that governs the body's activities at the molecular scale. Rob is seated in a chair next to Larry when he looks up into the camera to demonstrate the mechanism of his molecule. He leans right in to animate the mechanism with his body (see figures 6.2 and 6.3). He clicks the switch "off" by bending his right arm out in front of his body and rotating it downward until his palm cups the clenched fist of his left hand. To turn his

6.2 and 6.3. Robert Townley animates the molecular mechanism of the AMP Kinase metabolic switch. Screenshots from *Naturally Obsessed: The Making of a Scientist*.

switch "on" he rotates his right arm upward in a swift mechanical gesture. "On. Off. On. Off," he punctuates each move with a crisp click of his tongue. The molecular machine he's built both looks and sounds like a mechanical switch.

Rob's haptic aesthetic appears to contrast sharply with the lively inflections Edward articulates in his "breathing molecule," and the ways that Diane anthropomorphizes the molecule as a body that can experience pain. Chapters 7 and 8 will explore these differences in greater depth. For now the key point is that practitioners can fold *semiosis into sensation*, and *sensation back into semiosis*, in renderings that allow them to effect the affects of mechanical devices with their bodies.

MATERIALIZED REFIGURATION

Nature is . . . about figures, stories, and images. This nature, as *trópos*, is jerry-built with tropes; it makes me swerve. A tangle of materialized figurations, nature draws my attention.

—Donna Haraway, "A Game of Cat's Cradle"

Haraway's remarkable insights into machines and metaphors in the history of the life sciences make her especially good to think with on the topic of rendering life-as-machine. She invites her readers to take a closer look at the tropes and stories that organize our optics and shape how and what we know about the living world. She asks: "How do we learn *inside the laboratory and all of its extended networks* that there is no category independent of narrative, trope, and technique?"[42] She explains that in Greek, "*trópos* is a turn or a swerve"; tropes spin new significances; in so doing they "mark the nonliteral quality of being and language."[43] To bring attention to the figures that shape how and what we see and know is not to reduce the world to text or language. Tropes, metaphors, and analogies are not immaterial utterances or "mere textual dalliances": they are "material-semiotic actors," agents with the power to make new meanings and new bodies.[44]

Technoscience is a site where things and words can be made to "implode."[45] The machinic renderings so central to the life sciences today are the products of what Haraway would call "materialized refiguration." This is a concept that makes palpable the ways that tropes are sedimented or "corporealized" in objects and forms of life.[46] Materialized refiguration is about more than "metaphor and representation"; it is a way of grappling with the immense power of metaphors to rend the world. For Haraway, "Not only does metaphor become a research program, but also, more fundamentally, the organism for us is an information system and an economic system of a particular kind. For us, that is, those interpellated into this materialized story, the biological world *is* an accumulation strategy in the fruitful collapse of metaphor and materiality that animates technoscience."[47] Machinic renderings embody this "fruitful collapse of metaphor and materiality"; they are materialized refigurations that corporealize and resignify the molecular world in some ways, if not others. Indeed, machinic renderings not only constitute molecular visions, but also entire research programs, and, crucially, they inform precisely who is recruited to do the work of modeling life as machine.

It turns out that the work of rendering cells as factories and proteins as machines hinges on the productive meeting of biologists and engineers. During an interview with Joanna, a postdoctoral fellow trained in the sciences of protein folding, I learned about one such meeting. Jim, whose demonstration of the α-helix in his protein folding class illuminated the kinesthetic dimensions of molecular modeling (chapter 2), was the principal investigator in the lab where Joanna completed her PhD. Jim invited Geoff, a mechanical engineer based at the same institution, to join their weekly lab meetings.[48] Jim initially sought out Geoff's expertise for help in working on the lattice structure of certain viral proteins. According to Jim, "No wet biochemist could deal with a lattice. Who knows about lattices? Engineers know." On his first day at the group's lab meeting, Geoff completely refigured how Joanna thought about a protein with which she and long-term members of the lab were already quite familiar:

> Just as we were all sitting around the table describing [the protein], and talking about [it], all Geoff did was take a paper clip that was sitting at the table, . . . and he took the paper clip and he's like "Okay. I need to understand what you guys are talking about." And he just folded the paperclip into a three-dimensional representation of what we were talking about. And we were all sitting there going, "Wow." It was just kind of so bizarre that after twenty-some-odd years of working on [this protein], [we] had never thought about it in this way.
>
> And Geoff, as soon as we started talking about the structure, he wanted to see a model of it. It was like, immediately, "Let's make a model." He came back the next day with more elaborate wire that he had taken and molded at his house. . . . And he came in and said, "Oh! There's a clamp! This is a lock. I mean this is locking that molecule right in place." And now there're all these papers they've published on the molecular clamp. It's a lock! It's a clamp! And it's so exciting. And it's funny, his whole approach was entirely different. . . . I'd looked at that structure a million times. You know. And I was like, "Oh yeah. It's a lock! That's a clamp!" . . . I wasn't working on that project, but it opened a whole door of experiments that would not have happened otherwise. An entire postdoc was hired to work on this. And it hadn't been called a clamp until Geoff came to the meeting.

A modeler by training, this engineer needed to hold the structure in his hands. Simple modeling tools would suffice: a paper clip or wire that he could turn and twist. But he also had other tools at his fingertips that were integral elements of his visualization apparatus: he had a facility and a familiarity with clamps as mechanisms, and rendering the protein as a molecular clamp came easily to him. As Haraway suggests, a metaphor can even drive an entire research program. In this sense, the livelihood of a postdoc, the scientific significance of a protein, and the productivity of a research laboratory all turned around the figures of the clamp, the engineer, and his jerry-built model.

CULTIVATING A FEELING FOR MACHINES

It takes work to build machines into the bodies of organisms. Haraway's call to pay attention to the materiality of language is thus also a call to make visible how tropes are put to work, for whom, and at what cost. Today, protein modelers use a vast range of interactive molecular graphics media and a variety of stylistic conventions for depicting molecular structures (such as wire frame models, ribbon structures, space filling models, and so on). These distinct media produce numerous opportunities for modelers to express their own haptic aesthetic. Modelers can render proteins in a range of forms: in ways that make them appear to have glinting, metallic architectures; or as globular, gooey bodies that wriggle when they are animated on-screen. Thus, while the rhetoric of the molecular machine is pervasive, it is by no means the only way proteins are figured. When, how, and for whom do proteins cease to oscillate between the figures of the lively body and the machine? In what contexts do life scientists fixate on the molecule as a machine? What happens to other, more wily figurations of life? Chapters 7 and 8 explore how anthropomorphisms and animisms induce a range of anxieties in practitioners. Perhaps it is no surprise that the "mechanomorphisms" that most prominently inflect their models don't seem to cause as much concern.

In one of our several interviews I asked Fernando, a fifth-year PhD student in Diane's lab, if he ever used metaphors other than machines to talk about proteins. I mentioned that I heard protein modelers talk about proteins as wily, lively bodies. This suggestion put him on edge a little, and his response was firm: "A protein by itself is not a living thing," he tells me. "It is . . . it is a machine. And it will break down, just like machines do. Okay? And if something is not there to repair it, another machine, another piece of machinery," the whole system will "break down." At the suggestion that proteins had lively qualities, Fernando clamped down firm on the metaphor of molecular machines.

Machine analogies are not just pedagogical devices for Fernando. He likes to use the metaphor in part because he has a particularly nuanced feel for machines and their parts. He is a latecomer to science, and at forty, he is significantly older than most of the graduate students in his cohort. He grew up in a working-class Hispanic family and spent his twenties working as a plumber and manual laborer, and he took much pleasure in building cars. He later went back to school and eventually started teaching computer-aided design (CAD) to architecture and engineering students at a community college.[49] Machines are familiar to Fernando: they are, like Beale's clock, "actual visible works" into which he can see and intervene. He understands how they work, how their parts fit together, and what keeps them ticking. In the middle of our interview he actually got up to show me how door hinges work, how you can look at them, handle them, and know which way the door should swing. Our conversation produced dizzying Alice-in-Wonderland effects of scale as we zoomed along what seemed to be a continuum of tangible machines, from human-scale machines, down to the scale of molecules and back again.

As a crystallographer he builds models of proteins to figure out what the "machinery" of the cell looks like, and how it works. For him, X-ray crystallography is a visualization tool that he uses to get a "snapshot of the machine." He described his job as a protein modeler through an allegorical tale that took us to the factory floor of a robotics-mediated automotive assembly line:

> So you know, you are talking about the machine that screws in the fender at the Ford car plant. We're studying that machine because we are trying to find out what it does. And without [the X-ray crystal] structure we are just *feeling* it, just tentatively, sometimes with big thermal gloves. So we can't really get to *feel* the intricacies or the nuances of the drill bits. And all of a sudden crystallography is a snapshot of the machine. Okay. It [the machine] can even be in multiple states. Standing still, turned off. In a state when there is a screw being drilled into the fender. You know, it can be somewhere in between. Alright?
>
> But because we've seen a similar machine in another company, we kind of have an idea of what the machine does. We've seen the individual parts and stuff like that. I'm not going to mistake the machine for drilling for the machine for welding. Okay. What crystallography allows you to do is to say, "Hey that is a drilling machine, not a welding machine." Okay. And by looking at certain parts of the machine you can tell whether the drill bit is six inches long or two inches long or whether it has a neck that moves up and down, or whether the neck is static. That's the sort of stuff you get in a crystal structure that you don't have before.

Intense in his delivery, Fernando successfully sustained the analogy of the cell as the factory floor of the Ford car plant throughout his story. He had such a strong grip on the analogy that there was eventually a slippage from the machine as a metaphor for the molecule, to the molecule that had actually become a machine; in this case a (robotics-mediated) machine that could do highly specialized kinds of work in (a capital-intensive) cell.

Fernando's image of a human worker whose tactile and visual acuity is dampened by wearing "big thermal gloves" is an effective analogy for how hard it is for protein modelers to make sense of molecules without both the resolving power of X-ray crystallographic vision and an elaborate figural vocabulary to make sense of the substances they draw into view. Crystallographic modeling gives Fernando both a three-dimensional visualization of the spatiality of the molecule and a "nuanced" "feeling" for its "intricate" structure. As he made clear during another interview, modeling with interactive molecular graphics interfaces is, for him, a craft practice through which he has been able to develop what he calls a kind of "touchy-touchy-feel" for the molecular model as he builds it on-screen: "I don't want to say touchy-touchy-feely like that, but that sort of holding on to something and getting a feel of it." Taking off his thermal gloves, so to speak, he uses the interactive molecular graphics interface to bring models into haptic sensation, as well as into view. And yet, he goes a step further: he draws on his feeling for machines to inflect his rendering of molecular configuration with a new layer of significance that helps him tell a particular kind of story.

Though he is quite taken by the tools X-ray crystallography affords his curiosity, Fernando is ambivalent about his future in the field. "You can get so fascinated by the intricate gear work of a particular piece," he told me, "that you never learn how to operate the whole machinery." He finds he's very attracted by developments in biological engineering today, which do promise the possibility of "operating" the "whole machinery" of the cell. These are the same kinds of promises that have enticed a new generation of students to sign up for the new biological engineering major at his university. Molecular machines are thus powerful recruiting devices for rallying would-be engineers into the life sciences. Rendering molecules as machines gives engineers something they can get their hands on, something they can literally grasp. And it is through this metaphor that biology has become quantifiable, manipulable, and redesignable in ways that have enabled engineers to *rework biology*; literally and figuratively they have *put life to work* at the molecular scale.

ENGINEERING BIOLOGICAL ENGINEERS

Like proteins, metaphors left to their own devices don't automatically crystallize into forms that can produce new meanings and material effects. It takes effort for crystallographers to coax proteins to form "living, breathing" crystals. Similarly, metaphors must be nourished and sustained within the context of a practice and a culture that can keep them alive. This raises concerns for educators in the emerging discipline of biological engineering. While machines are prominent companions in daily life, as technical objects they tend to fall outside of the expertise of the classically trained biologist, and more to the point, outside the technical competence of many freshmen and sophomores taking a seat in introductory biological engineering courses. Few would likely have had as much experience with machinery as Fernando.

In order to learn how to *work the machine into the cell*, life scientists' expertise must be reconfigured. In order to properly build and deploy these machines, biological engineers must enlist and train a new generation of scientists to think and work like engineers. They do this by attracting new recruits with the technological promise of "direct access" to biological worlds at the molecular scale. In order to make the machine analogy do work, however, they must *engineer* practitioners who not only have a feeling for the molecule, but who also have a feeling for the machine. Biological engineers' kinesthetic and affective dexterities must be attuned to machinic life.

In 2005 I observed students and faculty in a semester-long undergraduate biological engineering laboratory course. They taught me that cultivating a feeling for machines is not an easy task. This was a required laboratory course for sophomores majoring in biological engineering. The third of this four-module course focused on what the instructors called "systems engineering." Over the course of six labs, students helped fine-tune a "bacterial photography" system that could produce images on the surface of bacterial colonies using the simple principles of pinhole camera technology. The goal was to optimize the conditions under which engineered light-sensitive bacteria would change color when exposed to light. In order to analyze this system they needed to understand the cellular "circuitry" that controlled the engineered bacteria's light sensitivity. Their laboratory manual offered this insight: "Biology is particularly well suited to model building since many natural responses appear digital. . . . The digital responses of cells to perturbations, combined with lab techniques for moving DNA parts around, allow logic functions and circuits to be constructed in living cells."

But what is a circuit? Electrical engineers are of course quite familiar with circuits. The concept of a cellular circuit is drawn directly from electrical engineering and uses the same iconography and nomenclature as engineers' electronic circuit diagrams.[50] In the lab, the students were introduced to an analogue—what could be considered a materialized figuration—of their photo-sensitive bacteria. This analogue was itself a circuit board. The students were expected to apply electrical engineering concepts, and to complete a light-sensitive electronic circuit in order to demonstrate their understanding of the circuitry already built into their bacterial cells. Each pair of students had at their desk a partly assembled electronic "solderless breadboard" that

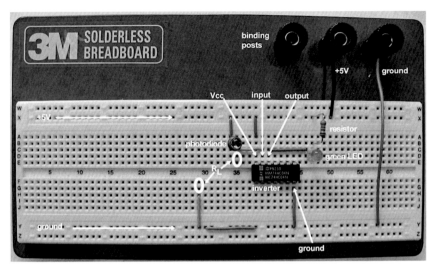

6.4. A "breadboard" wired up as an analogue of a bacterial photography system. Used with permission.

included a photodiode and an LED, which, if the circuit was completed correctly, would turn on and off in response to changes in light intensity sensed by a light-sensitive photodiode (figure 6.4). Students were asked to complete the circuit by connecting resistors of varying strength to appropriate sites on the breadboard.

Some of these biological engineering students, however, struggled with basic electrical engineering concepts. Meera, who was trained in computer science before coming into biological engineering, was the TA for this module of the lab. She had to run a remedial tutorial in electrical engineering several times over for small groups of students. Looking rather confused, they gathered around her at the white board. The laboratory director, herself not trained as an engineer, but as a molecular biologist, also joined the lesson. Current flow, resistors, converters, photodiodes, signal matching, and ground all had to be explained. Meera, who had assembled all the circuit boards herself, seemed a little surprised by how hard it was for the students to get the concepts: "Inverters . . . you all know what that is? . . . Okay? . . . Does it make sense when I say current flows through a wire? . . . Does that make sense?"

The students' blank stares and repeated questions gave the lie to the excited statement in their lab manual: "Notice how much easier it is to assemble electrical circuits as compared to biological circuits. It takes seconds to swap in a new resistor into your circuit but a few days to assemble" a couple of "biological parts." Apparently it was not that easy: not only did they not have facility in electrical engineering concepts, what they were being asked to do was to make sense of what was a rather dense material-semiotic tangle. Modeled on a circuit diagram, the bacteria had been engineered from "standardized biological parts"; that is, the proteins and genetic sequences were themselves modeled as input, output, and signal-matching devices, resistors, inverters, terminators, and protein generators.[51] The bacteria were thus already an analog, or rendering, of an electronic circuit. The circuit they fumbled with at their lab benches then was an analog of a cell that had been built on the model of the circuit board. Their attempts to wrap their heads and hands around this veritably loopy set of renderings did indeed give them cause to swerve.

One might expect that analogies are most useful when they draw on knowledge of a familiar realm to illuminate another, less well-known realm. The bacterial system they had been using throughout the module depended on an in-depth understanding of electrical circuits. Cellular signaling and regulation were consistently rendered as circuits in classroom lectures, yet the students did not yet have an appreciation of the full import of the circuit metaphor. They were not yet fluent in the proper terminology and techniques. The circuit-building exercises were, in this regard, quite productive: they were diagnostic of where students' understanding came undone; and they also offered to remedy the situation by enabling students to cultivate a feeling for these machines.

One of the lessons learned by the students, their instructors, and their ethnographer was that circuits are not self-evident figures for cells, and that the skills required to render cells and molecules as machines must be cultivated. It takes work to build machines into organisms, and this work requires thorough training in a practical, conceptual, and material culture.

DEADPAN LITERALISM

In an article in *Nature* in 1986, cell biologist Henry Bourne remarked astutely that in the life sciences "argument by analogy, like gambling, was once practiced behind closed doors."[52] Bourne was claiming that by the mid-1980s, analogical reasoning in biology had finally been "elevated into respectability" with rich "payoffs."[53] In this article Bourne could be said to have "outed" analogy as an integral practice in the work of science. This was a remarkable declaration in a context that doesn't normally reward efforts to acknowledge the figurations and fantasies that shape scientific knowledge. What is curious, however, is that just as Bourne put the rich productivity of some analogies on display, he simultaneously obscured others. Most notably he made no reference to the machinic figures that most promiscuously populated his essay. One hundred years after Huxley first introduced the machine metaphor, it seems to have lost its punch. Bourne's molecular machines are no longer animating figures that enable the scientist to take a leap across the divide between the visible and the invisible; with atomic-resolution molecular vision, molecular machines have been forged as fact and become unremarkable things-in-themselves. With Haraway, I am wary of such moments in which the richly "tropic" and figurative nature of technoscientific vision is erased or denied.[54] By naturalizing and literalizing machines in the bodies of organisms and asserting the neutrality of visualization technologies, practitioners risk giving the impression that they are merely unveiling the underlying machinery of life. In so doing they disavow the power of their own renderings. In this light, their creative labor is at risk of being made invisible and drawn back behind closed doors.

Molecular machines are more often naturalized as nature's tools, rather than recognized as the elaborately constructed figural machinery of the investigator. This phenomenon can be seen clearly in the context of the high-stakes debates unfolding in the United States between advocates of evolutionary biology and those who argue for "intelligent design." Remarkably, proponents of both intelligent design and neo-Darwinian evolutionary theories deploy the metaphor of molecular machines with serious deadpan literalism.[55] The question of "who" (God) or "what" (nature) made these molecular machines is anything but a trivial matter. What is ironic is that both sides continually defer the responsibility for engineering these machines to higher powers, either natural or divine. Neither the creationists nor the evolutionists want to take credit for their crafty work. Neither are in a position to laugh at the absurdity of their denial; their silence on the matter reveals the depth of their investments in the work of either naturalizing or deifying molecular machines. But the joke is on them: these are neither God's clever little devices nor evolution's sometimes-clumsy concoctions. Once they are recognized as renderings, molecular machines can be seen as none other than modelers' own marvelous inventions; indeed, machinic renderings are carefully crafted devices that enable both evolutionists and creationists to materialize and manipulate the molecular world to their own ends.

In the end, it is the biological engineers who recognize the absurdity of this denial. For, though they might nervously muffle their laughter, biological engineers do get the joke: as they struggle to reassert the respectability of "design" and "designers" in the realm of life science, they do, after all, want recognition for their labors—those massive, microscale engineering projects that they have rigged up within living cells. They are keenly aware of how machinic renderings have been productive of new objects, meanings, lines of research, and forms of life. In the face of pressure to patent their designs and inventions, they understand well how these renderings sustain their very livelihoods.

CONCLUSION

If rendering is a practice of "worlding" that gives substance and significance, body, and meaning to emerging technoscientific objects, then how practitioners figure the stuff of life matters: this is a practice that materializes some kinds of bodies and meanings, if not others. Refusing to take responsibility for crafting these figures and models is what Haraway calls an "avoidance of the tropic . . . tissue of all knowledge."[56] To literalize machines in the body of the cell is to refuse to recognize the haptic aesthetics and semiotic creativity involved in rendering life machinic. Machinic renderings are profound achievements. This chapter documented practitioners' pleasure and skill building molecular machines. The machinic renderings described here suggest that there is, after all, "no fundamental, ontological separation in our formal knowledge of machine and organism, of [the] technical and organic."[57] Haraway's cyborg figure is a reminder that "the machine is not an *it* to be animated, worshiped, and dominated. The machine is us, our processes, an aspect of our embodiment."[58] And yet, machinic renderings must always be seen as artfully engineered devices that have worked effectively to secure some kinds of objects, meanings, and forms of life, to the exclusion of others.

The Inner Life of the Cell is a computer graphic animation that has propagated online since 2007 (see plates 9–12). It warps time, space, and scale to invite viewers on an Alice-in-Wonderland journey to get down inside a cell and take a look around. Once the viewers are inside, this seething cell pullulates with life. A musical score lilts and swells as viewers orient themselves in a spacious expanse of morphing bodies. Textured surfaces ripple in vibrant color. Gravity is abandoned; large, lumbering orbs hover as if in some underwater dream. Some bodies slide past one another and slither between membranes. Fibers appear to self-assemble through some magical attractive force that pulls nearby particles into elongating spirals. Awkwardly interdigitating shapes skirt past one another as they ride like surfers across the swell of undulating surfaces. Small bodies in the distance swerve toward massive, heaving forms that breathe with a kind of dark force. In scenes that recall simultaneously the sublime and lonely images of space flight and deep-sea exploration, the animators choreograph moments of dramatic tension and sheer elation.

The Inner Life of the Cell was produced in 2007 for teaching core biological concepts to Harvard undergraduates.[1] Scientists and educators partnered with professional character animators at XVIVO Studio to use some of the industry's most sophisticated techniques. Computer graphic animations like this one offer close-up views of the ways that some practitioners imagine molecular life today. David Bolinsky, formerly a medical illustrator at Yale, is a cofounder of XVIVO Studios and the director of animation on this project. In March 2007, he gave a TED Talk on the making of the film.[2] He and his team built on known protein structures and biochemical data to render molecular events with "empirical accuracy." Bolinsky aimed to use his skills as an animator to "spark" students' "passions" and "imaginations" and inspire them to "find out really how life works." As the animation unfolds on the screens behind him, he tells the audience that he wanted to give students a "mental image of the cell" as "a large, bustling, hugely complicated city that's occupied by micromachines. And these micromachines are really at the heart of life. These micromachines, which are the envy of nanotechnologists the world over, are self-directed, powerful, precise, accurate devises that are made out of strings of amino acids. And these micromachines power how a cell moves. They power how a cell replicates. They power our hearts. They power our minds."

These power-generating "micromachines" recall biological engineers' renderings of proteins. Bolinsky's rendering is resonant with Stan's display of a seething cell crawling across the screen as he articulated the precision of the cell's molecular machinery for his biological engineering class (see chapter 6). Like Stan's performance, there is odd dissonance between Bolinsky's iterative and insistent description of molecules as machines and the zoomorphic forms dancing on the screen behind him. The animation's action-packed, fly-through scenes are set to ambient, orchestral music and inflected with stylized rhythms, caricatured movements, and animal-like affects. As one reviewer comments, the molecules and organelles "move with bug-like authority, slithering, gliding and twisting through 3D space."[3] The forms undulating across the screen look nothing like the mechanical devices Bolinsky repeatedly invokes. What forms of life are rendered in the wavering dissonance between these two animations, the computer graphic rendering and his story?

Remarkably, as Bolinsky tells his story, he loosens his firm grip on the metaphor of molecular machines. His machine rhetoric takes a swerve when he draws the audience's attention to one scene that animates the "bug-like authority" of a molecular assemblage. This protein marches with deft intention along a taut fiber, dragging a large load behind it. This, Bolinsky announces, is the "FedEx delivery guy of the cell": "This little guy is called kinesin. And he pulls a sack of brand-new manufactured proteins to wherever it is needed in the cell, whether it is a membrane or an organelle, whether it is to build something or repair something. Each of us has about 100,000 of these things running around right now inside each of your 100 trillion cells. So no matter how lazy you feel, you are not really intrinsically doing nothing."

What kind of machine is this kinesin? A mechanical device robotically enslaved on the factory floor of the cell? No, in this rendering kinesin is humanized. Bolinsky animates kinesin as a self-actualizing "little guy" laboring hard inside a bustling cell-cum-city, working for a corporation that positions itself as a vital economic service. Here his machine analogy gets tangled up in a knot of twenty-first-century economic desires and subjectivities. He remodels the molecular practices of cells on the imperatives of properly aligned subjects bent on keeping the economy in motion. The message: don't worry if you don't feel like you are working hard enough, your proteins are keeping pace with the self-actualizing demands of a neoliberal economy.[4]

What forms of life are taking shape in Bolinsky's animation of cellular physiology? His figural vocabulary oscillates between the registers of the animal, the human, and the machine. Held together, his story and animation engender ambi-valent renderings. It is this wavering, this inability to fully clamp down on mechanical analogies that this and the final chapter explore. Where and when does mechanism falter and give way to other kinds of stories? What happens when animators render proteins in livelier registers? Bolinsky and others featured in this chapter offer surprising insights into how practitioners "do mechanism" in the twenty-first-century life sciences.

Molecules move. Recall Diane's comparison of the movements of crystalline proteins in an X-ray diffraction experiment to the excitability of her dog each time she tried to take his picture (chapter 2); and how freeze-frame crystallographic techniques don't cope well with molecular movements like the sliding "backrub" motions of a polypeptide backbone when its amino acid side chains are subject to energetic bombardments (chapter 5). Molecules don't just vibrate with random Brownian motion; they also fold and unfold rapidly and continuously in the watery milieu of the cell. Crystallographic structures are in this sense quite deceptive: a protein may spend just nanoseconds in the folded state represented by a model. In order to address this limitation, protein modelers in a variety of research fields are making use of the spatial and temporal possibilities of time-based media to animate their models on-screen.[5] In the process they are giving form and movement to phenomena that can otherwise only be hypothesized and intuited. These animations are proudly displayed and available to be downloaded from laboratory websites, frequently projected to awe audiences in conference presentations and undergraduate classroom lectures, and they circulate widely through informal networks on the Internet. Some, like *The Inner Life of the Cell*, make use of high-end graphics, while others use much simpler imagery. Yet, what are the challenges modelers face rendering molecular time? And what forms of knowing are in the making as educators and researchers pull molecular phenomena into perceptible temporalities? This chapter looks at the phenomenon of protein folding in order to explore how modelers negotiate the challenges of rendering molecular dynamics.

Molecules acquire a range of affects as they are animated. Drawing attention to the oscillation between machinic and lively stories, this chapter explores the affective dimensions of molecular animations. More than other kinds of renderings, animations make practitioners anxious. Where molecular machines and machine inflected renderings are so readily incorporated into mechanistic discourse and naturalized as nature's marvelous tools (chapter 6), livelier renderings put many practitioners on edge. It appears that time-based media open up space for modelers to inflect animations with a range of animisms and anthropomorphisms that are otherwise anathema to the mechanistic logics that ground the life sciences. This chapter builds on issues raised in previous chapters, including questions about how modelers confront the shifting limits of molecular vision and their anxieties about misplaced concreteness. The issues they face are both epistemic and ontological: it's not just that the visualization techniques are limited and offer partial views; the nature of the substance being animated is also called into question. Animators like Bolinsky must confront the performative effects of their lively renderings. How do his lively renderings stand alongside the deanimated mechanistic ontology he and his colleagues are supposed to avow?

If mechanism is deployed to eradicate animistic stories and secure deterministic knowledge, these practitioners demonstrate the failure of mechanism to ensure complete disenchantment. This chapter explores how readily animations morph mechanism and lure practitioners toward livelier ontologies. Moreover, while it tracks the unease that this *lively mechanism* induces in those who feel compelled to adhere to neo-Darwinian views of evolution, this chapter also pays close attention to the opportunities animations generate for restaging reductive narratives of molecular life. In the process it explores how animism and mechanism are thoroughly entangled in the twenty-first-century life sciences.

ANIMATIONS

Researchers in the life sciences have engaged a wide range of media to produce animations of living phenomena. Before cinematic and time-lapse techniques, animated cartoons, and computer graphic animations, remarkable animations were crafted using the dynamic properties of organic materials.[6] Molecular animations are, like many of the other renderings described in this book, craft productions that hinge on the animator's creativity, intuition, and aesthetics. Moreover, as renderings, animations do not just describe the world of molecules; they are enactments that both express and sediment ways of seeing and knowing. And yet, the introduction of time, in the form of temporally dynamic media, appears to amplify these renderings' effects and affects.

Animations have a temporal structure and a telos, with a beginning, a middle, and an end. They are remarkable vehicles for telling molecular stories. Take, for example, the infamous 1966 film *Fantastic Voyage*, which can be read as an animation of the cellular and molecular physiology of the human body. It restages cellular life on film using textured fabrics and low-tech special effects to massively amplify cells, tissues, and molecules in a kind of physiological puppetry. This film renders a very different body politic than that narrated by Bolinsky's *The Inner Life of the Cell*: a nuclear-powered submarine, along with its crew, is miniaturized and injected into the bloodstream of a Cold War spy in a drastic attempt to perform microscale laser surgery and save him from a brain injury that threatens both his life and the critical state secrets that he holds. The viewer joins the brave surgeon and his crew as they are swept through the spy's circulatory system and into his cells and tissues. In one scene viewers swim alongside the laser surgeon's assistant, played by Raquel Welch, as she leaves the ship to investigate the spy's lymphatic tissue. Alarm sounds on the ship as the patient's antibodies sweep in to attack this "foreign invader"; their strangling hold on her body threatens to delay the mission. If Bolinsky narrates molecules in *The Inner Life of the Cell* as hard-working neoliberal subjects, then *Fantastic Voyage* dramatizes physiological and molecular forces through the violent and sexualized imagery of the Cold War.[7]

Not all molecular animations enjoy such wide circulation. Most are developed in-house for laboratory research and pedagogy. Time-lapse film is one medium widely used to animate biological processes. Time-lapse images, like the movie of the seething cell Stan presented to a captivated audience of biological engineering students, are part of an enduring genealogy of microcinematography in the life sciences. In the early twentieth century, researchers joined techniques of cell culture with film technologies to devise a form of animation that brought cells to life on film screens.[8] Time-lapse cameras can record a phenomenon changing over time. They capture static images at equidistant points in time and then reanimate the phenomena by playing still images in sequence to produce the effect of a moving image. Time-lapse photography can speed up or slow down a process to make subtle changes perceptible as movement.[9] Time-lapse cameras attached to microscopes can, in this way, be used to pull imperceptible phenomena, such as cellular processes that move too fast or too slow for us to see, into human time.[10] Subcellular molecular movements, however, cannot be visualized with a microscope. For this reason, molecular events are more often rendered in the form of cartoon films or computer graphic animations.

Chris Kelty and Hannah Landecker have proposed a "theory of animation" that examines how a range of moving image technologies have shaped knowledge production in the life sciences.[11] For Kelty and Landecker, media forms that use time-lapse and other animating techniques to keep pace with life forms "not only demonstrate the life of the organism in question, they also *animate* it in relation to other, often dominant, modes of static representation."[12] Kelty and Landecker are particularly interested in the "*status*" of animations "*as images in relation to knowledge.*"[13] They show how animations play biological theories and models forward in time, and so restage ways of seeing that have already been systematized in scientific research. Their approach is particularly salient for examining how molecular animations can be used to play mechanistic theories forward in time. And yet, what I find is that animations not only embed ways of seeing: by pulling static models into time; they also refigure our concepts and ways of seeing. Here I treat animations as affectively charged lures: they vectorize modelers' kinesthetic imaginations and activate viewers' haptic sensibilities. In so doing, they open up new spaces for molecular storytelling. Animations of all kinds inflect machinic renderings with a range of affects that have the potential to destabilize mechanistic theories of life. What stories of life do such animations engender?

FOLDING AND UNFOLDING

Proteins fold and unfold so fast their movements are nearly imperceptible. Protein folding researchers have developed a suite of biochemical techniques to give them at best indirect access to this ephemeral phenomenon. I gained a deep appreciation for protein folding dynamics and the limitations of static crystallographic structures while conducting ethnographic research in Jim and Geoff's undergraduate course on protein folding. Recall Jim's evocative demonstration of the haptics of seeing and the physical effort required to "grasp" molecular structures when he presented the α-helix to the class (see chapter 2), and Geoff's rendering of a molecular clamp with a paperclip for members of Jim's laboratory (see chapter 6). Jim and Geoff's collaboration extended beyond the laboratory and into the classroom in this course on protein folding.

The course was cross-listed in the departments of biology, chemistry, and chemical engineering, and over fifty students crammed into the classroom for the first meeting of the semester. In his first lecture, Jim recalled that when he had first started teaching the course just fifteen years earlier, only ten students signed up. "What has changed?" he asked the class. "Why so much interest in protein folding?" According to some, protein folding is "one of the most important problems in biology," and research has attracted funding and a multidisciplinary group of researchers.[14] Indeed, protein folding has been the "subject of immense scientific interest spanning the decades since the first structures of proteins were elucidated."[15] The publication of John Kendrew's models of myoglobin in 1958 (see figures 2.2 and 2.3) marked a defining moment for Jim: "The moment that image hit the press, people wondered: 'How does this [polypeptide] chain know where to go?'" What gave this strange molecule its special fold? How does a long chain of amino acids *know how* to fold?

Proteins don't always fold into their "proper" configurations. Indeed, there are a number of protein folding diseases identified, including Alzheimer's and Huntington's.[16] If misfolded proteins "escape the cellular quality-control mechanisms" they begin to aggregate in cells, which leads to "a wide range of highly debilitating and increasingly prevalent diseases."[17] Major efforts are under way to develop treatments based on knowledge of protein folding pathways. The problem as Jim defines it is that we still do not understand how "healthy" proteins "know how" to fold, or what triggers them to misfold. When protein folding fails, Jim is concerned that "we can't efficiently correct the problem."

In a vivid example of the protein folding problem on an industrial scale, Jim described Eli Lilly's Indianapolis fermentation plant, which produces genetically engineered, "biosynthetic" insulin for the massive influx of diabetics worldwide. To help resolve problems associated with animal sources for insulin production, including allergic reactions and concerns about animal cruelty, Eli Lilly collaborated with the biotechnology company Genentech to genetically engineer *E. coli* bacteria with a modified human gene for insulin to produce a product they call Humulin. Bacterial cells, however, don't provide the right conditions for these polypeptides to fold into their correct conformation; rather insulin chains accumulate in these cells as an insoluble mass. The synthetic peptides must be extracted from the bacteria, and then purified, made soluble, denatured, and processed in vitro to get them to fold into their active conformations. Jim amplifies the issue at stake: "If insulin is not folded correctly, it forms scrambled eggs." Though researchers at Eli Lilly have, through trial and error, developed a series of steps to go from "scrambled eggs" to active insulin, the biochemical processes involved in getting insulin to fold into the correct conformation are not fully understood. There is currently no coherent theory on which they can draw for the precise engineering of this process.[18] The indeterminacies of protein folding thus present serious obstacles to those hoping to render proteins visible, tangible, and workable.

It might seem that the dynamic phenomena of protein folding would be best rendered in time-based media. I encountered numerous animations of protein folding processes featured on laboratory websites (see, for example, plates 13–18).[19] Yet I was puzzled. Given that protein folding is so ephemeral and fast, how do researchers know how they move and change over time? A conversation with two postdoctoral fellows helped illuminate this problem. Joanna, a protein folding researcher, was Jim's former PhD student. She was hired along with her colleague Lynn to work with faculty members in the biology department and transform the introductory biology curriculum for undergraduate students at their university. In the midst of a wide-ranging conversation, I asked Joanna how she felt about computer graphic animations of protein folding pathways. Her response, and the conversation that unfolded between her and Lynn, were quite telling:

JOANNA: I've always hesitated [to make animations of protein folding]. . . .
Maybe it's because I'm from Jim's lab. I have always really hesitated trying to put what I see in my head onto paper. I was always the one in the lab up there who was able to make beautiful models for people. Not 3-D models. But I was the one who was easily able to take the data, and make the 2-D cartoon to see that "Okay this step goes first, then it's got to be this step, then this step." I could do that very easily. But I always hesitated to actually put anything amorphous, [like] the simulations together. Because from day one joining the lab, Jim was like "You can't make a simulation of protein folding. You can't do it! It's not going to be accurate!"

I can give you the steps that I know happen. I have no problem describing or creating representations that make people understand that very easily. But making an animation that goes from one to the other. . . . It's that ingrained Jim thing that says: "No you can't do that; it'll be wrong!"

For instance the molecule I was studying when I was there: It's got two sections. And we—myself and another grad student—we determined that yes, in fact, without a question, what happens is that one section folds before the other. We could see it and we know it happens. And we have lots of data that shows that's how it happens. But I would never try to model a . . .

LYNN: A continuous process?

JOANNA: Yes, a continuous process out of that. I could say yes, I can give you a model that shows that this guy is solid, this guy is loose. But I would never [animate it]. It's the Jim in me. . . . It's very difficult because so little is known. That's what makes protein folding so hard to describe to people. Cause, everyone seems to want to put a time sequence on things, to make a simplified animation. There are so few cases where we can say a simplified animation, oh, that's right. We just don't know.

LYNN: That's interesting. I just realized that's what irks me about the tRNA folding movie that [we] show in class.[20] It's precisely that. You don't know that's what happens.

JOANNA: You don't know the directionality, you just don't know . . .

LYNN: . . . that that's the order of the steps.

JOANNA: I have a protein folding background. So for me it is infinitely frustrating when people do that. I've seen more graduate students spend their careers trying to make animations and simulate folding when there's no experimental basis for what they are doing. . . . And it's very difficult as a true protein folder—I can't buy it. I don't believe it. I can see the desire to represent that, or to get at those steps. But there is just not enough data to support [it].

Throughout her delivery, Joanna kept repeating that her resistance to the use of animations (the visualizations produced from simulations) is the Jim "in her." Just as her kinesthetic imagination has been honed through her laboratory work with proteins, it appears her intuitions and judgments have also been shaped by the sensibilities and values of her mentor. Joanna and Lynn follow Jim in their concern that animations concretize the temporal sequence of what is for them a dynamic process. Joanna "knows" how a protein folds, and she can "see" it; and as she expressed evocatively, she can also feel it. She even understands the "desire to represent" the forms and movements of the folding process. However, she is concerned that computer graphic animations fix processes she knows to be amorphous, ephemeral, and impossible to capture. Joanna is particularly anxious about the ways these animations *rend* time: they impose a tempo and a direction; they put a time stamp on a process. An animation that runs time in a set direction like this effects a kind of closure; it directs how others see and experience a process. A viewer's sensory field is flooded with data; and the imagination is saturated with one vision of molecular life. This forecloses other interpretations and other ways of seeing. Animations thus performatively choreograph and concretize otherwise indeterminate ontologies.

POLICING ANIMISTIC PEDAGOGIES

Molecular animations provoke further anxieties about the ontological status of the stuff of life. Animations breathe life into phenomena. When animations render mechanical phenomena animate, they offer a potent reminder of the instability of mechanism and the uncanny and disruptive potential of animism. Molecular animations and their associated animisms are risky, particularly in the context of a culture that carefully polices accounts of living processes. Lively machines can quickly undermine the deterministic ontologies that ground neo-Darwinian evolutionary views. Though evolution remains a founding logic of the life sciences, dictating the order of things in such diverse fields as developmental biology, molecular genetics, animal behavior, and ecology, the public life of evolutionary views, especially in the early twenty-first-century United States, is remarkably fraught. Practitioners and educators in the life sciences often find themselves caught between proponents of neo-Darwinism, those of intelligent design, and other forms of creationism. This is especially so in the context of undergraduate biology classrooms in the United States, where educators see it as their responsibility to keep their student's understandings of evolution in check. The stakes are high for those who see introductory biology courses as part of a public relations effort to produce properly aligned supporters of evolutionary theory. And this means not only defending neo-Darwinian views from creationism, but also from other theories of evolutionary history.

Neo-Darwinism hinges on a mechanistic approach to evolutionary change. It proposes a mechanism for evolution, one that is directed by natural selection acting on random genetic variation in populations of organisms. The central premise is that the environment imposes selective pressures on genetically diverse populations, and only those variants that "fit" that particular environment will survive. Heritable genetic variation is generated by random mutations in an organism's DNA. In this view, organisms are the effects not the agents of evolutionary novelty. From the perspective of neo-Darwinism, other approaches to evolution, such as the theories developed by Jean-Baptiste Lamarck (1744–1829), are just as blasphemous as creationist views. Giraffes are a favorite example to distinguish neo-Darwinism from Lamarckianism.

According to neo-Darwinists only those giraffes whose genes endow them with long necks are able to reach the nourishing leaves high up in the trees; these are the ones that will survive to pass their genes on to their offspring. It is not the organism that changes their genome (nor some god designing it); it is an organism's randomly mutating genome that opens up opportunities for it to "exploit" specific environments. A Lamarckian approach would narrate this adaptation differently: giraffes are able to transform their physiologies to meet their needs and desires; they grow long necks to reach the most nourishing leaves. Neo-Darwinian logic is also applied to the molecular scale. Novel proteins and cellular functions arise only through random mutations in DNA. In this view proteins are mechanical actants, subject to the forces of physics and chemistry, and determined by the scripts coded in DNA. To suggest that proteins have any agency in directing their own actions within cells is just as blasphemous as suggesting giraffes grow long necks in order to reach into the trees.[21]

Defending evolutionary views in undergraduate biology classrooms thus often translates into educator's efforts to expunge agency, animisms, and anthropomorphisms from the stories they tell. Educators express serious concern that their animations may inadvertently ascribe agency to proteins. Lynn and Joanna's conversation on anthropomorphisms in the classroom was fascinating in this regard:

LYNN: The thing that is the number one problem with how we talk about biology, the expert scientists, I mean, is that we anthropomorphize our molecules. And when we do that, you know, we literally go, "If I was a DNA polymerase what would I want to do?" [Laughing.] And that's fine between us, because we have this underlying, deep, ingrained appreciation of the fact that we're talking about what is energetically favorable for a molecule to do! [Laughing.]

JOANNA AND LYNN [IN UNISON]: As opposed to what a molecule has a desire to do!

JOANNA: That's what's so difficult about evolution.

LYNN: And our students just so don't get that. And it's just . . . and I try not to ever say what the molecule wants to do, anymore. And if I catch myself, I always stop and ask my students to tell me exactly what did I just mean. And this year I am better at it than last year. [Laughs.] I have not used it nearly as many times.

NATASHA: So you're talking about the attribution of agency?

LYNN: Yeah. But our professors do it all the time. *All the time.*

JOANNA: Yeah. And they don't . . .

LYNN: . . . catch themselves . . .

JOANNA: . . . catch themselves *at all.* Yeah! I mean, and it's very difficult. And it's very problematic. I was going to say, like for instance, [in the case] of evolution, it's *very* problematic, because then the students will really believe that this molecule or *some other thing* decided that this molecule needed to change . . . and it's . . .

LYNN: Right, in terms of evolution, it's important for them to understand that the order isn't the environment changed and then therefore something arose that could deal with it; but the other way around, something was there that happened to exploit this changing environment. That's a huge, huge concept in there that they don't necessarily get and can't get, unless they are entirely free of any *illusions* of the . . .

JOANNA: [interrupts] Human characteristics of their molecule—that it's selecting to change itself! And it's just . . . I mean it's a little thing. As experts we can say, well "it's the chemistry in the molecule." But as "intro to biology" students, these guys don't see that it's . . .

LYNN: George [the professor] tells this story in class every year. Where—he likes to make the scientists human, which is a good thing. But then he tells a story about when he was at a conference and got asked a question. . . . He literally stopped and said, "Well if I was this molecule, *what would I want to do*?" And you know, it's funny, you know, and it's a characteristic of a person. But it is a great disservice to an introductory biology student to hear that. Because it doesn't even . . . from George, it never comes with a disclaimer. What he actually meant to say was, "What would be evolutionarily advantageous for an organism to have in terms of the system."

LYNN: It sounds little but it makes a huge difference.

Lynn and Joanna recognize that anthropomorphisms and animisms are potent pedagogical devices, but they worry that they produce risky, unwanted effects and affects for their impressionable students. As they explained later on in the conversation, they don't want their students to come away with the idea that molecules are miniature machines; they're not comfortable enough with their own knowledge of mechanical engineering to develop machinic analogies with any level of precision. But they also don't want students to come away with the "illusion" that molecules have "human characteristics" and "desires." For them, problematic anthropomorphisms include animations that endow molecular movements with human (and animal) forms and desires. As they see it, the risk in anthropomorphizing molecular behaviors is not that they will necessary instill creationist ideas in their students' minds; equally problematic is that by suggesting that molecules have desires—that they "want" to do things—they may be invoking Lamarckian views of evolutionary progress. What is so fascinating is that while they ardently police their own anthropomorphisms among undergraduates, they do recognize that experts—those who have cultivated a feeling for the affinities and energetics of "the chemistry of the molecule"—practice a kind of "molecular vitalism."[22]

DISCIPLINING THE ANIMATORS

This unease about molecular animisms and anthropomorphisms is particularly palpable among online respondents to David Bolinsky's TED Talk presentation on *The Inner Life of the Cell*.[23] Both his talk and his animations sparked an intense debate on the TED Talks website. Initial posts offered Bolinsky exuberant applause for the work that he had done. However, several vocal contributors to the thread were less than enthralled. Their critiques are evocative of a range of anxieties that undergird a general discomfort with the ontological implications of molecular animations. Here Bolinsky's critics chastise him for "glamorizing" molecular processes. Bolinsky's response is remarkable for its insistence on the limits of knowledge, the indirectness of perception, and his commitment to rendering the ontological indeterminacies of molecular events with modesty:

> FRANK MASON [APRIL 8, 2009]: What David is really doing is "Hollywood-izing," i.e., glamorizing or applying advertising techniques to biology. Everything has perfectly adjusted colors . . . glows, etc. All of which is fake. Much like most advertising.
>
> DAVID BOLINSKY [MAY 18, 2009]: Frank! Truth will ALWAYS be subjective! Were I able to send a camera to record the processes we depict here, and were you the viewer, able to accommodate the femtosecond frame rate and lack of color, parse out the threads of order and functional systems inside a densely packed molecular zoo, that is antithetical to an open vista allowing for clear visualization of structures and proximal context, and were we able to avoid the total disruption of the functions and structures being scrutinized

by our intrusive tools, I would welcome the release from the labor we endure to show you those processes. I would buy that camera in a heartbeat. Most of what you and I have learned in our lifetime was gleaned from stylized concoctions of narrative, distilled from a cacophony of conflicting content and perspectives, and filtered through a lens ground by the controlling zeitgeist of our time. My hope is that we enhance the number of neutral sources of our knowledge and learn to tell the difference.

Bolinsky defends his work in response to a charge that his animation is "fake." He would, he insists, jump at the opportunity to witness cellular phenomena play out before his eyes. But, as he points out, all he has to work with is conjecture drawn from "intrusive tools," "stylized" stories, and "conflicting" views. He is well aware of the politics of vision and extends an optical analogy to insist that his knowledge is "filtered" through a "lens ground by the controlling zeitgeist of our time." Rather than eliding his handiwork in the making of this animation, he foregrounds the labor involved in rendering this view. And at the same time he acknowledges that his knowledge has already been aestheticized and distilled. His rendering—part fact, part fabulation—is grounded in a situated objectivity that insists on the partiality and incompleteness of knowledge.

Frank Mason accuses Bolinsky of fetishizing molecules rather than providing a factual rendering. What is the distinction between a fetish and a fact? Bruno Latour offers an etymology of fetishes and facts that is helpful in parsing Mason's accusation.[24] At the root of "fetish" are the verbs "to do" and "to make." As an adjective, "fetish" signifies that which is "artificial, fabricated, factitious, and finally, enchanted." Mason's charge of fetishism, however, neglects to take into account that "facts" also have at their root an association with "things made," and so facts and fetishes share the same "ambiguous etymology."[25] Their linkage is crucial for Latour, because "each of the two words emphasizes the inverse nuance of the other." "The word 'fact' seems to point to external reality, and the word 'fetish' seems to designate the foolish beliefs of a subject. Within the depths of their Latin roots, both conceal the intense work of construction that allows for both the truth of facts and the truth of minds."[26]

It is Bolinsky who is able to acknowledge the fabricated nature of facts. He takes a risk and offers us the perfect example of a "factish"; neither pure fact nor pure fabulation, a factish celebrates the work involved in crafting partial truths.[27] For Bruno Latour and Isabelle Stengers, a factish stands for "the robust certainty that allows practice to pass into action without the practitioner ever believing in the difference between construction and reality."[28] Bolinsky's animation is a factish in this sense: by avowing his labor, creativity, and authorship in rendering molecular facts, he embraces his participation in crafting this view of life. Construction and reality are not polarized for Bolinsky in the same ways they are for Mason. But Bolinsky's critics, bound to a vision of science that refuses the situated partiality of the factish, find this approach reprehensible. A new voice wades into the mix:

ISAAC YONEMOTO [MAY 21, 2009]: Yeah, but what you are doing is actively lying. In the case of the animations where you are showing microtubule/microfilament polymerization, you have made it look like the monomers' units zip onto the growing tubule, magically "knowing" where to go. You have similar moments with the assembly of ribosomes onto plasmids, more subtle problems with the walking of kinesins, etc. This is completely bogus. Molecules do not magically know where to go. You have subsumed truth to display something of "beauty," creating a fiction in the process. That is not the same as deleting details to simplify. What is really dangerous . . . is that you have basically allowed ideology (and perhaps convenience) to trump what, arguably is a way more interesting phenomenon, dynamic assembly and disassembly being an emergent property from simple, stochastic kinetic processes . . .

BOLINSKY [MAY 26, 2009]: Sorry Isaac, I think you may have missed the point of the entire exercise: we were just exploring, with Harvard, ways to educate visually. We of course tried to stretch the limits of our technology, budgets and schedule to create a new way to imagine a tiny corner of science education. *Inner Life of the Cell* was never meant to be seen in public, so we never anticipated the need to develop a back-story or rationale to keep absolutists at bay! Your defense of "reality" is admirable though no magic is implied. No, molecules do not "know" where to go, but in their natural milieu, they do things and go places with an alacrity and precision that is admirable and worth visualizing in the context of studying the underlying science. No intelligent design is implied either, for those readers on the opposite end of the critical arc.

YONEMOTO [MAY 30, 2009]: You see, David. That is exactly my point. Molecules do not "go to places with an alacrity and precision that is admirable." Any given set of interacting molecules probably collide with each other thousands if not millions of times before "getting it right." . . . When the interaction is extremely critical, biological systems have evolved to compensate by saturating the system with . . . enough copies of the molecule so that the right interactions happen the right number of times.

Pay attention to Isaac Yonemoto's stern response to renderings that suggest that molecules are up to stuff in the cell, that they have the "know-how" to direct their activities with "alacrity and precision." Note his insistence that molecular events are not directed; they are, rather, the stochastic effects of random motion. In other words, molecules collide randomly in the cell, and only a small proportion of those interactions give rise to biochemical functions. Note the similarities with neo-Darwinian explanations that rely on random, directionless change. Yonemoto insists that biological systems "evolve" to "compensate" for this randomness by "saturating the system" with large numbers of colliding molecules, ensuring that the "right" biochemical events do take place.[29] The "right interactions" are an effect of the stochastic nature of the system, not some a priori goal. It is in this sense that Bolinsky's animation is a direct affront to neo-Darwinian principles, which do not allow for agency and intention at the scale of organisms, cells, or proteins.

Given the prominence of "abolutists" like Yonemoto, Bolinsky and his team have taken a big risk in animating molecules as if they had direction and intentionality. Is the risk worth it? What do viewers gain from this kind of rendering of molecular life? What does the liveliness of his rendering do to the machinic ontologies life scientists seem to be so invested in? What, other than failing to provide a "neutral" rendering of the molecular world, can an animation do? What do these animisms and anthropomorphisms do to conventional conceptions of molecular life?

AMBIVALENT FIGURATIONS

An organism . . . has greater latitude of action than a machine. It has less purpose and more potentialities. The living organism acts in accordance with empiricism, whereas the machine, which is the product of calculation, verifies the norms of calculation, that is, the rational norms of identity, consistency, and predictability. Life, by contrast, is experience, that is to say, improvisation, the utilization of occurrences; it is an attempt in all directions.

—Georges Canguilhem, *Knowledge of Life*

Biological engineers are proud of their inventions, those molecular machines that have proven to be so effective in mediating the reengineering of life.

French historian of the life sciences Georges Canguilhem, however, decries the use of mechanistic reasoning in the life sciences. There is, he insists, a fundamental gap between machines and organisms; and for him, mechanism's penchant for reducing organisms to machines is anathema to the kind of inquiry that life's "potentialities" and "improvisations" demand. For him, machines are "products of calculation" that obey "the rational norms of identity, consistency, and predictability." Yet, if we look closely we will find that the molecular machines propagating through the twenty-first-century life sciences are perhaps less calculating and deterministic. As Bolinsky's narration demonstrates, machinic renderings seem to waver between mechanical, animal, and human forms. Indeed, it is especially when modelers animate their molecular machines that the steadfast boundary that Canguilhem asserts between machines and organisms appears to break down.

Perhaps the boundary between machines and organisms was never that well guarded. In a study of eighteenth-century automata, historian Jessica Riskin suggests that "materialist and mechanist accounts of life worked in both directions. Not only did they shape how people thought about living creatures, but reciprocally, they also changed how people thought about matter and mechanism. If life was material, then matter was alive, and to see living creatures as machines was also to vivify machinery."[30] Riskin shows how, once set in motion, mechanical analogies loop recursively between the lively and machinic. In this view, animism is not necessarily stifled by mechanical analogies; rather, what comes into being are what could best be described as forms of lively mechanism and machinic liveliness.[31] Once mechanistic theories successfully level the stuff of life alongside inorganic matter, both matter and machines, by association, acquire new kinds of vitality. In this light, Canguilhem's attempt to distinguish life from machinery can be heard as a call for an impossible purification and separation of categories that have long ago folded in on one another. Perhaps protein modelers will help us to see how animism and mechanism are intimately imbricated in twenty-first-century life science.[32]

Attention to machine analogies in the rhythm of their articulation shows that mechanism does not resolve fully deterministic objects in the inner recesses of cells. Indeed, the biological engineers who designed the laboratory course described in chapter 6 recognized the limits of their electrical circuit analogies. After describing the "easy" features of the circuits they had engineered in bacterial cells, their laboratory manual provided this humbling caveat: "In practice, spatial and temporal factors hamper even simple designs. The cell is a messy circuit board without the static physical separation you could find between electronic circuit elements. Proteins are made and roam the cell, invariably interacting with nucleic acids and with other proteins in unpredictable and unspecified ways." Proteins in this "messy circuit board" have the agency to "roam the cell," and in their meanderings they escape full characterization and predictive analysis. Molecular phenomena take up space and move through time; the "spatial and temporal" limits of modelers' experimental tools "hamper even simple designs." Molecular machines are anything but determinate and tractable devices; they frequently fail to reduce otherwise messy systems to the deterministic logic of physical and chemical laws. In spite of efforts to clamp down on the figure of the machine, modelers produce renderings of molecules that are undeniably *lively*. Molecular animations restage machine analogies to engender what can best be described as a menagerie of *lively machines*.

Recall how Fernando tried hard to stick to mechanical analogies to describe living systems (chapter 6). And yet, even for him, cellular machinery has dynamic properties. In the course of the same conversation in which he sustained the analogy of the cell as the factory-floor of a Ford car plant, he also offered a series of more dynamic animations of cells and organisms (see the introduction of this book). It seems there is a widespread tendency among practitioners in this field to mix their metaphors as their stories oscillate rapidly between deterministic and lively articulations of life:

> I like to think of a cell as a house . . . or better yet, an office building. You have receiving, you have shipping and packing. You have your accounting office. You have your central distribution. You've got your CEO's office. In the middle of all these things, you have all these little offices working. It's more like an old building than a new building. In those old buildings you would have a vacuum system and you put a memo in and press the button. And shoowoo! There it goes! Instead of e-mail you've actually got vacuum mail! But that's just a single cell. And so you can look at a multicellular organism and you can look at a city with an underground rail system to deliver all the mail. Cells sit on a matrix, inside of an organism. Some of them travel on the matrix. Some of them stand. Some of them sit around because they are so close together they have no place to go. Some of them actually form interlacing channels that can pass stuff back and forth.

In the course of a single conversation, Fernando's figurations waver from cells as factories, to cells as office buildings, and organisms as bustling cities. Fernando's story of cellular mail delivery resonates remarkably with Bolinsky's analogy of the FedEx delivery system of the cell. In Fernando's rendering, the cell is kept alive through an economized division of labor organized to maximize the circulation of information. Information is exchanged not in the form of some body-less code, but in bundles of physical materials. In this rendition, the "offices" inside cells appear to have the capacity to act within their own being. Though he is reluctant to engage what he called an "overused metaphor," he goes on to animate cells as "communities" involved in the reciprocal exchange of materials. Even as he tries to clamp down on deterministic, mechanical accounts of protein molecules, he animates a much more lively scene.

Rob's molecular switch is another potent example. Recall that he could click the switch on and off with remarkable precision (see chapter 6). The clean lines of this little machine, however, seem to get blurred as the documentary shows him ramping up his rhetoric for his PhD defense. The camera pans through a large crowd gathered in a small lecture theatre at Columbia University. Rob is at the helm. "Today," he declares, "I'm going to tell you about visualizing metabolic control." Rob explains his efforts to "get at the mechanism of activation and repression of this metabolic control switch," and he delivers his message with intensity and aplomb: "What I want you to know and remember about this discussion is that the AMP protein kinase is really the Federal Reserve Board chairman of the cellular economy. It really does make decisions about how supplies are going to be used, and how the demands of the cell are going to be allowed. And so, this is much more than a fuel gauge." As he figures it here, this metabolic switch is not some inanimate machine, like a "fuel gauge" or mere mechanical switch; it is *Homo economicus* in the extreme, a rationally acting chairman who makes executive decisions to distribute resources through the cellular economy. This analogy takes the form of what might best be described as a hybridized mechanoanthropomorphism.[33] The protein comes to embody a rational, masculinized "master molecule" governmentality that rules over the energetics of life processes.[34] Note the ambi-valence in Rob's renderings. The figure of the chairman animates the molecular mechanism with an agency that Rob refuses to reduce to mechanical gear work. "Molecule as metabolic switch" and "molecule as chairman" are differently inflected figures that make molecular life waver materially and semiotically in his hands. In a clip posted online Rob is emphatic that his goal is to get at the "physical determinants" of cellular processes; but his chairman exceeds a mechanistic or deterministic analysis.

Another perhaps more puzzling ambi-valence haunts ways that Larry

Shapiro renders molecular machines. Recall Larry's rendering of drugs as monkey wrenches designed to jam the molecular machinery of the cell (chapter 6). A cafeteria scene featuring Larry and two of his students complicates this rendering and disrupts assumptions we might have about Larry's relationship to the molecular world. This scene does not make the final cut of the film, but it is posted on the *Naturally Obsessed* website under the heading "Examining Values." This is not one continuous stream of footage, but a sometimes sloppily spliced together rendering of voices, gestures, and expressions. The editors have rearranged the order in which each person speaks. For example, the scene opens at a moment when the cafeteria is quiet and Larry and his students have already finished eating. The final shot of the scene shows them just about to tuck into their meals. The filmmakers effectively reorient the direction of the conversation, altering the significance of the statements, and inflecting the characters' affects, gestures, and expressions with new meaning. This splicing in effect produces an animation of the event, telling a story that the directors find salient.

The scene opens with Larry at the table struggling to find the words to express to his students just what he sees when he looks at a molecular structure.

> LARRY: Every time that you see one of the basic microscopic machines that is necessary to our function as living organisms. . . . You . . . uh, this is corny. You see God? Is that right? I mean. Ummm . . . but. But. Um. *But evolution is God*, that's the . . . the horrible truth of it. And that's why there's so much fear of it. Because, that . . . that is what made all of this.

Just after this utterance, the editors splice in a segment from earlier in the conversation when the cafeteria was bustling. In the scene, Kil's contribution is framed as a direct retort to Larry:

> KIL: If of course evolution is actually happening. Let's not rule out alternate explanations.

Another cut brings us to Julius, a member of Shapiro's lab who is not actually featured in the film. He also appears to be one of the only nonwhite graduate students in the lab. He offers a different take:

> JULIUS: I was brought up in an Orthodox Christian family. [Shrugs.] I don't know. Science is based on observation. Religion is based on belief. That's it I think. You can't prove or disprove God by scien-

tific experimentation. So you know. That's what I believe. It's a belief. [In the background you can hear Kil in agreement, "I believe that too."] Right? Whether the belief is right or wrong, is a different question. If you believe, you just believe. You can believe. You think that's right, that's what it is. You know.

The scene cuts to an earlier moment when Larry is just about to start eating. He chuckles nervously, while shaking his head dismissively, "I don't know. I'm not getting this."

Kil's carefully spliced intervention makes a swerve that serves as a potent reminder of the fraught twenty-first-century U.S. context in which neo-Darwinian explanations of life are openly and hotly debated by creationists. Here Kil outs himself as an advocate of intelligent design. Julius, on the other hand, doesn't seem to need to fold gods into molecules or reinterpret scientific evidence for or against evolution. He seems to have come to terms with living in the contradictions between faith and fact by assigning them to separate domains. For him experimental proofs and mystical belief can exist comfortably side by side. While Kil nods avidly in affirmation of Julius's argument, this spliced rendering gives Larry the final word, and the filmmakers quickly bring to a close what appears to be an uncomfortable conversation for them, if not also for Larry.

Larry's invocation of God, however, deserves closer attention. This statement may produce some serious cognitive dissonance, particularly if one is expecting a particular kind of coherence or consistency in a scientist's approach to the stuff of life. This utterance seems counter to his articulation of drugs as "monkey wrenches" that intervene in the functioning of molecular machines (chapter 6). In this separate clip Larry offers up a barely articulable description of the kind of sublime rapture he experiences when he holds a molecular structure in his hands. Evelyn Fox Keller suggests that such forms of captivation in the face of "nature's secrets" are highly gendered; they hearken back to a time when scientists like Francis Bacon framed their inquiry as an attempt to wrestle secrets from a veiled, feminized nature.[35] Yet, Larry's invocation that God might actually *be* the molecule also produces a momentary disruption of the hegemonic mechanistic logic on which neo-Darwinian explanations of life and evolution rest. What Larry offers here is different from both the neo-Darwinists and the creationists: this is not a machine that does the work of a transcendent God or mechanized nature; rather, it is the embodiment of an immanent force. By folding God into the molecule, Larry has reorganized the logics of evolutionary processes. Where neo-Darwinists orient their explanations of speciation around random events of genetic mutation and nature's harsh and economizing selective pressures, here Larry recalibrates these transformational forces as the acts of an immanent (rather than transcendent) God. Here God is manifest inside the very theories that were conjured to banish him for good.[36] What is particularly significant is how powerfully these ambi-valences (molecule as machine or as God) can unhinge and reorient a modeler's relation to the molecular realm. The dissonance generated between these figures of the molecule shows how such vacillations are integral to the very logics of life science.

CONCLUSION

Practitioners in the life sciences are heavily invested in mechanistic reasoning, especially to ground their neo-Darwinian explanations of evolutionary change. Machinic renderings of molecules gear well into this deanimated mechanistic ontology. And yet, machines move: that is how they do their work. It turns out that animating molecular machines produces all kinds of unwanted effects and affects. Practitioners continuously attempt to tame their animistic and anthropomorphized renderings; and yet molecular mechanisms are continuously quickened. Modelers' firm grip on mechanistic logic and reasoning can be seen to slip and slide.

The ambi-valent inflections that haunt these molecular animations teach us that mechanism isn't what we thought it was. In practice, mechanistic theories of life do not adhere to a conventional script. Ethnographic attention to the wavering registers in which modelers render molecules makes it possible to see the strange ways that practitioners "do mechanism": mechanism takes shape through an improvised bricolage of animisms, affects, inflections, and stories. Indeed, practitioners demonstrate that a kind of liveliness is immanent to mechanism. What opens up when modelers animate proteins in these livelier registers? How might articulating molecules as if they were full of desires, wants, and needs actually be a generative practice for researchers and educators? And, how might these renderings be the most effective means of luring students and colleagues into new ways of grasping dynamic molecular phenomena? The final chapter explores these questions.

S tudents in Jim and Geoff's protein folding course were required to recall the fine details of molecular phenomena on tests; and to do this they were encouraged to learn the atomic configurations of proteins "by heart." Jim told the class: "We want you to have it in your head. You need to know it cold." Throughout the semester, Jim's commanding performances aimed to impart the skills students would need to get protein structures in their "heads" and "hearts." In one of his early lectures Jim drew attention to students' confusion around a homework assignment. Some apparently had trouble with the wording of one of the questions. The students were directed to a ribbon diagram of a protein structure found in the textbook (see plate 19 for an example of a ribbon diagram). They were asked to "draw, copy, or trace a version of Figure 2 (e) with the alpha carbons and nitrogen atoms clearly labeled or colored."

The difficulty it seemed was in their interpretation of the directive "copy." Jim clarified: "This means *hand copy*! If you Xerox it, you don't assimilate it!" "You have to trace it!" he implored. He demanded that the students get involved in the structures: in order to learn the fold *by heart*, they had to trace the direction of the polypeptide chain *by hand*. He demonstrated for the class. Against the backdrop of projected ribbon diagram, Jim swept his entire body up in the act of tracing the elaborate curvature of the fold. He caught the curve of the winding backbone and traversed the full visual space of this amplified model. With accelerating momentum, he hitched a ride on its folds. As he followed the polypeptide backbone, he told the class: "You have to signal actively" in order to "get" the structure of the fold. He demonstrated how, in his words, "you can't not learn something" if you get your body involved. Here Jim clued his students into their bodies as resources for learning the fine structures of complex three-dimensional molecular forms.

Jim showed his students a way to re-member the specificities of the fold. By leaning into the curves of the ribbon diagram, Jim showed his students how to look and learn. They too could articulate their memories and imaginations by *emulating* the fold. Jim modeled for the class what could be called a *kinesthetic diagram*. As he traced the peptide backbone, his entire body got swept up the fold. This act of tracing was nothing like photocopying. Rather than delegating the task to a replicating machine, Jim *transduced* the direction of the winding polypeptide chain through his tissues. This is a kind of rendering that not only manifests a representation of molecular structure; the person who follows the fold also gains a newly articulated sensorium, one that can register molecular differences in new ways. Kinesthetic diagrams thus produce both a representation and a sensation: Jim "signals" the folds indexically through a kind of gestic semiosis that tries to stay true to the model on-screen; and as he *leans into* the fold, this active and energetic tracing sensitizes his body to molecular form.

Affects, energies, and forms of knowing propagate through the dense thicket of entanglements that take shape between modelers and their models. Chapters 6 and 7 drew attention to the ways that modelers *inflect* their renderings with particular affects, aesthetics, animisms, and anthropomorphisms. In those contexts it appeared as if modelers were the ones directing the inflection with their culturally salient symbols and aesthetic forms. This chapter turns attention to the ways that modelers animate their molecular insights through the marvelously articulate medium of their own bodies. In this context, the source of the inflection becomes less clear. Kinesthetic animations, it seems, blur easy distinctions between the animator and the animated. Who is the animator and what is being animated when a modeler gives their body over to modeling molecular movements?

This chapter explores the concept of *agencement* to account for the ongoing distribution and redistribution of agencies in the "rapport of forces" that set modelers, models, and molecules in motion with one another.[1] Approached as an assemblage of agencies, as an *agencement*, the questions "who is animating what?" and "what is animating whom?" cannot be resolved; the source and direction of the impulse are indeterminate. Animacy, as I frame it here, is an intra-active phenomenon, in Barad's sense of the term. This chapter develops the concept of "intra-animacy" by observing how, alongside mechanism, these modelers "do animacy." Their practices suggest that animisms and anthropomorphisms are not entirely the effects of human ingenuity. The modelers here are not the sole directive force inscribing the relation: they don't simply impose animistic or humanistic qualities on molecules; rather, theirs is a "process of attunement" or emulation through which they become receptive and responsive to the subtleties of molecular energies and movements.[2] Perhaps it is by giving themselves over to the labor of making models and animations that they learn how to *move with and be moved by* molecular phenomena.

This final chapter takes a close look at the simultaneously *mimetic* and *transductive* dimensions of the kinesthetic renderings performed in classrooms, in laboratories, and at professional meetings. It expands the scope of the analysis to include the public life of protein models animated through what can best be called forms of "molecular calisthenics." These are contexts where scientists make their movement modalities explicit, sometimes even staging their kinesthetic animations as dance performances. This chapter examines the role that movement plays, not only in the popularization of science, but also the integral part it plays in scientific inquiry. In some circumstances, however, forms of expressive body-work can exceed the bounds of what is considered proper conduct in the laboratory. This chapter homes in on contexts that discourage body-work to see when, where, and how kinesthetic animations are policed. Throughout, this chapter considers what kinesthetic animations do, and how they rend imaginaries of molecular life.

KINESTHETIC PEDAGOGIES

In order to teach concepts associated with the phenomenon of protein folding, Jim and Geoff moved through a wide array of media including data, charts and graphs from biochemical assays, crystallographic structures, and computer-generated models and simulations. They also deployed a range of vivid analogies that could get students affectively entangled with molecular phenomena.[3] In class, Jim often bumped up against the limits of language for articulating the qualities of three-dimensional things. He struggled to find the right words to describe protein structures and movements: "It is clear from the X-ray diffraction patterns that proteins are objects with space in them. This is very different from packed polymers. So, we can ask: What is the character of the interior? Is it oily? Is it patchy with regions of solvent? But patchy is a two-dimensional word. I can't think of a three-dimensional word that gets at this." Where words fell short, and where two-dimensional images fell flat, Jim articulated the otherwise elusive texture, tensions, forms, and movements of proteins with his own body. This was the most ready-to-hand and articulate modeling medium he had available. He made his body over into a pedagogical tool: he became a dynamically morphing proxy that could stand in for a suite of other models. When describing the folding of a globular protein, he often drew his arms into the core of his body. Curving over and tucking inward to create a concave form, he used the shape of his arms to mimic a protein's internal organization of helices and sheets. When describing the packing of two helices in a protein, he repeatedly drew his arms in toward one another and crossed them at the forearms to specify the precise angle at which they were associated. He demonstrated the flexibility or inflexibility of this association through the energetics, affects, and tensions he held throughout his tissues.

Several studies have taken up the question of the role of gesture in scientific reasoning.[4] Elinor Ochs and her colleagues have approached the study of gesture in the performance of scientific concepts.[5] They describe the gestures that mediate communication among physicists who attempt to convey their research to one another in weekly lab meetings. They apply ethnomethodological conversation analyses to video recordings in order to track how bodily gestures help physicists narrate and dramatize their scientific stories. They observe what they call "understanding-in-progress" and show how "scientists can take seemingly immutable transcriptions such as published graphic displays, and, over narrative time, transform them into highly mutable, highly intertextual and symbolic narrative spaces through which they verbally, gesturally, and graphically journey."[6] In their example, physicists' gestures are seen as explicitly discursive: physicists' gestures generate a "dynamic grammar" that supports their spoken language, helping them to make statements about mathematical relations and two-dimensional graphic displays.[7]

By contrast, Jim's gestures reenacted molecular forms and movements. Rather than using gesture to amplify language, Jim emulated the forms and textures of substances, qualities that are hard to relate through words. His kinesthetic animations were in this sense kinds of mimetic renderings that attempted to convey a likeness to the thing modeled. Jim's gestural choreography was thus less a "grammar" than a form of *mimesis* that rendered the form and movements of the molecule through the form and movements of his body. Sociologist Pierre Bourdieu has likened such mimetic performance to a "rite or dance" in which there is "something ineffable," something that "communicates, so to speak, from body to body, i.e. on the hither side of words or concepts."[8] Practitioners like Jim articulate their kinesthetic animations through a kind of iconic and indexical "gymnastics."[9] And as they propagate "from body to body" they become lures, vectorizing the kinesthetic imaginations of both the modelers and, if they perform it well, their interlocutors.

Jim's former PhD student Joanna and her colleague Lynn in the biology department developed a series of workshops and lesson plans using specially designed three-dimensional models and interactive molecular graphics technologies to help students learn to visualize the structures and movements of biological molecules. Through this work, Joanna came to recognize that teaching these concepts places extra demands on her body to perform the multidimensionality of biological phenomena. Even though she was wary of "anthropomorphizing the molecule" (see chapter 7), she confessed to me and Lynn that in class she regularly animates molecular movements with her body: "I probably like the dancing and movement so much [in the classroom] because I do see these things rolling around in 3-D in my head. And yeah, it's like, if I could get my body to do this [she curves her body around an imaginary fold, voicing the movement with a "Schwooo!"], and have this little arm flapping in the breeze. I don't know. It just makes more sense." As she talked, her body came to life, and I could see her delight in transducing the details of the fold. She also appeared to be transducing Jim's penchant for molecular calisthenics. Apprenticeships, like those between graduate students and their advisors, appear to be sites where tacit knowledges, habits, and sensibilities are readily transduced.

In his study of gesture, science studies scholar Brian Rotman approaches mimetic gestures as the most "primitive" gestural form.[10] Yet, the kinesthetic animations I've observed are far from primitive or simple: they form an integral aspect of the rhythms of experimental reasoning. Moreover, they are precise in their articulation of molecular phenomena. Jim's in-depth demonstration of the packing of helices during processes of protein folding demonstrates this precision well. Jim held his arms out in front of him, crossed at the forearms, to demonstrate how it is that the "side chains are talking to each other." With his stance strong he explained to the class: "Now when two helices are packing against each other they form a junction." All of a sudden he paused and looked up. He called on his coteacher Geoff. "I want you to stand up." "Alright," Geoff nodded, and he stood up. Jim gave him explicit instructions: "Now, point to the junction." Geoff pointed vaguely at Jim's crossed arms. "No, not there!" With his "helices" still packed together, Jim gestured insistently with his eyes and head to redirect where Geoff was pointing: "Right between . . . yeah, okay."

Jim carried on with his description while Geoff stood by his side, pointing at the junction. In this moment, Jim made explicit that *he* was the model. By asking Geoff to point to the junction where his arms met, he demanded and received Geoff's confirmation that his arms really were helices. Moreover, Jim required Geoff to point, not to any place where his arms met, but to a specific site on his body-as-molecular model. He maintained a specificity in his body-cum-model that demanded Geoff locate the exact site of the junction. Giving his body over to the model, and enlisting Geoff as an indexical arrow to highlight his kinesthetic diagram, Jim figured out a way to transduce this molecular phenomenon for everyone in the room. A fuller examination of the propagating, transductive dimensions of such kinesthetic animations requires that we first take a closer look at the remarkable phenomenon of mimesis.

MIMESIS

The relation of emulation enables things to imitate one another from one end of the universe to the other without connection or proximity: by duplicating itself in a mirror the world abolishes the distance proper to it; in this way it overcomes the place allotted to each thing. But which of these reflections coursing through space are the original images? Which is the reality and which the projection? It is often not possible to say, for emulation is a sort of natural twinship existing in things; it arises from a fold in being, the two sides of which stand immediately opposite to one another.

—Michel Foucault, *The Order of Things*

Diane's "feeling for the molecule" offered insight into modelers' intimate entanglements with their models. Diane's response to the pain of a mangled molecule was sympathetic. She acquired this sympathy by *emulating* the molecule with her body, imitating the conditions that gave rise to its strain (see chapter 3). Alongside sympathy, one of the four similitudes, or forms of resemblance, that Foucault articulates in *The Order of Things*, is "emulation." Like sympathy, emulation brings far-flung phenomena together, "abolishing the distance" between things.[11] It arises out of a "fold in being" that articulates a "twinship" in things. Emulation is a kind of imitation in which it is not possible to say "which is the reality and which is the projection."[12] It is an intra-active mimetic exchange. Michael Taussig's multisensate approach to mimesis foregrounds this remarkable infolding that takes shape in acts of emulation.[13] Reading Walter Benjamin, Taussig develops the notion of an "optical tactility" that entangles movement, sensation, and perception in the mimetic act of becoming other. Taussig suggests that our mimetic faculty nourishes and sustains shared understanding and knowledge within a larger cultural milieu. On this, he suggests that the mimetic faculty is "the nature that culture uses to create second nature." It is "the faculty to copy, imitate, make models, explore difference, yield into and become Other."[14] Mimesis has two layers for Taussig: it contains both an element of copy or imitation, as well as the "palpable, sensuous connection between the body of the perceiver and the perceived."[15] For Taussig, "to ponder mimesis is to become sooner or later caught in sticky webs of copy and contact, image and bodily involvement of the perceiver in the image."[16] In this sense mimesis is not merely an optical relation, but a corporeal habit, guided by tactility.[17] This "palpable, sensuous" faculty describes well protein modelers' full-bodied efforts to craft "second nature" models of molecular life.

Mimesis is a "sensuous moment of knowing" that involves a "yielding and mirroring of the knower in the unknown, of thought in its object."[18] Mimesis resonates well with the concept of rendering. One of the meanings of the verb "to render" is "to yield," to "give oneself over to." In this sense, mimetic renderings involve a form of "copying" that does not rely on an a priori subject and object: model and modeler are mutually articulated in the act of mimesis.[19] For protein modelers, this yielding of the knower in the unknown opens up a space for an intimate dance between modelers and their media, models and machines. It is their desire to render imperceptible substances visible that takes modelers "bodily into alterity."[20] In their efforts to stay true to molecular form and movements, they entrain their kinesthetic imaginations to molecular phenomena.

Self and other, modeler and model, are intimately imbricated; they fold in on one another in acts of rendering. A modeler's willingness to give her body over to the practice of rendering demonstrates the "ineffable plasticity" of our morphological imaginaries "in the face of the world's forms and forms of life."[21] Yet, the kinesthetic animations explored here participate in a form of mimesis that does not hinge on generating mirror-image likenesses. Protein modelers articulate molecular forms within the range of motions available to their bodies. And while their contortions never actually look like the computer graphic models they project on-screen, this doesn't mean they are ineffective. Computer graphic animations appear to cause much anxiety among practitioners for the ways that they enliven and anthropomorphize molecules. Jim and Joanna's kinesthetic animations, on the other hand, demonstrate the multidirectionality of anthropomorphism. By giving their bodies over to the work of rendering molecular forms, they show us that what might appear as an anthropomorphism—the humanization of a molecule—actually takes shape in a mimetic exchange. This "yielding" of self in other simultaneously effects a *molecularization* of the human. Like Diane, whose articulated sensorium was sensitized to molecular strain (chapter 3), Jim and Joanna have allowed their bodies to be inhabited by their models. They situate themselves inside molecular phenomena in order to reach toward insight. It is the molecular phenomena that set their bodies into motion, and in the process, they have become effective proxies for their molecules.

INTRA-ANIMACY

Protein crystallographers produce static data forms; they must fix and freeze excitable molecules in crystal form in order to get a workable model of molecular structure. How then do modelers bring movement back into their models? How does Edward, for example, come to know that his molecule "breathes"? What happens when he emulates and animates his molecule by pulsing his hands as if he was holding a breathing lung? In an essay on the intimate association of animation, automation, and liveliness, film scholar Jackie Stacey and science studies scholar Lucy Suchman look to the *Oxford English Dictionary* to explore various inflections in the term "animate": "The term 'animate' is suspended between its transitive and qualifying meanings; 'to animate' is to impart life, vitality and motion, as a sign of life, to something that has previously been seen as inert; while 'animate' (as an adjective) identifies an entity that 'is endowed with life . . . having the power of movement, like an animal."[22]

In this frame, Edward would be seen as an *animator* who brings life to his inert model. He engages his moving body to enliven what is otherwise a static model. Stacey and Suchman cite anthropologist Alfred Gell and astutely identify how agency is presumed to operate in this relation: "As the 'action of imparting life', animation conventionally presumes an agent, a force that is antecedent to and independent of its object. In the secular domains of art and design, the agent is commonly taken to be human, its object a form of artifice endowed with the appearance of life."[23] In this formulation the liveliness inheres in Edward, the animator; the animacy of his model is merely the effect of his affective labors. From this perspective, his breathing molecule is "a transformative deception" that requires his audience to believe "in the illusion of movement."[24] If one were to follow this argument, it becomes clear that animation is the very enchantment that our critical analyses are meant to demystify.[25]

Brought back within the frame of mimesis, however, it is necessary to ask: is Edward animating his model, or is it setting him in motion? Stacey and Suchman ask: "Who is animating what?" and "what is animating whom?" They caution science studies scholars to "hold on to [their] skepticism when encountering the enchantments of technoscience"; yet they also encourage cultivating richer accounts of the distribution of agencies that take shape among humans and nonhumans in the dense tangles of bodies and meanings that constitute laboratories.[26] Indeed, the kinesthetic renderings documented here suggest that animations and animisms are an effect of a larger constellation of participating agencies. As practitioners lean in to make contact with molecular phenomena through the technological prosthesis of their modeling tools, their emulations are inflected and informed by the entire experimental configuration, not just by their own aesthetics, stories, and intentions.

Philosopher and animal studies scholar Vinciane Despret builds on Deleuze's concept of *agencement*, a French term that she prefers to "assemblage," its common English translation. Where "conventional" definitions of agency are "based on subjective experience and autonomous intention," Despret turns to the concept of agencement as "an assemblage that produces 'agentivity.'" For her, agency is not about autonomy or self-actualization, rather it is a phenomenon that is "extensively shared in the living world":[27] "An agencement is a rapport of forces that makes some beings capable of making other beings capable, in a plurivocal manner, in such a way that the agencement resists being dismembered, resists clear-cut distribution. What constitutes the agent and the patient is distributed and redistributed incessantly. . . . There is, in each agencement, co-animation, in the literal sense of the term, that is, in the most animist meaning of the term."[28]

Despret's formulation of agencement as a "co-animation" resonates well with the concept of "intra-animacy" that I am developing here, where animacy is the effect of the participation of an entire constellation of agencies in an experimental configuration. The inclusion of Barad's formulation of "intra-action" in this concept foregrounds the crucial insight that subjects and objects do not preexist their encounters; rather, they are precisely what precipitates out of their intra-actions. Modelers, models, and molecules *move with and are moved by* one another in the very moment that they are all in-the-making. "Intra-animacy" is in this sense not some immaterial vitalism that imbues inert matter with some external force, nor is it built up from a networked collection of autonomous individual agencies modeled on liberal notions of human subjectivity. It is an agencement produced through the affects and movements of intra-acting bodies, human and non. In this frame, there is no animator: Edward and his breathing molecule intra-actively animate one another. The resulting animisms are thus more than effects generated by modelers' aesthetic flourishes or their imposition of lively figurations on otherwise inert matter. These animate renderings transduce the energies, affects, and sensibilities that take shape in an ecology of laboratory entanglements.

These enlivened models and bodies go on to animate others' imaginations, shaping research questions and pedagogical interactions. In this sense they are lures that can tug on other modelers' kinesthetic imaginations, "vectorizing" their bodily experience to produce new forms of knowing. Transduction, as I use the term here, signals the energies, excitations, and affects that propagate through such renderings of molecular life. "Transduction" is a term in wide use in molecular biology. Proteins are frequently figured as working machines that *transduce* force and energy within the cell.[29] "Signal transduction" describes the transmission of extracellular signals into a cell and the propagation of this signal as a branching series of molecular events.[30] Once activated in a signaling network, molecules propagate chemical energy and mechanical forces through the cell. The characteristics of a signal depend on the specificity of the medium through which it is transduced. In the viscous and variegated milieu of the cell, a signal morphs as it moves. Pulling signal transduction out of the deterministic registers that inflect its use in molecular biology, it is possible to render it otherwise. Consider signal transduction, rather, as the propagation of energies in a contact improvisational dance among tightly packed molecules.[31] In this approach, a propagating signal cannot be reduced to information or code, and transduction cannot be reduced to information transmission or translation. Rather it becomes a relay of energies propagating through an excitable assemblage of responsive bodies. What gets transduced is in this sense not the outcome of a singular intention or agency, but an aggregate effect of all bodies, affects, and energies: an *agencement*, in Despret's terms.

Henri Bergson's *Matter and Memory* animates a theory of perception based on nineteenth-century physiology. He offers some insight to think through this transduction of affects and energies through modelers' bodies. He writes: "living matter, even as a simple mass of protoplasm, is already irritable and contractile; . . . it is open to the influence of external stimulation, and answers to it by mechanical, physical and chemical reactions."[32] In this work Bergson bundles perception and movement together in the nervous tissue of the body, and explores how affect and responsive action are produced through a "kind of motor tendency in a sensory nerve."[33] Bodies, in Bergson's formulation, are transductive; they propagate energies, affects, and excitations. Living bodies become fleshy antennae whose physiologies act as a kind of resonating medium that oscillates between conduction and resistance, and manifests energies as perception, affect, and action. From this vantage point, bodies

8.1 and 8.2. Albert Szent-Gyorgyi photographed while making opening remarks at the 1972 Cold Spring Harbor symposium on the mechanism of muscle contraction. Courtesy Cold Spring Harbor Laboratory Archives, Cold Spring Harbor Symposia on Quantitative Biology Collection.

can be seen as *excitable tissues* emulating and transducing the energetics and movements of the world.[34]

What do modelers' kinesthetic animations transduce? What is it that they propagate? Figures 8.1 and 8.2 show Hungarian protein scientist Albert Szent-Gyorgyi presenting his work on muscle proteins at the 1972 symposium, "The Mechanisms of Muscle Contraction," at the Cold Spring Harbor Laboratory. In these pictures he can be seen animating the sliding filament theory of the molecular forces involved in the contraction of muscle tissues in order to describe the interactions between the proteins actin and myosin. These images evoke the excitation in his tissues as he transduces these molecular forces and movements. He has attuned his sensorium to subtle molecular forces, and here he can be seen hitching a ride on movements he can imagine and intuit. In this sense, it is not just the molecules that are excitable: his tissues have become *excitable media* with the capacity to collect up and relay nuanced molecular affects. Just as modelers extend the contours of their kinesthetic imaginations to get a feel for the tensions and movements of molecules, they can also try on others' molecular gestures. Kinesthetic animations are thus lures that can entangle others. They are enticing for others to try on, and in a back-

and-forth relay of gestures and affects modelers can expand their repertoire. This opens up another dimension of the transductive nature of these renderings: by propagating their forms of knowing through affects and gestures, modelers can excite others into action. In this sense, pedagogy is a transductive practice. Just as Jim transduces the model through his tissues, he shows his students how they too can become transducers of molecular forms and movements. Such forms of molecular calisthenics demonstrate how the visual cultures of science are part of wider cultures of performance.

BODY EXPERIMENTS

The kinesthetic animations that enliven classroom pedagogy and scientific talks are mimetic and transductive agencements. In other words, they are part of a practice in which modelers learn to emulate the forms and forces of molecules in order to propagate their insights among others. At least once during my fieldwork, I was enlisted to participate in a collaborative kinesthetic animation. In the summer of 2005, anthropologist Michael Fischer and I were both pursuing research in the field of protein modeling.[35] We were invited to jointly conduct a series of interviews with modelers based in a group of laboratories at the privately funded cancer research institute where Zeynep works. There we met with Andrés, a protein crystallographer doing his postdoctoral research in an immunology lab. During our interview, Andrés demonstrated how a molecular mechanism he had hypothesized functions in intercellular adhesion. He wanted to understand how molecules bind cells together, and he postulated a mechanism that makes use of interlocking proteins to maintain the structural integrity of developing tissues. His study of a group of cell surface proteins showed that these molecules are long and straight. One part of the protein is embedded in the cellular membrane, while the other extends out into the extracellular environment, where it is available to connect to similar molecules on adjacent cells. The binding end of the protein has three short protrusions that give it a ratcheted structure. He hypothesized that this ratcheted structure provides a molecular mechanism to strengthen binding between adjacent cells.

Andrés, Mike, and I were seated facing one another on tall stools next to a workbench in the lab. Andrés was explaining how his protein works, and I was busy scribbling notes in my notebook while he was talking, with barely enough time to watch how he was demonstrating the structure. "Here, take my hand," he said. With this, I looked up. "As if we were shaking hands." I had to drop my notebook and pen in my lap, so that I could reach out my hand. He wanted to convey the strength of the associations made between molecules whose binding holds two adjacent cells together. We clasped hands in a firm handshake, and then without warning he leaned back. Unprepared, our hands slipped apart. "How would we make our grip stronger?" he queried. "Suppose we are climbing a mountain, what kind of grip would we need?" Still holding hands, he eased me into an answer by gripping me at the wrist. I followed along, and clasped his wrist in turn. We both leaned away. Our grasp was decidedly stronger. "Right," I confirmed. Molecules binding at their first and second hooks would form a stronger bond. "And how would we make it even stronger?" He extended his grip further up my arm, clasping me at my elbow. I followed suit and we tested our combined strength. Together, Mike, Andrés, and I acknowledged the augmented stability of this third hold.

By enlisting my participation in this performance, Andrés interrupted my note taking and redirected my attention toward the body-work of reasoning in protein modeling. Ratcheting up the grip, from binding at the hands, to the wrists, to the elbows, Andrés was conducting a collaborative *body experiment*. He was testing both his model and my understanding by enlisting our jointly corroborated physical intuitions to test the strength of molecular associations. In this striking example of haptic creativity, his body had become a key resource for him to be able to experiment with and make arguments about molecular mechanisms. His body was invested in his interpretation of protein structures, and his dexterities and intuitions played a significant role in his experimental reasoning. His "empirically derived 'practical reason'" had become a form of "scientific reason."[36] In a resonant study of the role of kinesthetic knowledge in the practices of measure and geometry, phenomenologist of touch Mark Paterson explores how "body parts" can become "an investigative aid to the perception and measurement of external space," in such a way that they become "components of the mechanisms of measure; the body itself becomes instrument."[37] Body experiments such as this one were for Andrés "vehicles for materializing questions";[38] that is, they propelled him toward new kinds of conceptual and corporeal understandings that shaped his hypotheses. Body experiments can rehearse, renew, or radically transform how molecular phenomena are understood. In Andrés's case, his kinesthetic intuitions had committed him to several years' worth of research with the hope of visualizing this particular intermolecular phenomenon. Despite little evidence to support his theory that these proteins bind to each other using all three hooks, he still held out hope that he might one day find the crystal structures that could validate his feeling for the strength of these molecular associations.

Protein modelers perform their kinesthetic knowledge of molecular forms and movements frequently, and in all kinds of settings. In Diane's lab, many of the advanced graduate students and postdocs (particularly those who had successfully built crystallographic models) were delighted to tell me all about the molecular movements they could intuit, but couldn't otherwise see. In interviews where I asked them to explain how they conducted their experiments and how their proteins worked, they performed the vibrations of molecules embedded in growing protein crystals and waved their arms about to emulate the floppy ends of polypeptide chains that would come out blurred in crystallographic snapshots. They also contorted their bodies into sometimes-awkward configurations to demonstrate the conformational changes of the molecule, and to show how it does its work mechanically and chemically in the cell. Graduate students, postdocs, and principal investigators would sometimes fumble and correct their kinesthetic animations, occasionally realizing mid-gesture that they had the structure wrong. In these situations they were quick to correct their kinesthetic models as a means to correct the model that they were simultaneously figuring out in their heads. They learned new things while they used their bodies to play through possible molecular configurations and movements.

Protein modelers are constantly trying to find ways to share their kinesthetic knowledge with others. Their body experiments can communicate only if others in the space are willing to participate. Research groups make active use of body experiments as they attempt to communicate their findings with one another. Brian, the biological engineer who co-taught the course on protein dynamics that I observed (see chapter 6), runs a protein simulation laboratory that specializes in designing novel protein structures. At one of their meetings, Kabita, a fifth-year PhD student, presented her recent progress. The protein she works on is a complex protease: it forms a dimer, which means that it is made up of two similar polypeptides folded up together. It also has intracellular and extracellular domains, with parts of the protein that must traverse the lipid bilayer of the cell membrane. In the midst of her presentation she was repeatedly interrupted with a constant stream of questions from her colleagues asking her to clarify the structure of the protein. Even with intricate computer graphic renderings of the molecule projected on the screen behind her, the group demanded more detail, and she was compelled to articulate the structure with her own body.

To convey this structure to the group she transduced a ribbon diagram of the molecule. She lifted both her hands over her head and traced the winding backbones of the twinned molecules, one with each hand, following them as they traversed extracellular and intracellular spaces. Her gestures were large and sweeping: she brought her arms from high up over her head, all the way down in front of her body. Her molecular dance ended with her fully bent over, hands touching the floor. Yet, questions still surfaced from the group, and Kabita was asked to describe the mechanism that bound the molecules together. "I like to think of it this way," she said and repeatedly crossed her arms at the forearms, fists clenched, demonstrating with the tension in her musculature the binding energy between the molecules. A visiting professor, still confused, leaned over the table and repeated this gesture over and over as he asked questions, inquiring and confirming with her that this was indeed the form of the molecular interaction she was describing.

Kabita later told me that I had had the "misfortune" of attending the "most contentious group meeting in at least two years": "Brian still refers to it as 'that disastrous group meeting,' because I didn't succeed in conveying my concepts, apparently." What is interesting is that even in such a tense situation, she was comfortable to fully animate her hypothesis with her body, and moreover, others engaged their bodies to relay back to her both their understandings and their misunderstandings. In this transductive, mimetic exchange, Kabita's kinesthetic model was reenacted until misunderstandings gave way to partial confirmations. This kind of relay can often resemble an improvisational dance. Here it is possible to see these researchers trying to tune their bodies in to one another's renderings as a means to enable fuller communication of the forms and functions of particular molecules. In the process they learn how to *move with* and *be moved by* one another's kinesthetic sensibilities.

In pedagogical situations, such as Jim and Geoff's protein folding class, kinesthetic animations are often made explicit. Jim instructed his students how to conduct a body experiment, how to give their bodies over to the folds. Yet, most often, body experiments remain tacit. They are part of the rhythm of communication among practitioners working at laboratory benches, gathered in group meetings, engaged in conversations at conferences, and performing scientific talks. There are, however, a number of moments where practitioners draw attention to their body-work. In some cases, protein modelers fashion themselves as dancers and choreographers animating the molecular realm.[39] It is by tracking this phenomenon of dancing scientists that it is possible to explore contexts in which movement is explicitly recognized as integral to scientific research.

Only rarely is there an opportunity to participate in a molecular happening. You are going to have that opportunity, for this film attempts to portray symbolically, yet in a dynamic and joyful way, one of nature's fundamental processes: the linking together of amino acids to form a protein.

—Paul Berg, in *Protein Synthesis* by Robert Alan Weiss

Kinesthetic renderings can take on epic proportions. In 1971 a football field at Stanford University became the scene of a large-scale "molecular happening." Over one hundred dancers and musicians gathered to perform *Protein Synthesis: An Epic on the Cellular Level.* This was a collaborative kinesthetic animation that massively amplified the intricate molecular interactions involved in protein synthesis. This reenactment of protein synthesis was directed by Robert Alan Weiss, choreographed by Jackie Bennington, and set to a proteinaceous revision of Lewis Carroll's poem "Jabberwocky." This "happening" was recorded on film and has since circulated widely as a pedagogical tool, entertaining undergraduate biology classrooms across the United States for decades. More recently it has become a major hit on YouTube, and university campuses around the United States have staged their own renditions.[40]

The film opens with a mini-lecture by molecular biologist and Nobel Prize laureate Paul Berg. In his "Protein Primer," Berg addresses his audience from within a classroom. He stands in front of a blackboard diagram that indexes the molecular pathways involved in protein synthesis (see figures 8.3–8.4). Before him is a colorful ball-and-stick model of a protein. His sweeping gestures instruct his audience how to make sense of this model:

We know now that the three-dimensional structure and the function of a protein is determined by the order of the amino acids along the backbone of the molecule. So protein synthesis involves programming and assembly. And this film—with people portraying molecules using the dance idiom—tries to animate these two processes: the programming and assembly of a protein (see figures 8.5 and 8.6).

Our genes carry the instructions for ordering the amino acids of each protein. Those instructions are encoded in a messenger molecule, in RNA, depicted in this film as a long snaking chain. Each of the message units is played by three adjacent people in the chain. Colored head balloons indicate the bases. Green for Guanine, blue for Uracil, yellow for Adenine.

If there is a message, there must be a way to translate that message. And

Molecular Calisthenics 219

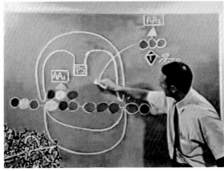

8.3 and 8.4. Screenshots from "A Protein Primer" in Robert Alan Weiss's *Protein Synthesis: An Epic on the Cellular Level.* Paul Berg uses his blackboard diagram to index the sequence of molecular interactions involved in protein synthesis.

that's the job of the ribosome and the transfer RNAs. The ribosome is composed of a large and small subunit, and these are depicted in the film as tumbling, rolling clusters of bodies, amorphous by themselves, but organized and structured in the act of translating the message.

Berg continues on with his description, indicating how "puffs of smoke" in the performance indicate the use of energy. He concludes with this proviso: "My diagram is of necessity static. But protein synthesis is a dynamic process. This movie tries to bring those dynamic interactions to life." The phenomenon of protein synthesis defies the flat or static representational conventions Berg has available in his classroom. In his narration, moving bodies are the ideal animating media for visualizing molecular dynamics. The "dance idiom" enables participants to collectively articulate the "programming" and "assembly" of a polypeptide chain. Molecular structures and interactions are brought to life through "tumbling, rolling clusters of bodies" that weave across the field. In this large-scale reenactment of protein synthesis, dancers' bodies stand in as proxies for individual amino acids and other molecular assemblages. These bodies-cum-molecules coalesce and converge across the massive surface of a football field, forming higher-order aggregates. This "molecular

8.5 and 8.6. Screenshots from Robert Alan Weiss's *Protein Synthesis: An Epic on the Cellular Level.*

happening" thus animates and amplifies Berg's two-dimensional blackboard diagram of protein synthesis, playing his diagram forward in time. This "epic" thirteen-minute reenactment is full of life: joyful, excitable bodies laugh and shout, careening and cartwheeling across the field. If this is indeed a reenactment of the cybernetic metaphors of code and information flow, it is at least a cybernetics on LSD (see figures 8.5 and 8.6).

DANCE YOUR PHD

Dance and science have recently converged again. On February 14, 2008, the *New York Times* reported the results of the world's first "Dance Your PhD Contest" held in Vienna, Austria. In his article "Dancing Dissertations," John Tierney, a science reporter for the *Times*, writes: "We've been warned that America's lead in science is in jeopardy. Now it looks as though Europe has definitely forged ahead in one field: interpretative dancing of PhD theses."[41] Twelve contestants responded to a widely distributed call for submissions sent out by science journalist John Bohannon. The event was hosted by the American Academy for the Advancement of Science (AAAS) and *Science Magazine*. Mounting enthusiasm for the contest resulted in a second call for submis-

8.7, 8.8, 8.9, and 8.10. Screenshots from *A Molecular Dance in the Blood, Observed* by Vincent LiCata, winner of the 2009 Dance Your PhD Contest in the category of Professor.

sions just eight months later. The Second Annual Dance Your PhD Contest attracted thirty-six entries from scientists in the United States, Canada, Australia, Europe, and the UK. In 2010 a third contest ran, this time attracting forty-five submissions from PhD students and recent graduates in physics, chemistry, biology, and the social sciences. In 2011 the fourth annual competition drew fifty-five contestants, and the 2012 competition attracted a further thirty-six entries.[42] To date hundreds of scientists have participated in choreographing and performing the results of their doctoral research. The submissions, which after the first live event were uploaded as YouTube and Vimeo videos, were widely accessible on the Internet and caused a serious stir in the blogosphere as well as in print and radio news.[43]

One of the winning entries for the second annual Dance Your PhD Contest is an example of a collaborative molecular animation. It engages four dancers to animate and amplify the multidimensional movements of a single hemoglobin molecule as it binds and releases oxygen. Vince LiCata, a tenured professor in the Department of Biological Sciences at Louisiana State University, won in the category of Professor. His choreography, *A Molecular Dance in the Blood, Observed*, was set to Laurie Anderson's "Born, Never Asked" (see figures 8.7–8.10).[44]

Hemoglobin is a protein that transports oxygen in the blood. It is made up of two pairs of identical subunits that interact with one another to endow the protein with structure and chemical function. In LiCata's staging, two pairs of dancers wear red T-shirts and light blue jeans. One pair dons purple gloves and matching goggles, while the other set is clad in white gloves and goggles. The two pairs rotate around a single axis in tight formation, curving, swaying, and reaching out into the space. They rotate as an aggregate, and as they move, they keep two small white balls of oxygen in play among them.

One of the compelling features of this choreography is that the movements of this animated molecule resonate with a repertoire of movements that come alive among practitioners in research and teaching contexts. More specifically, LiCata's dance re-members a well-rehearsed gesture that has propagated widely among protein modelers since Nobel Prize laureate Max Perutz solved the structure of hemoglobin in 1962 (see chapter 2). Hemoglobin is one of the primary structures that life science students must learn, almost as a rite of passage, early on in their studies of biochemistry. Because its conformation changes each time it binds and releases oxygen, it is often imagined and performed as a "breathing molecule." Here, the origins of the gesture that Edward enacted with this breathing molecule become clear. The undergraduate biochemistry and biological engineering students I interviewed would often re-enact this breathing molecule by making a gesture as if they were holding a pulsing substance breathing with life in their hands. Such movements allow students and practitioners at all levels to *lean into* this phenomenon in order to feel through the kinetics of molecular motions. If the "epic" protein synthesis performance and LiCata's staging of hemoglobin are choreographed forms of kinesthetic animation, then these more mundane, everyday gestures are more like improvised body experiments. These dances open up new ways of thinking about forms of knowing and experimental inquiry in the life sciences. While it is true that the Dance Your PhD Contest submissions inhabit a carnivalesque modality that explicitly breaks from the mundane context of day-to-day research practice, they do gesture at the ways movement is crucial to laboratory research and pedagogy.[45]

TEMPORAL ELASTICITY

Time-lapse imaging and computer graphic animations are two prominent techniques for visualizing phenomena in the contemporary life sciences. Kinesthetic animations, like LiCata's hemoglobin dance and the epic reenactment of protein synthesis, appear to offer a third modality. These animations are choreographed, scripted, and staged reenactments of dynamic molecular processes. And while there is nothing inherently limiting about these media forms—each can enable modelers to engage their objects, concepts, and data in open-ended explorations—they all rend time in the very ways that cause Joanna and Lynn such concern (see chapter 7). These renderings put a time stamp on otherwise indeterminate and ephemeral processes, and in so doing they set the tempo and the direction in a way that concretizes the dynamic temporalities of molecular events.

Joanna's anxieties about animating protein folding resonate with concerns that philosopher Gilles Deleuze raises in his meditations on cinematic movement. In *Cinema 1*, he examines the ways that moving images mechanically recompose movement. The movement that is perceived when watching films is the result of a "mechanical succession" of "snapshots" (what he calls "immobile sections") that cut moving phenomena into equidistant moments of time (what he terms "any-instant-whatever"). Snapshots section and immobilize movements, generating "closed sets"; and these are laid out in series to produce the effect of movement on screen. He associates this mechanical recomposition of movement with the achievements of the scientific revolution, which succeeded in abstracting time as a measure and an independent variable in experiments.[46]

For Deleuze, both time-lapse techniques and cartoon films rely on this frame-by-frame mechanical recomposition of time. And by this account, any rendering that fixes the temporal flow, such as the epic protein synthesis reenactment or LiCata's choreography, also fall into this category. Body experiments can be distinguished from these other forms of kinesthetic animation. Body experiments are not mechanical reconstitutions of movements, and so in Deleuze's definition, they are "foreign to the cinema." Rather, they are forms of movement that "cannot be divided without changing qualitatively at each stage of the division."[47] In this sense, body experiments, like Andrés's collaborative model of molecular hooks and Kabita's emulation of the helices winding through her protein, are resonant with the forms of movement Bergson described with his concept of the "open whole" of "duration";[48] that is, they are open, improvised explorations that unfold in the rhythm of hypothesis formation, communication, and experimentation. They cannot be isolated from a larger constellation of affects and energies. They are improvised rather than choreographed, or fully composed. And while they can be repeated, it is always with a difference. In their rehearsals and reenactments, these renderings are also continuously revised and can be refuted mid-gesture. What's crucial here is that these animations do not fix the temporal flow of a process: their temporality is elastic.[49] Where an animated movie traces a narrative arc that runs directionally in time, modelers conducting a body experiment can hesitate and prevaricate as they test out the attractive and repulsive forces and tensions between atoms in a molecule. By shifting the rhythm of their movements, they can expand and contract molecular time. Where Bolinsky's *Inner Life of the Cell* induced significant backlash, body experiments don't seem to encounter the same kinds of resistance. Molecular forms and temporalities are left open and available for reinterpretation, rather than foreclosed. And yet, this does not mean that modelers' expressive movements are exempt from scrutiny. It turns out that kinesthetic animations are frequently policed. The following section examines when and where body experiments get shut down. It poses the question: If body experiments keep open what it is possible for protein modelers to see, say, feel, and know about the molecular world, how might practitioners find ways to nourish and sustain this practice?

DISCIPLINED BODIES

Practitioners confront significant pressures to contain their exuberant body-work. Tacitly and explicitly they are conditioned to conform to norms of proper conduct in the laboratory and the classroom. Take another look at Nick Downes's cartoon featured in the introduction (figure 1.1). Michaels, depicted in the cartoon as a helically wound-up scientist, gets chastised for his elaborate kinesthetic model. Hold this image together with the media reports that weighed in on the first Dance Your PhD Contest. They took up a position similar to the curmudgeonly scientist who, hunched over his work at the bench, barks at his colleague telling him to "get back to work." As the buzz of the Dance Your PhD contests circulated through radio, newspapers, and most prominently through the blogosphere, they were rendered as light-hearted, fun-loving means for hard-working scientists to distract themselves from their heady work in the supposedly serious and sterile space of the lab. The media pitched the contest as an opportunity for "stressed students" to "relax and distract their minds from what can be a long, arduous project."[50] Bohannon, "clad in a white disco leisure suit" at the American Association for the Advancement of Science (AAAS) meetings for the finale event of the second Dance Your PhD contest, exclaimed to a reporter from the *Chicago Tribune*: "How amazing that scientists around the world busy with lab work took a break to do something as bizarre as this. I love that."[51] These reports frame participation in the contest as a diversion from more serious work.

The discussions about the contest that have propagated through the media reveal that popular conceptions about scientific labor, and what scientists are supposed to be up to in their laboratories, are rather constrained. Apparently, there is a moral imperative for scientists to engage in rigorous, disciplined labor in the form of a kind of disembodied cognition. This imperative is not just policed from inside the laboratory, but also by members of the public. One of Bohannon's interviews on NPR makes this clear:

> BILL LITTLEFIELD: You gave grad students, postdocs and professors six weeks to create and film a dance based on their scientific research. And since scientists are pretty busy people and not known for their dancing ability, you couldn't have gotten more than, what, two or three entries?
>
> JOHN BOHANNON: Well, that's what I feared. The amazing thing is that in the end about a hundred scientists took part producing three dozen dances.
>
> BILL LITTLEFIELD: Well obviously our scientists are not busy enough doing science! [Laughter.][52]

What counts as scientific productivity appears to be rather narrowly defined. Dance—figured in this context as leisure—is seen as an inappropriate distraction for busy scientists. In the interview and elsewhere, Bohannon was forced to defend his contest as a form of public outreach. And yet, it is not only the scientists who are chastised for getting distracted from their important work; the science reporters who covered this story were also held culpable. In the comment section of his online article, *New York Times* reporter John Tierney is blasted by one of his readers for expending his valuable labor to report on the contest. In this reader's mind, the contest is clearly a grand waste of time: "So instead of John Tierney writing about something worthwhile like America's lost power in the sciences, he showcases something completely irrelevant. I used to believe John Tierney cared about science. His last two articles, on this and NASCAR physics, [have] proven otherwise."[53]

This reader taps right into the widespread assumption that science is all work and no play, and that play itself has no proper place in the lab. The dance contest gets figured as excessive and wasteful of scientists' time, talent, and energy: the kind of excess that must be reined in to keep the machinery of publicly funded science well greased, and the "pipeline" of graduate students moving smoothly. Tierney's critic, anxious about "America's lost power in the sciences," reveals a pervasive biopolitical imaginary about the proper relationship between bodies, pedagogies, power, and knowledge, one that tethers the labors of well-disciplined scientists to the scientific prowess of the nation-state.[54]

As publicly staged performances, the Dance Your PhD contests and the "epic" protein reenactment are both explicit in their use of dance as a medium for rendering dynamic phenomena. Most of the time, in both pedagogical and professional settings, practitioners moved their bodies without inhibition, but they rarely if ever referred to what they were doing as dancing. When conducting interviews, I had to be careful to avoid directing a researcher's attention to their own movements. The few times I directly called attention to their elaborate movements, they became self-conscious or froze mid-gesture. For these reasons I avoided using video to record interviews or scientists at work. What follows is an account of a series of encounters in which modelers' bodies became the focus of attention. In each of these events, the modelers' body-work was interrupted. What are the forces that shut kinesthetic animations down? When do their kinesthetic renderings get read as excessive? This line of inquiry explores how and why kinesthetic animations are such precarious phenomena.

During my extended conversation with Lynn and Joanna, I told them about "one crystallographer" who contorted her body in pain when students presented her with misshapen protein models. They laughed as I told them how the modeler corrected the model by realigning her own body. They really got how one could feel the strain in the model. Lynn interjected, "Well, you see, she feels the pain of the molecule. She anthropomorphizes it, which is fine, for *her*, as long as she doesn't imply . . . you know." Lynn was referring here to the kind of Lamarckian intentionality this might signal for a nonexpert (chapter 7).

And yet, this story elicited accounts of how Lynn and Joanna act out molecular movements in class. Joanna confessed to taking delight in dancing her molecules and got right into it, demonstrating for us what she does in class to teach her students about the ways molecules move through a gel in an electrophoresis apparatus: "You know, this little molecule needs to move through the gel. And it's a net, and it's stuck like this. And now I'm a big long molecule," she laughed as she wriggled along with all the other molecules moving through the gel. The effect was vivid, and I came away with a remarkably clear sense of these molecular interactions. Yet after reflecting on what it was they were doing in class up in front of their students, Lynn was visibly embarrassed: "We're making fools of ourselves is what we are doing!" When I asked them to focus on their body movements they became self-conscious. It seems as if, upon reflection, their lively performances were excessive, as if they were breaching some code of proper conduct. As junior women in science they perhaps felt doubly anxious about staying within acceptable bounds.

Andrés, who had unabashedly enlisted me in his collaborative animation of molecular hooks, also experienced discomfort when I drew attention to his body-work. I was at an annual meeting for protein modelers with Michael Fischer. I was off looking at student posters when Mike told Andrés that I was studying how protein modelers "danced" their molecules. Andrés confessed to Mike that he had choreographed "a little dance" for one of the molecules that he had modeled. When I rushed up to him having just heard the news, he balked and told me: "I hate dancing, but there was just no other way to communicate the mechanism. I had to dance it." He looked positively mortified when I asked him to show me his dance. I quickly learned that to call out a modeler's movements as the focal element of interest, without acknowledging the content of what she or he is trying to communicate, is blasphemous. I had skipped over the part that was crucial. What I should have asked was: How does this molecule move? While he refused to show me his "secret" dance as we stood in the bustling lobby at the scientific meeting, he admitted that he had performed it before for a small group of colleagues. In this moment, he taught me that kinesthetic animations are a means not an end to the work of communicating molecular insights.

One last example shows how differential relations of gender and power can severely constrain how modelers use their bodies. Zeynep, the postdoc whose work with engineered peptides was described in chapter 1, conducts her research in a lab at the same institute as Andrés. She recounted a story that brought these issues to the fore. We were at a cocktail party held by a mutual friend when she told me about an incident that occurred when she was talking with the principal investigator of her lab. She was describing the molecular mechanism of a particular protein she was working on. We were standing in the corner of the kitchen, and she cleared space to show me the elaborate choreography she had used to convey the molecular movements to her professor. The mechanism involved one part of the protein making an upward, twisting, piston-like movement into another part of the protein. She demonstrated this with zeal, making a large upward gesture with one arm to enact the piston, while using her other arm to hold the space occupied by the rest of the protein. Apparently her professor read her molecular dance as an overtly sexual performance. He taunted her and called her out for making such rude gestures in public. In this moment he sexualized her body-work, and so rendered it as an excessive and improper expression of knowledge. She took this warning seriously and told me that she'd become much more self-conscious since then about how she moved when relating molecular mechanisms.

Kinesthetic animations extend and expand what counts as a model or as data. Protein models are embodied and performed in ways that propagate more than structural information. The effects and affects engendered through these renderings are not extrascientific phenomena; rather, they appear to be intrinsic to forms of knowing and inquiry in this field. Kinesthetic diagrams and animations are tacit and ephemeral transductions, and they are subject to constraints that make them impossible to sustain in certain power-laden contexts. And yet, in spite of ardent attempts to eradicate animisms and anthropomorphisms, and to discipline excitable bodies, it is clear that modelers don't always keep their bodies in line: they continue to rely on their bodies to articulate the wiles of molecular life. Documenting these sometimes fleeting and precarious forms of knowing is thus crucial. Body experiments, kinesthetic pedagogies, and large-scale kinesthetic animations appear to defy the normative moral economy of the laboratory, where scientists' cognitive labor is at a premium. In the face of normalizing forces, it becomes crucial to pay attention to the ways that kinesthetic animations continuously disrupt the norm. Where and when, in spite of their better judgment, do educators slip up and enact molecular desires? Practitioners' affective entanglements with their models and molecules are never fully suppressed. Modelers are constantly reinventing ways to articulate their intimate knowledge of molecular phenomena. The enthusiasm with which scientists have embraced the Dance Your PhD contests, for example, suggests that practitioners are on the look-out for ways to circumvent conventions that limit what is possible for them to see, say, imagine, and feel. Perhaps it is through movement that they are best able to avow, rather than disavow, the entanglements that condition what and how they know. I would go so far as to suggest that it is protein modelers' capacity and willingness to *move with and be moved by* their molecules, models, and animations—to *emulate* and *transduce* the excitability of proteinaceous forms—that enable lively narratives to thrive, even inside of the mechanistic logic life scientists are supposed to avow. It is from this vantage point that it is possible to ask what it is that modelers' excitable tissues transduce? What forms of life do their renderings propagate?

> . . . life itself destroys beings.
>
> —Michel Foucault, *The Order of Things*

What is life becoming in the hands of these twenty-first-century protein modelers? What forms of life are in the making in their laboratories? Life, according to Michel Foucault, was an invention of the nineteenth century. Prior to that "life itself did not exist": "All that existed was living beings, which were viewed through a grid of knowledge constituted by natural history."[1] In his classic text *The Order of Things*, Foucault digs deep into the archive to observe mutations in the human sciences from the classical episteme to the nineteenth century. Caught between vitalism and mechanism, the sciences of life in the nineteenth century manifested what Foucault calls an "untamed ontology."[2] At this time, "life itself" came to be figured as both an "inexhaustible force" that propels "all existence" and as the very force behind the undoing of living beings: "For life—and this is why it has a radical value in nineteenth-century thought—is at the same time the nucleus of being and of non-being: there is being only because there is life, and in that fundamental movement that dooms them to death, the scattered beings, stable for an instant, are formed, halt, hold life immobile—and in a sense kill it—but are then in turn destroyed by that inexhaustible force."[3] Life, in this formulation, is a force that cannot be held still without dissipating its vitality: to "hold life immobile" is to destroy it and the beings that were its transient manifestation. In this sense the "itself" of life, figured as a force, could be named only once it was extracted from rhythms of life and its manifest beings. Any attempt to capture its essence would extinguish it and the beings whose lives it propelled. "Life itself" thus would remain always a secret force lurking behind living beings.

As a secret force, "life itself" has long been a conceptual lure for experimental inquiry in the life sciences.[4] The promise of its capture continues to entice practitioners today. And yet its contours have transformed significantly since the nineteenth century. Indeed, the "untamed ontology" of nineteenth-century thought appears to have been disciplined by twentieth-century mechanistic visions of life: this "inexhaustible force" could not, it seems, evade the techniques and instruments that brought life into view at higher and higher magnification. By the mid-twentieth century, practitioners could claim that they had captured "life itself" in the form of high-resolution models of DNA and proteins.[5] It would seem as though "life itself" lost its power to escape. Writing in the 1970s, at the cusp of the informatics revolution in the life sciences, French molecular biologist François Jacob could assert: "Biology has demonstrated that there is no metaphysical entity hidden behind the word 'life.' The power of assembling, of producing increasingly complex structures, even of reproducing, belongs to the elements that constitute matter. From particles to man, there is a whole series of integration, of levels, of discontinuities. But there is no breach either in the composition of the objects or in the reactions that take place in them; no change in 'essence.'"[6] For Jacob, there was "no breach" in the series of links that tethered molecules to "man": the "essence" of life remained intact, from the atomic particles that make up matter, to the complex organization of the organism, and there was no need to invoke a special force to make sense of these phenomena. He broke with the view that "life itself" was an enigmatic force, and posited a fundamental continuity between inorganic materials and vital forms. By his account, the untamed ontology of nineteenth-century thought was reined in, and a deanimated mechanism gained ground.

Rhetorician and historian of the life sciences Richard Doyle explores this history to examine how a demystified life science reconfigured "life itself" in the form of a molecule.[7] He documents the 1960s rise of a molecular biology that flattened living bodies into a one-dimensional code. In this period, the primary object of biological interest, deoxyribonucleic acid, was fetishized as information.[8] In Doyle's formulation, twentieth-century molecular biology reduced living bodies to a kind of thinness and transparency with nothing left lurking secretly beyond or behind DNA's codes. It was in the flatness of the rhetoric of "body as code" that the enigmatic force of "life itself" was simultaneously squeezed into a molecule, and spread out into the thinness of legible text. In the case of the helically coiled DNA molecule, life got unraveled, and unzipped.

For Doyle, the emblematic moment in this history was marked by a profound ennui: this was the moment when one molecular geneticist would look down at the transparent body of a genome-mapped and cell-fate-mapped laboratory worm (*Caenorhabditis elegans*) and shrug and say, with maps at hand, "That's all there is."[9] Doyle narrates a moment in the history of molecular biology where a kind of "boredom" settles in among scientists involved in genomics projects: once "life itself" was captured, the exhilaration of the chase appears to have evaporated. There was "nothing more to say": this was "a story of resolution told in higher and higher resolution."[10] Reduced to legible codes, it turned out that "life itself" was not so lively. In Doyle's reading, "the great unsaid of the life sciences, of a molecular biology that sought and found 'the secret of life,' is the fact that life has ceased to exist. Or, rather, it never did exist, that the life sciences were founded on an embarrassing but productive ambiguity, the opaque positivity called 'life.'"[11] In the late-twentieth-century, molecular biology had rendered the organism "postvital"; that is, "beyond living."

But what if the "untamed ontology" of the nineteenth century was never fully captured or contained? Evelyn Fox Keller has shown that a kind of vitality continued to lurk just below the surface of mechanistic theories throughout the twentieth-century life sciences.[12] She has examined mechanistic explanations in the history of genetics and finds that geneticist J. H. Muller and his colleagues figured the gene as an agential "entity embodying the capacity to act within its own being."[13] Early concepts of the gene, it seemed, "betray[ed] a subconscious adherence to 'the ancient lore of animism.'"[14] Does a disavowed liveliness still linger in spite of the rise of presumably deanimated solutions to the problem of life?

This book has shown that twenty-first-century protein modelers "do mechanism" in surprising ways. Their ambi-valent renderings demonstrate that their mechanistic theories of molecular life are rather "untamed." The stuff of life does not resolve for them into fully deterministic machines. In practice, they figure proteins in registers that waver between the machinic, the human, and the animal. While practitioners may boast that they have captured and put "life itself" to work in the cell, they also simultaneously animate molecules as wily creatures who continually evade such capture. They show us that mechanism is not the hegemonic logic that would suppress the animisms and vitalisms of the earlier sciences of life. Rather, they show how the disenchantment of the life sciences is incomplete. Modelers continually fail to contain the animisms and anthropomorphisms that keep erupting through their renderings. Could it be that the "postvital" apathy of late twentieth-century molecular genetics was just a passing phase? Or perhaps this ennui was just an effect of the ways that the life sciences have been observed and documented? Regardless, it is through an attention to the affective entanglements of life science inquiry that a lively mechanism can be seen pullulating below the surface. This book has documented scientists tracing the contours of an ontology so untamed it is veritably *excitable*. What is this liveliness that continuously erupts inside of and alongside mechanistic descriptions of life? And what effect does such a lively mechanism have on the life sciences' grounding logics?

LIVELINESS

Life itself is the psychic, cognitive, and material terrain of fetishism. By contrast, live-liness is open to the possibility of situated knowledges, including technoscientific knowledges.

—Donna Haraway, *Modest_Witness*

This book has aimed to make strange the idea that "life itself" is a thing that can be captured. The twentieth-century concept that life has an "itself" that it could manifest in the form of a molecule should give us pause. Haraway identifies "life itself" as a fetish object. Indeed, it is a ruse. To have an "itself," life must be pulled out of time and out of relation. In this sense, "life itself" is a lonely concept: this abstracted "it" is on its own; unresponsive and unaffected, it is left for dead. The ambi-valent renderings of life propagating through the life sciences today suggest that "life itself" is not the only lure. Twenty-first-century life scientists, it seems, are unable to resist hitching a ride on what might be called the "unapologetic swerve of liveliness."[15]

Liveliness is not a kind of vitalism; it is not a vital force that infuses matter with life; and it is not the opposite of mechanism. Lively narratives are not positioned against machines. Machines, as I showed in chapter 7, can be quite lively. Lively machines and machinic life are hybrid forms made possible through stories that don't get mired in the divide between vitalism and mechanism that has haunted biology since its inception. Liveliness is a way of telling stories that refuses to make clean distinctions between organisms and machines, or between vitalism and mechanism. Lively narratives reach toward a world in which thrive barely recognizable forms of life. Liveliness is a relational concept. It hinges on an intra-active conception of agency or *agence-ment*. Intra-animacy is generated in contexts where bodies are open to move with and be moved by one another. It is in the kinesthetic and affective entanglements of inquiry that modelers, models, and molecules intra-animate. Liveliness is thus an effect and a range of affects that arise out of the labor of rendering life molecular. This liveliness is expressed as a kind of excitability transduced through modelers' tissues. And it is through their lively renderings that life scientists learn to transduce the excitable life of matter.

EXCITABLE MATTER

"Protein folding is a deep problem." This is a statement that Jim repeated over and over again in the course of his lectures on protein folding. Why is protein folding a "deep problem"? What gives this problem its "depth"? Part of what Jim was asserting was that this was a problem that would not likely be solved with improved algorithms or faster computers. He clearly recognizes the epistemic uncertainties involved in determining protein forms from an ad hoc combination of techniques that provide partial views. But it is not just that the experimental techniques aren't adequate to the task of visualizing this process. Rather, I hear Jim's insistence about the depth of this problem as an affirmation that the *molecular practices of cells* evade complete capture.

The recalcitrance of the protein folding problem has major implications for some of the core assumptions that life scientists have long held dear. One particular vision of cellular life, circumscribed by what practitioners call "the central dogma," has organized assumptions in molecular biology since the discovery of the structure of DNA.[16] The central dogma, which hinges on an informatic model of the cell, dictates that complete information for life moves from DNA to RNA (ribonucleic acids) to protein sequence. Indeed, this is the very process that was rendered in massively amplified form on a football field at Stanford University in 1971 (see chapter 8; for more on the central dogma, see the appendix). This view of life has been a fundamental tenet in evolutionary theories that postulate that random mutations in an organism's DNA are the sole source for heritable variation.

University of Chicago biologist James Shapiro has explored some of the many "surprises" researchers have recently unearthed about "unexpected" activities ongoing in cells that would be "forbidden" under the model of life laid out by the central dogma.[17] These "complications or contradictions of the central dogma" demonstrate that "there is no unidirectional flow of information from one class of biological molecule to another."[18] His "short (and partial) list" of these "forbidden" activities include such "epigenetic" processes as: reverse transcription, which allows DNA to be produced from RNA sequences; posttranscriptional RNA processing, which includes processes that can splice messenger RNA and profoundly alter the message; the recent discovery of catalytic RNA, which shows that some RNA molecules can behave like protein catalysts and spur metabolic activities in the cell; recent discoveries that the "non-coding regions" found in the genome, those long stretches of DNA commonly dismissed as "junk," do have functional properties; epigenetic posttranslational protein modifications that can transform a protein's cellular activities by the addition of methyl or phosphate groups; and DNA proofreading and repair processes that may be involved in directing specific changes in a DNA sequence.[19]

If in the last half of the twentieth century, the central dogma scripted genes as the primary actors determining the course of life, it seems that new cellular actors are emerging for twenty-first-century life scientists. In addition to the vast menagerie of proteins that make up each cell's "proteome," other molecules, including RNAs, are now being hailed as active participants in the story of *how cells do life*. One group of cell biologists has even made the call for a "molecular sociology" in order to keep track of "who is talking to who" in the dense thicket of molecular interactions ongoing in a cell at any given moment.[20] According to James Shapiro, "biological specificity" does not have a "rigidly deterministic character"; rather it obeys a "fuzzy logic."[21] Indeed the life sciences can now be seen articulating the fine details of a story that does not just inquire into the genetic determinants of life, but rather documents the *molecular practices of cells*. These stories reconfigure cells and molecules as active participants in the *agencements* that shape their growth, development, and reproduction. This new view of the cell and its molecular agents has implications for our understanding of evolutionary processes. James Shapiro tests out a new metaphor of "cellular cognition" to describe how cells actively sense their worlds and rework their biological activities in responsive relation to their environments:

> If we are to give up the outmoded atomistic vocabulary of 20th-century genetics, we need to develop a new lexicon of terms based on a view of the cell as an active and sentient entity, particularly as it deals with its genome. The emphasis has to be on what the cell does with and to its genome, not what the genome directs the cell to execute. In some ways, the change in thinking reverses the instructional relationship postulated by the central dogma. The two basic ideas here are: 1. Sensing, computation, and decision-making are central features of cellular functions; and 2. The cell is an active agent utilizing and modifying the information stored in its genome.[22]

Shapiro takes a bold leap to reconfigure the role of information transfer and the direction of its flow in cellular life. In this view, cells make use of their DNA as a tissue that not only remembers their evolutionary history but also records ongoing life experiences.[23] No longer a deterministic code, the genome can be seen as a materialized archive of memories left behind *in the wake of cells doing life.*

In James Shapiro's hands, cells become sensing and sentient. His formulation suggests that cells have a kind of know-how, an ability to sense and respond actively and dynamically to their worlds. He is not alone in this attempt to articulate smartness at the scale of cells.[24] And though he invokes the concept, this is a model of sentience that fundamentally reconfigures what we understand as "cognition." It is an approach to sentience that might best be described by Maurice Merleau-Ponty's insistence that there is "a carnal adherence of the sentient to the sensed and of the sensed to the sentient," where sensitivity, and the capacity to sense, is always already the promise of sentience.[25] Rereading Aristotle, literary scholar Daniel Heller-Roazen has described this relation between the sensitive and sentient as a kind of "inner touch" or "sense of aliveness" in things.[26] Cells in this view are not just subject to their environments; they can be seen to actively make and remake their worlds in responsive relation.[27] Perhaps Jim's insistent question—"How do proteins know how to fold?"—can be understood within this context. How might protein folding be understood as a *molecular practice* in a cell that is actively sensing and modifying its internal and external milieu?[28]

Advocates of neo-Darwinism would likely consider this formulation blasphemous. A neo-Darwinian evolutionary view of the cell is beholden to the central dogma: DNA carries all the information for life; and it is mutations in DNA that generate heritable variation within populations. In this view, organisms are figured as blind, mechanical actants driven by genetic determinants. Only those with the genes to economize their energy and maximize reproductive output will survive.[29] An organism's genes will determine whether or not it goes on to reproduce and whether its progeny will thrive. Cellular events in this view are random and stochastic, even as advocates simultaneously figure proteins as deterministic molecular machines whose efficiencies and economizing logics are perfectly calibrated to their function.

And yet, as we have repeatedly encountered in this book, life scientists regularly defy their good training and deviate from these evolutionary scripts. They describe cells and molecules as if they had intentions and desires. It is by tuning into their often-muted deviations that I have been able to amplify the contours of a life science not so thoroughly bound to mechanism or neo-Darwinian logics. Their renderings manifest life as a relational phenomenon playing out in an excitable medium of wily matter. Theirs is not an ontology of vibrant things that appear to shimmer by their own light. Excitability is not property not of bodies themselves, but of bodies in relation. Matter becomes excitable for these practitioners in an *ecology of practices* that includes both human and more-than-human practitioners.[30]

OPENINGS

If liveliness is a lure for the life scientists documented in this book, it is surely also the lure that got me entangled in this inquiry. This ethnography makes no attempt to mask how I move with and am moved by life scientists' lively renderings. My rendering transduces modelers' pleasures and passions, their winces and cringes, along with my own. Transduction of course does not promise complete or perfect translation.[31] Affects and gestures diffract through excitable tissues; signals get torqued as they are transduced. The rendering I've performed here selectively amplifies some stories of life and life science practice, if not others. I may in the process have crafted a story that many practitioners in the field will want to disavow.

My aim, however, has been to challenge accounts of science that make it seem as if the capture of "life itself" is the only kind of story scientists know how to tell. Stories about such forms of capture are powerful lures that have drawn both life scientists and their critics into their respective investigations. As stories told by scientists, narratives of the capture of "life itself" secure an image of the scientist's power over nature and model objectivity as a disembodied practice. In such stories, once captured, "life itself" can be put to work in the form of deterministic molecular machines and made to generate value for capital enterprise. Stories that follow the lure of capture thus may ever see only life scientists reproducing deanimated renderings of molecular life. The effect is that such stories keep the life sciences locked into the logics of both neo-Darwinian mechanism and market forces. If observers of science just follow the scripts scientists think they are supposed to follow, we would aid in entrenching and normalizing the hubris of stories of capture. We would be complicit in limiting the kinds of inquiry, and the modes of attention and relation that are possible in the life sciences. In this sense, stories told about science by those of us observing from a distance, which frame science as the capture of "life itself" for capital gain, risk reproducing the very conditions that constrain what scientists can see, say, feel, imagine, and know.

How might we open up, rather than foreclose what it is possible to see, say, feel, imagine, and know in today's sciences of life? This account has resisted the lure of capture narratives in order to imagine the life sciences otherwise.

It is precisely because they are so readily disavowed that the lively renderings documented here need to be amplified. This book thus has offered a tactical, aspirational account.[32] It amplifies otherwise muted phenomena in an attempt to disrupt the constrained discursive fields of science through interventions that might catalyze new ways of thinking, speaking, and feeling. Perhaps the lively stories rendered here can offer a way for scientists to recognize and so avow their affective entanglements with the living phenomena they draw into view and so keep open what it is possible to see, say, feel, imagine, and know about the stuff of life today.

What is a protein? Consider joining me in my kitchen for a little demonstration.

I pick out an egg from the carton and grab a bowl from the cupboard. Chicken eggs are simple living structures composed of just a couple very large cells, and the contents of these cells are easy to observe. Eggs are also potent analogs for life and living bodies; they are full of substance and overflowing with significance. The one in my hand is worth close examination.[1] I crack open the egg and in one half of the shell I catch the yolk, a massive, bright yellow single cell. The viscous whites of the egg drip over my fingers, until they plop down into the bowl below. Rubbing my fingers together, I can feel the slippery textures of the egg whites. What is this stuff made of?

When scientists divide this substance up into its molecular constitution, like they do the rest of our foods, they find that it is mostly water. In addition to the fats, trace minerals, vitamins, and sugars it contains, a mixture of proteins accounts for about 10 percent of its weight. One particular protein, ovalbumin, accounts for just over half of all the protein in the egg whites. If it had been a fertilized egg, this is the protein that would have nourished a growing chick.

When I look at the egg whites, however, all I see is a slippery, translucent mass. Protein molecules are so small they lie beyond the limits of our perception. Even if I placed a sample on a microscope slide and applied the highest power microscope lens, I still would not be able to see an individual protein molecule.

Plate 1 in this book shows a crystallographic rendering, a three-dimensional, atomic-resolution model of ovalbumin. What does it depict? What are those two blobs that seem to be joined together? And those ribbon-like spirals that thread through the molecule?

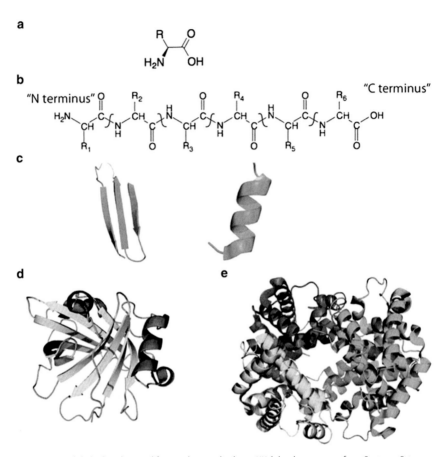

A.1. Amino acids linked end-to-end form polypeptide chains. With kind permission from Springer Science and Business Media.

Life scientists tell us that cells synthesize thousands of different proteins that are vital to their growth, structural integrity, and reproduction. Proteins are made of long chains of amino acids, linked end to end (see figure A.1). These long chains are called peptides or polypeptides. Amino acids are small organic compounds made of varying configurations of nitrogen, carbon, oxygen, and hydrogen. There are twenty amino acids commonly found in proteins. Each type of protein has a different sequence of amino acids. As cells synthesize new proteins, they rely on genetic sequences in their DNA to tell them which amino acids should be added to a growing polypeptide chain. Cells deploy a whole suite of proteins to do the work of transcribing and translating genetic sequences to make more protein.

But proteins aren't just long strings of amino acids. For a polypeptide to acquire its biological activity in a cell, it must fold itself up into specific configurations. Polypeptide chains form secondary structures, like helices, coils, and sheets (see section "c" of figure A.1). The formation and structural integrity of these secondary structures depend on which regions of the peptide are hydrophilic (water loving) and which are hydrophobic (water hating), as well as chemical affinities between amino acids, including delicate hydrogen bonds and the stronger bonds occasionally made between sulfur atoms.

Take another look at the model of ovalbumin shown in plate 1. Those ribbon-like structures inside the model trace the backbone of the polypeptide chain as it winds through the molecule. Such ribbon diagrams indicate the secondary structures. Sometimes multiple polypeptides come together to form larger molecular assemblages. Ovalbumin, pictured here, is made up of two polypeptides that have folded themselves up together.

Folded proteins are ephemeral forms. This is easy to demonstrate with a hot frying pan. When I cook the egg whites, their translucence quickly gives way to opacity. Heat denatures proteins; that is, it breaks the delicate bonds that hold folded proteins together. Denatured proteins coagulate and clump together. They lose the specificity of their chemical properties and their biological activity. Once denatured and coagulated, cooked ovalbumin also loses its transparency.

How are proteins made? Processes of DNA transcription and translation are frequently narrated through the metaphors of information and code. The "central dogma" of molecular biology has long dictated that the complete information for life moves in the form of an informatic code from DNA to protein sequence. In this view, polypeptide chains are considered the end product of a complex process that involves the reading and writing—the "transcription" and "translation"—of "information" stored within the genome. DNA is transcribed into an RNA molecule, which, in turn, serves as a "messenger" that transports an RNA "transcript" of the nuclear DNA into the cytoplasm. It is in the cytoplasm that ribosomes, tiny macromolecular organelles, "read" the ribonucleic transcript, and "translate" it into a polypeptide chain of amino acids linked end to end. In this view of life, the genetic code is the agent and determinant of life and its processes.[2]

Yet, it is precisely when the polypeptide chain is released into the cytoplasmic matrix that protein folding emerges as a practical problem for the cell. This is also the point at which metaphors of information and code break down and cease to do productive work for illuminating molecular events in the cell.

Indeed, one of the primary phenomena for which the central dogma cannot account is that of protein folding. Amino acid sequences do not communicate the full information required for a protein to acquire its active form. For example, many different amino acid sequences can produce similar tertiary structures, and highly similar sequences can produce different folds. Moreover, some proteins can fold only in the presence of other molecules called chaperones, which support these proteins in achieving their final form.[3]

Protein folding is a complex and poorly understood process. Where the sequence of nucleic acids in DNA tells the cell what the amino acid sequence of a peptide should be, it is only in the crowded, watery milieu of the cell that polypeptides figure out how to fold, unfold, and refold into their active conformations. Because these processes take place on a rapid timescale, the specific folds of each protein are especially difficult to detect. Protein structures cannot be calculated or otherwise deciphered directly from amino acid sequence; the specific configuration of each protein must be determined empirically. The protein crystallographers and other protein modelers documented in this book invest incredible time and energy to figure out the structures of protein molecules.

PREFACE

1. Readers who are not well versed in molecular biology or protein biochemistry will want to consult the appendix, which offers a primer in protein structure.

2. See Dumit, *Drugs for Life*. Molecular conceptions of life can even propagate through conversations with one's naturopath or yoga teacher. See, for example, Martin, *Flexible Bodies*; Dumit, *Picturing Personhood*; Taussig et al., "Flexible Eugenics."

3. For a resonant argument about the making of the cell in twentieth-century life science, see Landecker, *Culturing Life*.

4. The ancient Greek philosopher Democritus (c. 460–c. 370 BC) postulated the first atomic theory of matter. On Democritus, see Hacking, *Representing and Intervening*. On the concept of "subvisible," see Sagan and Margulis, *Garden of Microbial Delights*.

5. On twentieth-century visions of molecular life, see, for example, Kay, *Molecular Vision of Life*; Rheinberger, *Toward a History of Epistemic Things*; de Chadarevian, *Designs for Life*.

6. See J. Butler, *Bodies That Matter*; Barad, *Meeting the Universe Halfway*.

7. On the ways we fashion ourselves around the received facts of science, see Joseph Dumit, *Picturing Personhood*. See also Emily Martin, *Flexible Bodies*.

8. For critiques of the metaphors and stories told in biology, see, for example, Martin, "The Egg and Sperm," and *Flexible Bodies*; Haraway, "Biopolitics of Postmodern Bodies"; Keller, *Refiguring Life*, and *Making Sense of Life*.

9. On feminist approaches to materialism and materialities, see, for example, Haraway, *Simians, Cyborgs, and Women*, *Modest_Witness*; J. Butler, *Bodies That Matter*; Barad, "Meeting the Universe Halfway," and *Meeting the Universe Halfway*; Alaimo and Hekman, *Material Feminisms*; Murphy, *Sick Building Syndrome*; Bennett, *Vibrant Matter*.

INTRODUCTION

1. Many thanks to Hasok Chang for sending me this cartoon.

2. On kinesthetics, see, for example, Paterson, *The Senses of Touch*.

3. On ethnographic and philosophical approaches to affect more generally, see Stewart, *Ordinary Affects*; Massumi, *Parables for the Virtual*; Seremetakis, *The Senses Still*; Gregg and Siegworth, *The Affect Theory Reader*; Clough and Halley, *The Affective Turn*; Deleuze and Guattari, *A Thousand Plateaus*.

4. Other accounts tend to treat the affects and the senses separately, designating the senses as bodily, physiological phenomena, and the affects as sociocultural forms. This account challenges claims that the senses are innate functions of individual bodies or that they are biologically determined and so aims to treat the senses in ways similar to how affects are understood as social and cultural effects.

5. I have changed the names of all participants in this study and do not name the institutions where they work. This agreement was reached to ensure anonymity for the graduate students with whom I worked. It was important to them and their faculty advisors that I not disclose their names or the specifics of their projects, as many were concerned that their research might get "scooped," that is, that others might publish the structures before them. The names of those whose work I have encountered in public talks, in print, or in historical sources remain unchanged. In addition, I have retained the names of some of the historically significant figures whom I have interviewed.

6. Fischer cited in Clardy, "Borrowing to Make Ends Meet." See also Kay, *Molecular Vision of Life*, 165.

7. See Alač and Hutchins, "I See What You Are Saying."

8. See also Daston and Galison, *Objectivity*.

9. See also Alač, *Handling Digital Brains*; Prentice, "Anatomy of a Surgical Simulation"; Hopwood, "'Giving Body' to Embryos"; Myers and Dumit, "Haptic Creativity."

10. On affective entanglements, see Myers, "Animating Mechanism," and "Dance Your PhD"; Murphy, *Seizing the Means of Reproduction*.

11. On "doing" theory as practice, see Mol, *The Body Multiple*.

12. On the affectivity of matter, see Clough, *The Affective Turn*; Bennett, *Vibrant Matter*.

13. See, for example, Latour and Woolgar, *Laboratory Life*; Downey and Dumit, *Cyborgs and Citadels*; Fischer, *Emergent Forms of Life*; Dumit, *Picturing Personhood*; Franklin, *Dolly Mixtures*; Franklin and Lock, *Remaking Life and Death*; Gusterson, *Nuclear Rites*; Hayden, *When Nature Goes Public*; Helmreich, *Silicon Second Nature*, and *Alien Ocean*; Martin, *Flexible Bodies*, and "Anthropology and the Cultural Study of Science"; Rabinow, *Making PCR*; Sunder Rajan, *Biocapital*; Thompson, *Making Parents*; Traweek, *Beamtimes and Lifetimes*; Suchman, *Human-Machine Reconfigurations*; Shapin and Schaffer, *Leviathan*.

14. On the facts we live by, see Dumit, *Picturing Personhood*.

15. See also Knorr-Cetina, *Epistemic Cultures*; Keller, *Making Sense of Life*.

16. See Fleck, *Genesis and Development of a Scientific Fact*.

17. On "thinking with care," see Puig de la Bellacasa, "Matters of Care."

18. See, for example, Sunder Rajan, *Biocapital*; Nikolas Rose, *The Politics of Life Itself*.

19. See Geison's account of Huxley's views in "The Protoplasmic Theory of Life."

20. The history of life science abounds with the language of "excitation," "irritability," and "sensitivity." These are terms deployed in the life sciences in the late eighteenth through to the nineteenth century. See Geison, "The Protoplasmic Theory of Life"; and Steigerwald, "Figuring Nature," for a taste of the language of irritability and excitation in the history of the protoplasmic theory of the cell and early studies of nervous tissue.

21. Geison, "The Protoplasmic Theory of Life," 280. For more on Huxley's protoplasmic theory, see chapter 6 of this book.

22. See Kay, *Molecular Vision of Life*, and *Who Wrote the Book of Life*. See also Abir-am and Outram, "Uneasy Careers and Intimate Lives"; Abir-am, "The Politics of Macromolecules"; Cambrosio et al., "Erlich's 'Beautiful Pictures'"; Rheinberger, *Toward a History of Epistemic Things*; Creager, *The Life of a Virus*; de Chadarevian, *Designs for Life*. Tanford and Reynolds, *Nature's Robots*, offers a "scientist's account" of this history.

23. On Weaver's contributions to the formulation of molecular biology's mandate, see Kay, *Molecular Vision of Life*. On the early days of protein crystallography, see Law, "Development of Specialties in Science."

24. Stent, "That Was the Molecular Biology That Was."

25. Cited in Stent, "That Was the Molecular Biology That Was," 390.

26. Stent, 391.

27. Stent, 391.

28. Stent, 391; see also Kay, *Who Wrote the Book of Life*; Doyle, *On Beyond Living*.

29. See, for example, Kay, *Who Wrote the Book of Life*; Keller, *Century of the Gene*.

30. Haraway, *Modest_Witness*.

31. On metabolomics, see Levin, "Multivariate Statistics." On the limitations of genetic approaches to life see, for example, Kirschner et al., "Molecular Vitalism"; Lewontin, *Biology as Ideology*; Shapiro, *Evolution*.

32. In addition to X-ray crystallography, other structure determination techniques include nuclear magnetic resonance (NMR) and cryoelectron microscopy (cryoEM). Researchers in NMR apply their techniques to produce high-resolution models of small, soluble proteins. Electron microscopists use high-energy microscopes to resolve low-resolution models of larger protein complexes, such as the massive protein assemblages that form membrane pores and channels. These techniques account for approximately 10.6 percent (NMR) and 0.007 percent (cryoEM) of all the structures in the Protein Data Bank. On the moral economies of data exchange in the life sciences, see Kohler, *Lords of the Fly*; and on databases, see Edwards et al., "Science Friction."

33. Berman, "The Protein Data Bank," 88.

34. Berman, "The Protein Data Bank," 88.

35. Berman, "The Protein Data Bank," 88.

36. Berman, "The Protein Data Bank," 88.

37. Berman, "The Protein Data Bank," 88. Initially, the PDB received funding from the National Science Foundation. That small grant secured a home for the archive at Brookhaven National Laboratories in New York State. The data bank remained there until 1998, when it was clear that use of the archive was growing exponentially and new technologies and investments were required. That year alone 2,058 new structures had been uploaded, bringing the total to 8,610 protein structures. With a desire to expand this service, in 1998 a consortium of researchers from Rutgers, the University of California at San Diego, and the National Institute of Standards and Technology (NIST) applied for and won a $10 million grant to form the Research Collabo-

ratory for Structural Bioinformatics (RCSB). They received funds from the National Science Foundation (NSF), the Department of Energy (DOE), the National Institute of General Medical Sciences (NIGMS), and the National Library of Medicine (NLM).

38. By 2008 the Research Collaboratory for Structural Bioinformatics's (RCSB) first paper (published in the year 2000) was already one of biology's most cited journal articles, with over five thousand citations. Today, this single paper has amassed nearly seventeen thousand citations. See Berman et al., "The Protein Data Bank," and Berman, "The Protein Data Bank," 94.

39. The Research Collaboratory for Structural Bioinformatics (RCSB) Protein Data Bank can be accessed at http://www.rcsb.org/pdb/home/home.do.

40. See the RCSB *Newsletter*, no. 60 (winter 2014), http://www.pdb.org/pdb/general _information/news_publications/newsletters/2014q1/query.html. See also the RCSB 2013 Annual Report, which lists the top five countries that access this resource as the United States (29 percent), India (8 percent), the United Kingdom (6 percent), China (5 percent), and Germany (5 percent). To access the Annual Report see http://www .pdb.org/pdb/static.do?p=general_information/news_publications/index.html.

41. On the globalization of data in the life sciences, see Thacker, *Global Genome*.

42. See Berman, "The Protein Data Bank," 88–89.

43. On "cross-species" databases, see Leonelli, "When Humans Are the Exception."

44. These were the distributions listed on the PDB website as of January 2014. See http:// www.pdb.org/pdb/statistics/histogram.do?mdcat=entity_src_nat&mditem=pdbx _organism_scientific&numOfbars=50&name=Source%20organism%20%28Natural %20Source%29.

45. Structural genomics centers are contributing large quantities of data to the PDB. As of January 2014, the Riken Structural Genomics/Proteomics Initiative in Japan has uploaded the highest proportion of structures (2,739 data sets since 1999). The aim of this initiative, as explained on its website, is to make contributions to the pharmaceutical industry by analyzing protein structures to elucidate "protein functional networks," and determine which proteins are interacting with one another in the cell. See http://www.rsgi.riken.go.jp/rsgi_e/.

46. See Daston and Galison, "The Image of Objectivity," and *Objectivity*.

47. This book builds on an extensive genealogy of literature on the visual cultures of science. See, for example, Lynch and Woolgar, *Representation in Scientific Practice*; Latour, "Drawing Things Together"; Lynch, "The Externalized Retina"; Coopmans et al., *Representation in Scientific Practice Revisited*; Dumit, *Picturing Personhood*; Cambrosio et al., "Erlich's 'Beautiful Pictures'"; Daston and Galison, *Objectivity*; Galison, *Image and Logic*; Landecker, "Cellular Features."

48. Hooke quoted in Shapin and Schaffer, *Leviathan*, 36.

49. Shapin and Schaffer, *Leviathan*, 36–37.

50. On embodied sensorial modalities in live cell microscopy, see Myers and Dumit, "Haptic Creativity."

51. Shapin and Schaffer, *Leviathan*, 37.

52. Art historian Jonathan Crary's history of vision describes how observers' bodies and subjective perceptions were brought under scientific scrutiny and control. He exam-

ines how the senses were approached as a corruptive force to be disciplined, and how this discipline was enacted through scientific instrumentation. See Crary, *Techniques of the Observer*, 9.

53. Merleau-Ponty, *Phenomenology of Perception*, and *The Visible and the Invisible*.

54. Merleau-Ponty, *The Visible and the Invisible*, 133–34.

55. Merleau-Ponty has recently been taken up in studies investigating visualization technologies in the sciences. See, for example, Sturken and Cartwright, *Practices of Looking*; Hayles, *How We Became Posthuman*; Rasmussen, *Picture Control*; Myers, "Visions for Embodiment in Technoscience"; Myers and Dumit, "Haptic Creativity"; Ihde, *Bodies in Technology*; Haraway, "Crittercam."

56. In his history of the electron microscope, for example, Nicolas Rasmussen describes the intimacy electron microscopists experienced with their machines and brings phenomenological attention to the embodied nature of intuition and inferential reasoning, such that "experimental thinking" can be understood as "embodied action." Rasmussen, *Picture Control*, 223.

57. Haraway, "Situated Knowledges."

58. Haraway, "Situated Knowledges," 581.

59. On "body-lessness," see Haraway, "From Cyborgs to Companion Species." For a resonant approach to the body-fullness of the life sciences, see Deborah Heath, "Bodies, Antibodies, and Modest Interventions," 71.

60. For more on feminist epistemology see Harding, *The Science Question in Feminism*; Barad, "Meeting the Universe Halfway"; Hayles, "Constrained Constructivism."

61. For other accounts of body-work in other scientific fields, see Prentice, "Anatomy of a Surgical Simulation"; Alač, *Handling Digital Brains*; Vertesi, "Seeing Like a Rover." These accounts build on ethnomethodological approaches to documenting scientific practice, including Lynch, *Scientific Practice and Ordinary Action*; Suchman, *Plans and Situated Action*, and *Human-Machine Reconfigurations*; Goodwin, "Practices of Seeing" and "Action and Embodiment"; Lave and Wenger, *Situated Learning*.

62. See, for example, Cartwright et al., "The Tool Box of Science"; N. Cartwright, *Models*; Evelyn Fox Keller, "Models Of and Models For," and *Making Sense of Life*; Morgan and Morrison, *Models as Mediators*.

63. See, for example, H. M. Collins, *Changing Order*; Shapin and Schaffer, *Leviathan*; Lynch and Woolgar, *Representation in Scientific Practice*; Kohler, *Lords of the Fly*; Galison, *Image and Logic*; Haraway, *Modest_Witness*; Mol, *The Body Multiple*; Suchman, *Human-Machine Reconfigurations*.

64. The low status of models in the historical and social studies of science literature reflects a general lack of interest in pedagogy and training in science. The term "mneumotechnical device" is used in Cambriosio et al., "Erlich's 'Beautiful Pictures,'" 681, to describe models as tools used to aid the memory. Historian of physics David Kaiser's *Drawing Theories Apart* is a study of the dispersion of Feynman diagrams in postwar physics. Kaiser takes special interest in the lives of models, not just as research objects, but also as pedagogical tools.

65. See, for example, Francoeur, "Beyond Dematerialization"; Hopwood, "'Giving Body' to Embryos"; de Chadarevian and Hopwood, *Models*; Keller, *Making Sense of Life*.

66. See, for example, Francoeur, "The Forgotten Tool," and "Beyond Dematerialization and Inscription"; Hopwood, "'Giving Body' to Embryos"; de Chadarevian and Hopwood, *Models*; Cambrosio et al., "Erlich's 'Beautiful Pictures'"; Keller, *Making Sense of Life*.

67. Hacking, *Representing and Intervening*, 216.

68. Morgan and Morrison, *Models*, 25.

69. Sismondo, "Models," 258.

70. Sismondo, "Models," 247.

71. On the senses in science, see, for example, Puig de la Bellacasa, "Touching Technologies"; Latour, "How to Talk about the Body"; Roosth, "Screaming Yeast"; Mody, "The Sounds of Science"; Burri et al., "The Five Senses of Science"; Haraway, "Crittercam"; Helmreich, "An Anthropologist Underwater."

72. On craft, see Paxson, *The Life of Cheese*.

73. See Latour, "Drawing Things Together," Latour and Woolgar, *Laboratory Life*.

74. See Hopwood, "'Giving Body,'" 49; Francoeur, "Beyond Dematerialization."

75. See Hopwood, "'Giving Body'"; and Francoeur, "Beyond Dematerialization." On reading and writing in the laboratory, see Latour and Woolgar, *Laboratory Life*.

76. Griesemer, "Three-Dimensional Models," 435.

77. On the haptics of vision, see Marks, *Skin of the Film*; Hayward, "Fingery Eyes." On kinesthesia and proprioception, see Paterson, *The Senses of Touch*. For accounts of kinesthetic knowledge in other contexts, see Çelik, "Kinaesthesia"; Geurts, "On Rocks, Walks, and Talks in West Africa"; Samudra, "Memory in Our Body"; Schwartz, "Torque."

78. See Myers and Dumit, "Haptic Creativity."

79. See Austin, *How to Do Things with Words*.

80. On performance in science, see, for example, Schaffer, "Self Evidence"; Shapin, *A Social History of Truth*; Morus, "Placing Performance"; and the 2009 focus section of the journal *Isis* entitled "Performing Science." On performativity and critiques of performance in science, see Herzig, "On Performance, Productivity, and Vocabularies of Motive"; Myers, "Pedagogy and Performativity."

81. Rendering has been developed as a concept by others examining scientific visualization. See, for example, Lynch, "Discipline and the Material Form of Images."

82. Hacking, *Representing and Intervening*.

83. On modes of fabrication in the life sciences, see Roosth, "Crafting Life," and "Evolutionary Yarns in Seahorse Valley."

84. See Hacking, "Making Up People."

85. On material-semiosis, see Haraway, "A Game of Cat's Cradle" and *Modest_Witness*.

86. On performativity in science, see Barad, "Posthuman Performativity"; Haraway, *Modest_Witness*; Mol, *The Body Multiple*; Pickering, *The Mangle of Practice*; Doyle, *Wetwares*. On performativity more generally, see J. Butler, *Bodies That Matter*; Goffman, *The Presentation of Self in Everyday Life*.

87. See Kaiser, *Drawing Theories Apart*. Other studies have provided much insight into this arena. Studies of pedagogy in science studies have focused on various themes, including trajectories of training (e.g., Traweek, *Beamtimes and Lifetimes*); the "moral

economies" of laboratories and apprenticeship training (e.g., Kohler, *Lords of the Fly*); disciplinary formation (e.g., Bourdieu, *The Logic of Practice*; Fleck, *Genesis and Development of a Scientific Fact*; Foucault, *Discipline and Punish*; Kuhn, *The Structure of Scientific Revolutions*; Keating et al., "The Tools of the Discipline"; Maienschein, *Transforming Traditions in American Biology*); the stabilization and dispersion of knowledge and tools through pedagogical and institutional networks (e.g., Clarke and Fujimura, *The Right Tools for the Job*; Kaiser, *Pedagogy and the Practice of Science*); the physicality of scientific training and theoretical work (e.g., Kaiser, *Drawing Theories Apart*; Warwick, *Masters of Theory*); the inculcation of tacit and craft knowledge (e.g., Bourdieu, *Practical Reason*; H. M. Collins, *Changing Order*; Latour, "How to Talk about the Body"; Olesko, "Tacit Knowledge and School Formation"; Polanyi, *Personal Knowledge*; Warwick and Kaiser, "Conclusion: Kuhn, Foucault, and the Power of Pedagogy"); and the problems of generational reproduction (e.g., Gusterson, "A Pedagogy of Diminishing Returns").

88. By contrast the laboratories documented in Latour and Woolgar's *Laboratory Life* were part of a research institute dedicated not to pedagogy but to the production of facts.

89. On face-to-face apprenticeship, see Kaiser, *Drawing Theories Apart*.

90. On tacit knowledge, see Polanyi, *Personal Knowledge*; Collins, *Tacit and Explicit Knowledge*.

91. Jones, "The Mediated Sensorium," 8.

92. Jones, "The Mediated Sensorium," 8.

93. On efforts to move beyond the five senses model see, for example, Howes, *The Empire of the Senses*.

94. On the "distribution of the sensible" and sensory regimes, see Rancière, "The Aesthetic Dimension"; Buck-Morss, "Aesthetics and Anaesthetics"; Taussig, *Mimesis and Alterity*; and Foucault, *The Birth of the Clinic* and *The History of Sexuality*.

95. Latour, "How to Talk about the Body," 206. Latour's argument builds on contributions from Vinciane Despret ("The Body We Care For") and Isabelle Stengers (*Cosmopolitics*).

96. Latour, "How to Talk about the Body," 207.

97. Latour, "How to Talk about the Body," 210.

98. Latour, "How to Talk about the Body," 218, 210. On passionate involvement, Latour insists that "a disinterested scientist abstaining from any interference with uninterested entities will produce totally uninteresting, that is, redundant articulations!"(218). He continues: "It must be clear, according to this formulation, that abstaining from biases and prejudices is a very poor way of handling a protocol. To the contrary, one must have as many prejudices, biases as possible, to put them at risk in the setting and provide occasions of manipulation for the entities to show their mettle. It is not passion, nor theories, nor preconceptions that are in themselves bad, they only become so when they do not provide occasions for the phenomena to differ" (219).

99. This ethnography has benefited from two experimental field sites, what anthropologist George Marcus and others have called "para-sites" (see Marcus, *Para-Sites*).

These are sites in which I have had the opportunity to deepen this inquiry and hone the ethnographic questions I am asking in the more open, experimental medium of an art practice. The first para-site was a yearlong collaboration with visual and movement artist Clementine Cummer. *Cellular Practices and Mimetic Transductions: A Dance in Four Scores* was performed as a choreographed, live-feed video installation-performance at the 2006 European meeting of the Society for Science, Literature, and the Arts (SLSA) in Amsterdam. That collaboration gave me the space and time to raise new questions about the ambi-valent registers in which modelers perform molecular knowledge (for more on that project, see Myers, "Dance Your PhD").

A second para-site was a 365-day experiment in ethnographic methods launched in 2008 (see http://adanceaday.wordpress.com). I used the form of a daily movement practice as an experimental medium for crafting new methods for anthropological research. In that project, I approached ethnography as a practice of learning how to move with and be moved by other moving bodies. The aim was to cultivate dexterities for sensing and transducing movements and affects. I explored ways of documenting experiences with ready-to-hand materials, like pencil and paper, and producing short texts to hone my ability to write movement. Over the course of a full year I developed gesture drawing techniques for generating what could be called "kinesthetic graphemes," or "energy diagrams." I used these as mnemonic devices to help me re-member the energetics, movements, and affects of things or bodies I observed or events I participated in over the course of each day. Once tuned in to moving phenomena, these drawings helped me become a "transducer" in a field of affects. That research project deepened my dexterities perceiving, documenting, writing, and theorizing movement, and has informed the accounts of laboratory life I relay in this book. See Myers, "Anthropologist as Transducer in a Field of Affects," and chapter 8 of this book.

100. Richard Rifkind describes his approach to filmmaking on BiotechBlog, posted July 27, 2009: http://www.biotechblog.com/2009/07/27/naturally-obsessed-the-making -of-a-scientist/. For an online interview with the filmmakers, see Amy Charles, "The Making of *Naturally Obsessed*," Lablit.com, posted November 1, 2009: http://www .lablit.com/article/554.

101. See the documentary's website at http://naturallyobsessed.com.

102. Warwick and Kaiser, "Conclusion: Kuhn, Foucault, and the Power of Pedagogy," 402–3; see also Kaiser, "The Postwar Suburbanization of American Physics"; Foucault, *Discipline and Punish*.

103. See Foucault, *Discipline and Punish*. See also Warwick and Kaiser, "Conclusion: Kuhn, Foucault, and the Power of Pedagogy," 399.

104. On "promissory capital," see Thompson cited in Franklin and Lock, *Remaking Life and Death*.

105. For more on the metaphor of science as a competitive game or race in the film, see Myers, "Pedagogy and Performativity."

106. See Foucault, *History of Sexuality*, vol. 1; Rabinow and Rose, "Biopower Today."

107. On the relationship between biology and capitalism, see, for example, Sunder Rajan,

Biocapital, and *Lively Capital*; Helmreich, "Species of Biocapital"; Franklin and Lock, *Remaking Life and Death*; Waldy and Mitchell, *Tissue Economies*.

108. On life "enterprised up," see Sarah Franklin, "Science as Culture"; Haraway, *Modest_Witness*.

109. On "life itself," see Franklin, "Life Itself"; Franklin and Lock, *Remaking Life and Death*; Nikolas Rose, *The Politics of Life Itself*; Foucaut, *The Order of Things*. For more on "life itself," see the conclusion to this book.

110. On biological weapons design, see also Melinda Cooper, *Life as Surplus*, and "Preempting Emergence."

111. I received a surprise telephone call from a DARPA official early one morning to confirm my participation in the workshop. It was only at the very end of the conversation that I inadvertently outed myself as a "nonresident alien" Canadian living and working in the United States. I did this by verbally confirming the e-mail address of the woman I was talking with on the phone. Her e-mail address contained several "z's." Repeating her name back to her, I said the word "zed" instead of "zee." Joe Dumit, one of my PhD advisors at the time, wrote a Dr. Seuss–inspired poem to mark the occasion:

> Darpa huffed and puffed,
> that with a Zed, she had misled,
> the artist would cause inspire,
> the anthropop much perspire,
> but the Canuck, a vampire!
> For the sake of the empire,
> sadly the Director now fired.
> moral: don't inquire.

112. See Haraway, *Simians, Cyborgs, and Women*.

113. Nikolas Rose, *The Politics of Life Itself*.

114. For studies of the ongoing militarization of the life sciences and efforts to bring living bodies and populations into the realm of calculation in the name of state power, see, for example, Kosek, "Ecologies of Empire"; Cooper, *Life as Surplus*; Masco, "Mutant Ecologies"; Hartman et al., *Making Threats*; Shukin, *Animal Capital*. Eugenics can be considered one formation of such a molecular biopolitics, which raises crucial issues about the use of biology to racialize bodies. See, for example, Reardon, *Race to the Finish*; Tallbear, *Native American DNA*; Pollock, *Medicating Race*; Nelson, "DNA Ethnicity as Black Social Action?"

115. See especially Cooper, *Life as Surplus*; Murphy, *Seizing the Means of Reproduction*; Murphy, *The Economization of Life*.

116. This "thinking otherwise" is a foundational gesture for feminist scholars in science studies. See, for example, Haraway, *Simians, Cyborgs, and Women*; Leigh Star, "Distributions of Power"; and Puig de la Bellacasa, "Matters of Care." For efforts to imagine science otherwise, see also contributors to the radical science movement, including Rose and Rose, *The Political Economy of Science*; Hubbard and Wald, *Exploding the Gene Myth*; Rose, *Love, Power, and Knowledge*.

117. Da Costa and Phillip, *Tactical Biopolitics*.

118. On an affirmative biopolitics, see Esposito, "Community, Immunity, Biopolitics"; Hardt, "Affective Labor"; Negri, "Value and Affect."

119. Latour, "How to Talk about the Body," 209. See also Deleuze and Guattari, *A Thousand Plateaus*.

120. On other accounts of this wavering between vitalism and mechanism, or romanticism and mechanism, see Lenoir, *Strategy of Life*; Tresch, *Romantic Machine*.

121. For critiques of the metaphors and stories told in biology, see, for example, Martin, "The Egg and Sperm," and *Flexible Bodies*; Haraway, "Biopolitics of Postmodern Bodies"; Keller, *Refiguring Life*, and *Making Sense of Life*.

122. On this question of enchantment, re-enchantment, and disenchantment, see Despret, "From Secret Agents to Interagency," 36.

123. For a range of rich insights on the concept of rendering as a form of extraction (of capital, affect, and media) from animal bodies, see Shukin, *Animal Capital*.

124. *Oxford English Dictionary*, OED 3rd ed. (December 2009).

125. On haptic creativity, see Myers and Dumit, "Haptic Creativity."

126. *New Oxford American Dictionary*, 3rd ed. (2010).

ONE | CRYSTALLOGRAPHIC RENDERINGS

1. Diane is a pseudonym, and the name of the university where she works is withheld. The names of many of her colleagues and all members of her laboratory have also been changed.

2. See Landecker, "Living Differently in Time"; Rabinow, *Making PCR*; Rheinberger, *Toward a History of Epistemic Things*; Knorr-Cetina, *Epistemic Cultures*, for more on the history of laboratory apparatuses such as freezers and PCR machines in use in molecular biology labs.

3. See, for example, Star, "Craft vs. Commodity." For more on models in the history of the life sciences, see chapter 4.

4. See, for example, Kay, *The Molecular Vision of Life*; Keller, *Refiguring Life*; Haraway, *Modest_Witness*; Doyle, *On Beyond Living*.

5. See the RCSB PDB website for statistics on "Yearly Growth of Structure by X-ray," http://www.rcsb.org/pdb/statistics/contentGrowthChart.do?content=total&seqid=100.

6. Note the formulation here that "nature" has "tailored" these molecules. See chapters 6, 7, and the conclusion of this book for discussions of the evolutionary narratives that undergird concepts of molecular life.

7. Chapters 5, 7, and 8 further explore the significance of molecular movement and the limitations of protein crystallography.

8. See Stengers, *Cosmopolitics I* on the "achievement" of science.

9. On "bricolage of expertise" see Garfinkel et al., "The Work of a Discovering Science."

10. For more on the career trajectories of scientific training, see Traweek, *Beamtimes and Lifetimes*; Kaiser, *Drawing Theories Apart*, and "The Suburbanization of Physics."

11. On the concept of "ethos," see, for example, Geertz, *Interpretation of Cultures*; Ortner, "Theory in Anthropology Since the Sixties"; Rabinow, *Marking Time*.

12. See Csordas, "Somatic Modes of Attention," for this interpretation of habitus. In his

essay "Techniques of the Body," Mauss developed the concept of the habitus to analyze the range of bodily techniques, forms of training, and the propagation of traditions within distinct cultures. His work has inspired decades of research into the inculcation and propagation of habits, attitudes, and postures. Sociologist Pierre Bourdieu defines "habitus" as that system of "durable, transposable *dispositions*," which he calls "structured structures." The habitus for him is "objectively 'regulated' and 'regular,'" but this regulated schema is acquired without following explicit rules. He recognizes the habitus as a "collectively orchestrated" social form "without being the product of the orchestrating action of a conductor" (*Outline of a Theory of Practice*, 72). The concept of "habitus" has acquired an extensive genealogy and wide application in contemporary anthropology. See, for example, Lock and Farquhar, *Beyond the Body Proper*; Mascia-Lees, *A Companion to the Anthropology of the Body and Embodiment*.

13. See the *Naturally Obsessed* website for footage from this high school screening: http://naturallyobsessed.com/blog/post-screening-discussion-with-briarcliff-manor-high-school/.

14. See Haraway, "A Game of Cat's Cradle"; Martin, "Anthropology and the Cultural Study of Science."

15. See the Rifkinds' online interview with Amy Charles, "The Making of *Naturally Obsessed*," Lablit.com, posted November 1, 2009, http://www.lablit.com/article/554.

16. Richard Rifkind describes his approach to filmmaking on BiotechBlog, posted July 27, 2009: http://www.biotechblog.com/2009/07/27/naturally-obsessed-the-making-of-a-scientist/

17. For Carol Rifkind's comments, see Amy Charles, "The Making of *Naturally Obsessed*," Lablit.com, posted November 1, 2009, http://www.lablit.com/article/554.

18. See Weber, "Science as a Vocation."

19. This event at the CUNY Graduate Center was documented and the video recording was posted online in three parts. For part 1, see https://www.youtube.com/watch?v=kS1Ewxsnpjo. For part 2, see https://www.youtube.com/watch?v=qKbUrWxGrAY. For part 3, see https://www.youtube.com/watch?v=cVYqFHcsHFU.

20. On neoliberalism in science, see, for example, Lave et al., "Introduction"; Cooper, *Life as Surplus*.

21. See Ishani Ganguli's profile on Ben Ortiz posted November 7, 2005, in an online supplement of *The Scientist* at http://www.the-scientist.com/?articles.view/articleNo/16824/title/Ben-Ortiz/.

22. Herzig, *Suffering for Science*, and "On Performance." For more on self-fashioning, see Traweek, *Beamtimes and Lifetimes*; Dumit, *Picturing Personhood*.

23. Mahmood, "Feminist Theory, Embodiment, and the Docile Subject"; Foucault, *Discipline and Punish*.

24. On neoliberalism and "scientific entrepreneurialism," see Ong, "A Milieu of Mutations" (69).

25. On gender and the constrained masculinities performed in this documentary, see Myers, "Pedagogy and Performativity."

26. Tompkins provided comments on an early version of my account of *Naturally Obsessed* at the Value/Knowledge workshop, University of Chicago, June 2–3, 2011.

27. On affective labor, see, for example, Clough, *Affective Turn*; Murphy, *Seizing the Means of Reproduction*; Hardt and Negri, *Empire*; Hardt, "Affective Labor"; Negri, "Value and Affect."

28. Hardt, "Affective Labor," 90.

29. Hardt, "Affective Labor," 90.

30. Hardt, "Affective Labor," 96–97.

31. See Foucault, *The History of Sexuality*, vol. 1, and *The Care of the Self*.

32. For a similar argument on "care of the data," see Fortun, *Promising Genomics*.

33. On the distinction between requirements and obligations in scientific experiments, see Stengers, *Cosmopolitics I*.

34. On "matters of fact," see Shapin and Schaffer, *Leviathan*. On "matters of care," see Puig de la Bellacasa, "Matters of Care."

35. Caldwell, "Some Analogies between Molecules and Crystals," 89.

36. On the basic principles of diffraction and for an account of the significance of the concept to feminist theories of scientific knowledge and practice, see Barad, *Meeting the Universe Halfway*, 28.

37. Bragg quoted in Hodgkin, "The X-ray Analysis of Complicated Molecules," 71.

38. Hodgkin, "The X-ray Analysis of Complicated Molecules," 71.

39. See Glusker, *Structural Crystallography in Chemistry and Biology*; Glusker and Trueblood, *Crystal Structure Analysis*.

40. Note that the diagram identifies the scattering of light rays as "diffracted light." This is incorrect. Light moving through objects like amoebas is "refracted." On refraction, see Hayward, "Enfolded Vision."

41. See also Hacking, *Representing and Intervening*.

42. On "mechanical objectivity," see Daston and Galison, "The Image of Objectivity," and *Objectivity*; Galison, "Judgment against Objectivity." On human-machine entanglements, see Suchman, *Human-Machine Reconfigurations*. For more on the peculiar form of objectivity that this form of rendering engenders, see chapter 5.

43. See Shukin's (*Animal Capital*) account of industrial plants that render materials like cellulose from animal bodies. For an account of the ways rats are rendered into scientific objects in a neuroscience laboratory, see also Lynch's "Sacrifice and the Transformation of the Animal Body into a Scientific Object."

44. See de Chadarevian, *Designs for Life*.

45. In his Nobel lecture, "Myoglobin and the Structure of Proteins," Kendrew described his search for an appropriate animal source for myoglobin: "First of all it was necessary to find some species whose myoglobin formed crystals suitable, both morphologically and structurally, to the purpose in hand; the search for this took us far and wide, through the world and through the animal kingdom, and eventually led us to the choice of the sperm whale, *Physeter catodon*, our material coming from Peru or from the Antarctic, with some close runners-up including the myoglobin of the common seal" (677).

46. On love and violence in the life sciences, see Hustak and Myers, "Involutionary Momentum"; Myers, "Interrupting the Order of Things."

47. Fersht, "From the First Protein Structures," 651.

48. Fersht, "From the First Protein Structures," 651.

49. Fersht, "From the First Protein Structures," 651.

50. On the origins of experimental inquiry and the relationship between "seeing and believing," see Shapin and Schaffer's account of Robert Boyle's experimental form of life in *Leviathan*.

51. See Latour, *On the Modern Cult of the Factish Gods*, 42.

52. See Thompson, *Making Parents*.

53. Larry Shapiro and Rob Townley in the documentary *Naturally Obsessed: The Making of a Scientist*. For more on crystals and crystal analogies in the history of the life sciences, see Haraway, *Crystals, Fabrics, and Fields*.

54. For more on the entangled relations between instruments and objects in science, see Rheinberger, *Toward a History of Epistemic Things*.

55. For more on the structure of graduate student careers in science, see Traweek, *Beamtimes and Lifetimes*.

56. Protein crystallography fails to resolve the structures of many molecules. Other molecular visualization techniques, such as nuclear magnetic resonance, or NMR, have been developed and are in wide use. Nuclear magnetic resonance is especially useful for modeling the structures of smaller protein molecules in solution.

57. Mauss, "Techniques of the Body," 67.

58. Barley, "The Social Construction of a Machine," 513. On ritual in science, see also Mody's ("A Little Dirt") interpretation of Mary Douglas's approach to ritual purity in the context of a materials science laboratory.

59. On the "vibratory milieu," see Deleuze and Guattari, *A Thousand Plateaus*.

60. On magical thinking in technological contexts, see Barley, "The Social Construction of a Machine," 520.

61. For a critique of this practice of modeling just a single structure from the data, see Davis et al., "The Backrub Motion."

62. See Ahmed, *Cultural Politics of Emotions*.

63. See, for example, Suchman, *Human-Machine Reconfigurations*; Traweek, *Beamtimes and Lifetimes*; Downey, *The Machine in Me*; Turkle, *Simulation and Its Discontents*; Turkle et al., "Information Technologies and Professional Identity."

64. See, for example, de Chadarevian, *Designs for Life*; Francoeur and Segal, "From Model Kits to Interactive Computer Graphics"; Siler and Lindberg, *Computers in Life Science Research*; Tsernoglou et al., "Molecular Graphics."

65. See Doing, *Velvet Revolution at the Synchrotron*, for more on the use of synchrotrons by protein crystallographers.

66. Buck-Morss, "Aesthetics and Anesthetics."

67. For insight into experimental physicists' affectively charged relationships with their massive instruments, see Traweek, *Beamtimes and Lifetimes*.

68. Isabelle Stengers describes this relation between scientific capture and scientists' captivation as a form of "reciprocal capture" (see *Cosmopolitics I*).

1. Paxson, *The Life of Cheese*, 133, 151.
2. For a history of artisanal production, see Pamela Smith, *The Body of the Artisan*.
3. McCollough, *Abstracting Craft*, 22.
4. McCollough, *Abstracting Craft*, 213, 217.
5. McCollough, *Abstracting Craft*, 2.
6. On the crossing over between the visible and the tangible, see Merleau-Ponty, *The Visible and the Invisible*.
7. Paxson, *The Life of Cheese*, 131.
8. See Mark Paterson, *The Senses of Touch*, ix.
9. See Fisher, "Relational Sense," 5.
10. Smith, *The Body of the Artisan*, 21, 95.
11. McCullough, *Abstracting Craft*, 7.
12. Paxson, *The Life of Cheese*, 135.
13. Paxson, *The Life of Cheese*, 136.
14. On the haptics of vision, see also Marks, *Skin of the Film*; Hayward, "Fingery Eyes"; Prentice, "Anatomy of a Surgical Simulation."
15. See Myers and Dumit, "Haptic Creativity."
16. On bricolage, see de Certeau, *The Practice of Everyday Life*.
17. Paxson, *The Life of Cheese*, 152.
18. See de Chadarevian, *Designs for Life*; Francoeur, "The Forgotten Tool."
19. See Francoeur, "The Forgotten Tool," "Beyond Dematerialization and Inscription," and "Molecular Models and the Articulation of Structural Constraints in Chemistry"; Francoeur and Segal, "From Model Kits to Interactive Computer Graphics."
20. Francoeur, "The Forgotten Tool."
21. Francoeur, "The Forgotten Tool," 14.
22. Francoeur, "Beyond Dematerialization and Inscription," 6.
23. See also Nye, "Paper Tools and Molecular Architecture in the Chemistry of Linus Pauling."
24. Perutz, "Obituary," 670.
25. See Kaiser, *Drawing Theories Apart*.
26. On how models attract "curious hands," see Langridge, "Real-Time Color Graphics," 661.
27. See Hopwood, "'Giving Body' to Embryos," and Francoeur, "The Forgotten Tool," on the ways that molecular models challenge Latour's analysis about the immobility of three-dimensional facts.
28. See chapter 7 for more on this course and protein folding.
29. On "epistemic things" and "technical objects," see Rheinberger, *Toward a History of Epistemic Things*.
30. Rheinberger, *Toward a History of Epistemic Things*, 28.
31. For a description of embodied metaphors like "grasping," see Lakoff and Johnson, *Philosophy in the Flesh*.
32. See Francoeur, "The Forgotten Tool."
33. See Roosth, "Crafting Life."

34. See Watson, *The Double Helix*, 62.
35. Watson, *The Double Helix*, 123.
36. Watson, *The Double Helix*, 62.
37. Watson, *The Double Helix*, 122.
38. Watson, *The Double Helix*, 122.
39. I asked Mike Fuller whether he could identify the modeler pictured in the footage. He was certain that it was Anne Cullis. According to Fuller, "she was one of the secretaries, a computing girl–cum–. . . you name it . . . a jack of all trades. As everybody was. There was no demarcation in those days. You did everything, full stop. And that was it. One minute they'd be acting as a secretary or typist, at another minute they'd be making the models. And I think everybody worked the same. You just mixed in and did the jobs. There was no demarcation at all." Fuller's account is more generous than some other accounts of how the "girls" in the labs at the LMB were treated. In *Designs for Life*, historian Soraya de Chadarevian examines the history of women in protein crystallography laboratories, in particular the women technicians, or "computors," who were responsible for most of the labor of analyzing crystallographic data using punch card computers. They also measured atomic distances in the electron density maps and built the models by hand and by eye.
40. De Chadarevian, *Designs for Life*, 142; Perutz, "The Hemoglobin Molecule," 45.
41. De Chadarevian, *Designs for Life*, 142.
42. See Kendrew et al., "A Three-Dimensional Model of the Myoglobin Molecule."
43. Perutz, "The Hemoglobin Molecule," 45.
44. De Chadarevian, *Designs for Life*, 143.
45. Chapter 6 documents how a mechanistic logic and its accompanying mechanical aesthetic—which was made possible through the use of molecular modeling kits with machined metal parts—seems to have provided a tacit framework through which these researchers evaluated both the aesthetic value and the truth status of their models.
46. See also de Chadarevian, *Designs for Life*; Francoeur, "The Forgotten Tool."
47. Francoeur, "The Forgotten Tool."
48. See interviews with Max Perutz online. "Face to Face with Max Perutz," Vega Science Trust, accessed January 21, 2014, http://www.vega.org.uk/video/programme/1. Also shown alongside Perutz and his mechanical model is a video clip of an interactive molecular graphics screen animating the same movements on-screen. Perutz's handling of a hemoglobin model is also described in de Chadarevian, *Designs for Life*.
49. Cyrus Levinthal's evocative rhetoric, displayed in the epigraph at the opening of this section, echoed the discourse alive at MIT at that time. The MIT computer scientist and psychologist J. C. R. Licklider published his now-famous speculative article "Man-Machine Symbiosis" in 1960, and his vision informed the ethos of the work ongoing at Project MAC. Licklider anticipated a cybernetic tangle of man and machine, and crystallographers were among those who invested effort in cultivating such intimate associations in the form of the "human-computer lens." For more on human-computer entanglements, see Suchman, *Human-Machine Reconfigurations*.
50. This story is related in Francoeur and Segal, "From Model Kits to Interactive Computer Graphics," 412.

51. See Francoeur, "Cyrus Levinthal, the Kluge and the Origins of Interactive Molecular Graphics"; Levinthal, "Molecular Model-Building by Computer."

52. See Francoeur and Segal, "From Model Kits to Interactive Computer Graphics."

53. Thanks to Stefan Helmreich for this analogy. Interactive graphics does seem to offer a gravity-free, buoyant environment for protein modeling. Indeed, Donna Haraway has reminded me that interactive computer graphics may in this way reproduce the watery worlds that support protein structures in their cellular environments.

54. Langridge et al., "Real-Time Color Graphics," 661, emphasis added.

55. Francoeur and Segal, "From Model Kits to Interactive Computer Graphics."

56. Langridge quoted in Francoeur and Segal, "From Model Kits to Interactive Computer Graphics," 418.

57. Langridge quoted in Francoeur and Segal, "From Model Kits to Interactive Computer Graphics," 418.

58. Barry et al., "Evolving Macromodular Molecular Modeling System," 2368–69.

59. Tsernoglou et al., "Molecular Graphics," 1379.

60. Langridge, "Interactive Three-Dimensional Computer Graphics in Molecular Biology," 2333.

61. Langridge et al., "Real-Time Color Graphics," 666.

62. Langridge et al., "Real-Time Color Graphics," 666.

63. See Pique, "Technical Trends in Molecular Graphics," 13. See also Milburn, "Digital Matters," on videogame technologies used in nanoscience.

64. Diamond et al., "Three-Dimensional Perception for One-Eyed Guys," 286.

65. Diamond et al., "Three-Dimensional Perception for One-Eyed Guys."

66. Several visualization techniques in the life sciences use this tomographic logic to slice through the depth of organs and tissues, including the techniques used to produce virtual anatomical models in the Visible Human Project and in PET scans. See Waldby, *The Visible Human Project*; Dumit, *Picturing Personhood*.

67. See Kendrew, "Myoglobin and the Structure of Proteins"; de Chadarevian, *Designs for Life*.

68. Kendrew, "Myoglobin and the Structure of Proteins," 681.

69. Collins et al., "Protein Crystal-Structures," 1049.

70. Collins et al., "Protein Crystal-Structures," 1049.

71. Collins et al., "Protein Crystal-Structures," 1049.

72. Collins et al., "Protein Crystal-Structures," 1049. See also de Chadarevian, *Designs for Life*.

73. Collins et al., "Protein Crystal-Structures," 1049.

74. Collins et al., "Protein Crystal-Structures," 1049.

75. Collins et al., "Protein Crystal-Structures," 1049.

76. Tsernoglou et al., "Molecular Graphics."

77. Susan Fielding is a pseudonym.

78. For more on model building in classrooms, see Myers, "Performing the Protein Fold."

79. Similar phenomena are observable in other research contexts. See also Prentice,

"Anatomy of a Surgical Simulation"; Alač, *Handling Digital Brains*; Vertesi, "Seeing Like a Rover," on gestural communication at computer interfaces.

80. See Hayles, *How We Became Posthuman*.

81. For more on how the documentary stages graduate student research life as a race or game to be won, see Myers, "Pedagogy and Performativity."

82. Hodgkin, "The X-ray Analysis of Complicated Molecules," 75–76.

83. Polanyi, *Personal Knowledge*, 53, 62.

84. Barry et al., "Evolving Macromodular Molecular Modeling System," 2368.

THREE | MOLECULAR EMBODIMENTS

1. Keller, *A Feeling for the Organism*.

2. *Oxford English Dictionary*, 3rd ed. (December 2009).

3. Foucault, *The Order of Things*, 26.

4. See Ahmed, *Cultural Politics of Emotion*, on the movement of affects in a circuit.

5. Foucault, *The Order of Things*, 26.

6. On becoming with, see Haraway, *When Species Meet*. On involution, see Deleuze and Guattari, *A Thousand Plateaus*; Hustak and Myers, "Involutionary Momentum."

7. See Hustak and Myers, "Involutionary Momentum," for a resonant account of Darwin's experiments with orchids.

8. Hacking, *Representing and Intervening*, 216.

9. The concept of "mental image" has been a useful convention for describing elements the scientific imagination. For discussions of mental images in science, and accounts of the scientific imagination in general, see, for example, Cambrosio et al., "Erlich's 'Beautiful Pictures'"; Meinel, "Molecules and Croquet Balls"; Trumpler, "Converging Images." For an elaboration of the diversity and complexity of mental images, see Sacks, "The Mind's Eye."

10. Merleau-Ponty, *Phenomenology of Perception*.

11. For a resonant formulation of the kinesthetic imagination, see Roach, *Cities of the Dead*; Bernstein, "Dances with Things."

12. Hopwood, "'Giving Body' to Embryos."

13. Hopwood, "'Giving Body' to Embryos," 466.

14. Hopwood, "'Giving Body' to Embryos," 482.

15. Hopwood, "'Giving Body' to Embryos," 483.

16. Hopwood, "'Giving Body' to Embryos," 482.

17. Hopwood, "'Giving Body' to Embryos," 482.

18. See Mol, *The Body Multiple*; Barad, "Posthumanist Performativity."

19. Ahmed, *Cultural Politics of Emotion*, 24.

20. Ahmed, *Cultural Politics of Emotion*, 24–25.

21. In their account of cultures of objectivity and visualization practices across a range of sciences at various points in history, historians of science Lorraine Daston and Peter Galison explore contexts where visualization relies on practitioners' trained judgment and intuitions rather than mechanical apparatuses. These skills are crucial in contexts where practitioners cannot rely on "mechanical procedure" to "synthe-

size, highlight, and grasp relations" (*Objectivity*, 314). They propose the term "physiognomic sight" to describe a way of seeing and feeling that resembles the crystallographic kinesthetic imagination that I document here. Indeed, it is in part because there is no fully automated procedure that can render a model from crystallographic data, and because they rely so heavily on improvisation with tangible media, that protein modelers acquire such well-developed forms of "physiognomic sight." For more on trained judgment in protein crystallography see chapter 5.

22. On "truth to nature" as a modality of objectivity, see Daston and Galison, *Objectivity*.

23. Merleau-Ponty, *Phenomenology of Perception*; Polanyi, *Personal Knowledge*, and "The Logic of Tacit Inference."

24. Merleau-Ponty, *Phenomenology of Perception*, 142.

25. J. Butler, *Bodies That Matter*.

26. I attended a reunion of protein crystallographers who had all done their training under two well-known practitioners in the United States. In one talk, a male crystallographer joked that you couldn't trust anything that a particular female crystallographer said: she was too exuberant and prone to exaggeration. The sexism of his sentiment was loud and clear: he aligned expressiveness with femininity and maligned both as excessive. Her performance apparently threatened his view of what counts as objective knowledge. For more on the myths of body-lessness, distance, and neutrality that pervade such conventional accounts of objectivity, see Haraway, "Situated Knowledges."

27. On "infolding," see Haraway, "Crittercam," and *When Species Meet*; Hayward, "Fingery Eyes"; Merleau-Ponty, *The Visible and the Invisible*.

28. Perutz was an avid mountain climber. This quote is transcribed from an interview with Max Perutz conducted by the Vega Science Trust, "Face to Face with Max Perutz."

29. In *The Elementary Forms of Religious Life*, which includes a close study of totemic religions, Durkheim suggests that "it is most often on the body itself that the totemic mark is imprinted" (115). I found some evidence in an online search that some protein crystallographers have tattooed protein structures on their bodies. In a related context, the RCSB Protein Data Bank's outreach efforts for public education have included providing temporary tattoos of protein structures to young participants in their workshops. See Zardecki, "Interesting Structures."

30. See Marx, *Capital*.

FOUR | RENDING REPRESENTATION

1. Perutz, "The Medical Research Council Laboratory of Molecular Biology."

2. Hacking, *Representing and Intervening*, 140.

3. Hacking, *Representing and Intervening*, 140–41.

4. Hacking, *Representing and Intervening*, 137.

5. Barad, "Posthumanist Performativity," 806.

6. On material-semiosis, see Haraway, *Modest_Witness*, and "A Game of Cat's Cradle." On performativity, see J. Butler, *Bodies That Matter*; Barad, "Posthumanist Performativity," and *Meeting the Universe Halfway*.

7. The long and varied history of the construction and use of three-dimensional models

has captured the attention of scholars tracking material culture in the sciences. See, for example, Haraway, *Primate Visions*; Daston, "The Glass Flowers"; de Chadarevian and Hopwood, *Models*; Hopwood, "'Giving Body' to Embryos"; Francoeur, "The Forgotten Tool"; Star, "Craft vs. Commodity, Mess vs. Transcendence"; Keller, *Making Sense of Life*.

8. Star, "Craft vs. Commodity, Mess vs. Transcendence," 261. In her discussion of Carl Akeley's taxidermy expeditions to "darkest Africa," Donna Haraway offers a potent reminder of the ways that capitalism and colonial violence are implicated in the treacherous labor of rendering nature's bounty for the edification of the American public. See *Primate Visions*.

9. See Star, "Craft vs. Commodity, Mess vs. Transcendence," 262.

10. Griesemer, "Material Models in Biology," 80.

11. On similarity relations, see Morrison and Morgan, *Models as Mediators*, 29. On equivalence, see Galison, *Image and Logic*, 106.

12. See also Keller, *Making Sense of Life*.

13. Daston, "The Glass Flowers," 31.

14. Daston, "The Glass Flowers," 8.

15. Daston, "The Glass Flowers," 8.

16. Francoeur, "The Forgotten Tool," 12.

17. Francoeur, "Beyond Dematerialization and Inscription," 64.

18. In "Beyond Dematerialization and Inscription," Eric Francoeur develops Peter Galison's terms "homomorphy" and "homology" in relation to the representational function of molecular models. See Galison, *Image and Logic*.

19. We investigated domains as disparate as remote sensing in deep-sea oceanographic research, human-machine entanglements in the history of aviation and space flight, simulations used in nuclear weapons design, computer modeling techniques developed by architects, and imaging, modeling, and simulation techniques used by life scientists and biological engineers. See Turkle et al., "Information Technologies and Professional Identity"; Turkle, *Simulation and Its Discontents*.

20. See also Turkle et al., "Information Technologies and Professional Identity"; Turkle, *Simulation and Its Discontents*.

21. On the power of scientific images, such as PET scans, see Dumit, *Picturing Personhood*.

22. An angstrom is a unit of measurement. One angstrom is equal to 1.0×10^{-10} meters or 0.1 nanometers.

23. Barad, "Posthumanist Performativity," 802. For queer and feminist theories of performativity, see Haraway, *Simians, Cyborgs, and Women*; J. Butler, *Bodies That Matter*; Barad, "Meeting the Universe Halfway"; Herzig, "On Performance."

24. Barad, "Posthumanist Performativity," 802.

25. Karen Barad's theory of "agential realism" offers insight into the ontics and epistemics of scientific visualization (see "Meeting the Universe Halfway," "Posthumanist Performativity," and *Meeting the Universe Halfway*). Agential realism disturbs assumptions about the nature of experimentation. Barad meditates on a set of classical experiments in physics that appear to generate contradictory insights into the nature

of light: some experimental configurations manifest light as a wave, while others manifest light as particles. She insists that the properties of light depend on the precise configuration of the experimental apparatus. The "apparatus of observation" is for her an expansive concept and includes social, material, technical, and discursive features. In her formulation, the phenomena that are produced in a given experimental setup are expressions of an entire configuration of elements. From this perspective, a scientist is not the singular directive force organizing the experiment, and the object is not the primary determinant that engenders the observed experimental phenomenon. Agency is distributed among human and nonhuman participants, and subjects and objects are mutually constituted from within the scene of inquiry. In this view subjects and objects do not preexist one another; rather, they are what precipitate out of an experimental encounter. It is in the context of an experimental configuration that a "cut" is made between subjects and objects; and different cuts produce distinct experimental phenomena.

26. Barad's approach to "mattering" builds on a critique of Judith Butler's theory of performativity, which appears to give discourse the power to sculpt matter as if matter were somehow inert and inactive. Barad's intra-active performativity insists on the agency of the material world.

27. See also Haraway, *Simians, Cyborgs, and Women*, and *Modest_Witness*.

28. Rheinberger, *Toward a History of Epistemic Things*.

29. On material-semiosis, see Haraway, "A Game of Cat's Cradle," and *Modest_Witness*. For a further discussion on material-semiosis in protein modeling, see also chapter 7.

30. See J. Butler, *Bodies That Matter*; Barad, "Posthumanist Performativity," *Meeting the Universe Halfway*; Haraway, *Simians, Cyborgs, and Women*, and *Modest_Witness*.

31. Stengers, "A Constructivist Reading of Process and Reality," 97–98.

32. Keller, "Models Of and Models For," s82.

33. On enactment, see Barad, *Meeting the Universe Halfway*; Mol, *The Body Multiple*.

34. Stengers, "A Constructivist Reading of Process and Reality."

35. Stengers, "A Constructivist Reading of Process and Reality," 96.

36. Stengers, "A Constructivist Reading of Process and Reality," 96.

37. Charis Thompson, *Making Parents*.

FIVE | REMODELING OBJECTIVITY

1. Chang et al., "Retraction," 1875.

2. Dawson and Locher, "Structure of a Bacterial Multidrug ABC Transporter."

3. G. Miller, "Scientific Publishing," 1856.

4. C. Miller, "Pretty Structures, but What about the Data?"

5. C. Miller, "Pretty Structures, but What about the Data?"

6. C. Miller, "Pretty Structures, but What about the Data?," 459. "Reciprocal space" is a technical term in crystallography to refer to a mathematically calculated space inside a crystal.

7. G. Miller, "Scientific Publishing," 1856.

8. G. Miller, "Scientific Publishing," 1856.

9. Pathogens can readily evolve resistance to drugs like antibiotics just by learning how

to up-regulate their membrane pumps to clear antibiotics and other toxins from their cells. A drug designed to disable specific membrane pumps can thus make a pathogenic cell vulnerable: the cell would no longer be able to rid itself of toxins. For this reason, new compounds that can target and regulate these pumps have potential therapeutic applications because they may facilitate the uptake of drugs more effectively. Membrane pumps are thus clear targets for drug design and pharmaceutical development. See, for example, Cowen and Lindquist, "Hsp90 Potentiates the Rapid Evolution of New Traits," and Cowen, "The Evolution of Fungal Drug Resistance" for the role of membrane pumps in the evolution of drug resistance in fungi.

10. G. Miller, "Scientific Publishing," 1856–57.
11. Gawrylewski, "Retractions Unsettle Structural Bio."
12. G. Miller, "Scientific Publishing," 1857.
13. G. Miller, "Scientific Publishing," 1857.
14. C. Miller, "Pretty Structures, but What about the Data?," 459.
15. On visual persuasion, see Dumit, *Picturing Personhood*.
16. See Daston and Galison, *Objectivity*.
17. On "ontological indeterminacy," see Schrader, "Responding to *Pfiesteria*"; Barad, *Meeting the Universe Halfway*.
18. On situated objectivity, see Haraway, "Situated Knowledges"; on modesty, see Haraway, *Modest_Witness*.
19. Daston and Galison, *Objectivity*.
20. Daston, "The Moral Economy of Science," 6.
21. Daston, "The Moral Economy of Science," 6.
22. On social facts, see Durkheim, "What Is a Social Fact?" in *The Rules of Sociological Method*.
23. Scholars in STS and the history of science have long taken interest in the moral economies that shape scientific cultures. On the "moral economies" of science, see Kohler, *Lords of the Fly*; Daston "The Moral Economy of Science." On affective economies, see Ahmed, "Affective Economies." For an excellent analysis of ambivalences in the moral and affective economies shaped by biopolitics, see Murphy, *Seizing the Means of Reproduction*.
24. Daston, "The Moral Economy of Science," 4.
25. Daston, "The Moral Economy of Science," 11.
26. Daston, "The Moral Economy of Science," 11.
27. Daston and Galison, *Objectivity*, 40–41.
28. The concepts of ethos and habitus were explored in part I of this book, especially in connection to the ways that model making articulates modelers' modes of embodiment, affects, and kinesthetic imaginations.
29. Daston, "The Moral Economy of Science," 4.
30. Daston, "The Moral Economy of Science," 4.
31. Kohler, *Lords of the Fly*.
32. Kohler, "Moral Economy," 243.
33. Kohler, "Moral Economy," 243.
34. See Sarah Ahmed's analysis of affect on the model of Marx's theory of capitalist ex-

change values in the essay "Affective Economies." She makes the important point that affects do not inhere in bodies, but are in-the-making as they circulate between bodies. However, the values and affects that move through laboratories suggest to me that there are other topologies on which we can theorize affect. Elsewhere historian Carla Hustak and I explore the concept of "affective ecologies" to consider how entanglements among humans and nonhumans open up a space for theorizing affect on the topology of an ecology. See Hustak and Myers, "Involutionary Momentum." For more on affect see Gregg and Seigworth, *The Affect Reader*; Clough, *The Affective Turn*; Stewart, *Ordinary Affects*; Massumi, *Parables for the Virtual*.

35. See Daston and Galison, *Objectivity*.

36. Daston and Galison, *Objectivity*, 40–41.

37. See Haraway, "Situated Knowledges," and *Modest_Witness*.

38. For more on this generational transition and the new divisions of labor taking shape with increased reliance on computer power, see Turkle, *Simulation and Its Discontents*.

39. See, for example, Dumit, *Picturing Personhood*; Galison, "Judgment against Objectivity"; Daston and Galison, *Objectivity*.

40. Latour, *Science in Action*.

41. See Turkle et al., "Information Technologies and Professional Identity," for how researchers in a number of fields describe codes as the "guts" of the computer.

42. See the University of Alabama Reporter Archive for "UAB's Statement on Protein Data Bank Issues," http://www.uab.edu/reporterarchive/71570-uab-statement-on -protein-data-bank-issues.

43. Ajees et al., "The Structure of Complement C3b."

44. Janssen et al., "Structure of C3b Reveals Conformational Changes"; Wiesmann et al., "Structure of C3b in Complex with CRIg."

45. Janssen et al., "Crystallography," E1.

46. Janssen et al., "Crystallography," E1.

47. Ajees et al., "Crystallography."

48. See Stephen Curry's December 13, 2009, blog entry "Ego Tripping at the Gates of Hell" on his blog "Reciprocal Space": http://occamstypewriter.org/scurry/?s=gates+of+hell.

49. Borell, "Fraud Rocks Protein Community," 970.

50. Borell, "Fraud Rocks Protein Community," 970.

51. See the World Wide PDB's website describing these new validation procedures: http:// www.wwpdb.org/validation/; see also Read et al., "A New Generation of Crystallographic Validation Tools."

52. Barad, *Meeting the Universe Halfway*, 127.

53. For Karen Barad, "apparatuses provide the conditions for the possibility of determinate boundaries and properties of 'objects' within phenomena, where 'phenomena' are the ontological inseparability of objects and apparatuses" (*Meeting the Universe Halfway*, 127).

54. Schrader draws attention to *Pfiesteria piscicida*, the unicellular marine organism implicated in the deadly phenomenon known as red tide. She draws attention to the indeterminacies that keep tripping researchers up as they try to pin this organism

down. *Pfiesteria* are slippery and elusive. Experiments generate contradictory and discontinuous data. *Pfiesteria*'s ontology is precisely what must be settled in order to determine the causative agent in the toxic fish kills that threaten food webs, fishing communities, and consumers. But experimental inquiry into its beings and doings produce data sets that refuse to cohere. Rather than beginning and ending her analysis with an account of the uncertainties plaguing researchers who generate discordant data, Schrader takes the evasive ontologies of *Pfiesteria*'s form of life seriously. She points to indeterminacies "between *Pfiesteria*'s beings and doings," suggesting that "it is not only that their ontology is indeterminate; rather, ontology itself is put into question" ("Responding to *Pfiesteria*," 283).

55. Schrader, "Responding to *Pfiesteria*," 283.
56. Schrader, "Responding to *Pfiesteria*," 277.
57. Celikel et al., "Modulation of α-Thrombin Function"; Dumas et al., "Crystal Structure of the GpIbα-Thrombin Complex."
58. Sadler, "A Ménage à Trois in Two Configurations."
59. Kobe et al., "The Many Faces of Platelet Glycoprotein Ibalpha," 551.
60. Vanhoorelbeke et al., "The GPIbalpha-Thrombin Interaction," 33.
61. Sadler, "A Ménage à Trois in Two Configurations," 177.
62. Sadler, "A Ménage à Trois in Two Configurations," 178.
63. Sadler, "A Ménage à Trois in Two Configurations," 179.
64. Sadler, "A Ménage à Trois in Two Configurations," 179.
65. Sadler, "A Ménage à Trois in Two Configurations," 179.
66. See Adam et al., "Thrombin Interaction with Platelet Membrane"; Goh et al., "Conformational Changes Associated with Protein-Protein Interactions"; Vanhoorelbeke et al., "The GPIbalpha-Thrombin Interaction"; Kobe et al., "The Many Faces of Platelet Glycoprotein"; Ruggeri et al., "Unraveling the Mechanism and Significance of Thrombin Binding."
67. Such a claim would make it difficult for practitioners in this field to make any progress. It would be difficult for protein crystallographers to argue that these structures could be compared only if the experimental conditions were identical. Indeed, crystallographers need room to improvise with their protocols to ensure they get crystals and that those crystals diffract well; there is no standard protocol to which practitioners can adhere.
68. Ruggeri, "Unraveling the Mechanism and Significance of Thrombin Binding," 895.
69. Kobe et al., "The Many Faces of Platelet Glycoprotein Ibalpha," 556.
70. Gerstein and Chothia, "Proteins in Motion"; Echols et al., "MolMovDB"; Goh et al., "Conformational Changes Associated with Protein-Protein Interactions"; Bhardwaj and Gerstein, "Relating Protein Conformational Changes to Packing Efficiency and Disorder." See also Mark Gerstein's Database of Macromolecular Movements online at http://www.molmovdb.org/.
71. Goh et al., "Conformational Changes Associated with Protein–Protein Interactions," 108.
72. Bhardwaj and Gerstein, "Relating Protein Conformational Changes to Packing Efficiency and Disorder," 1236.

73. Wright and Dyson, "Intrinsically Unstructured Proteins," and "Linking Folding and Binding."
74. Wright and Dyson, "Intrinsically Unstructured Proteins," 321.
75. See, for example, Richardson and Richardson, "The Kinemage"; Block et al., "KinImmerse"; Davis et al., "The Backrub Motion"; Georgiev et al., "Algorithm for Backrub Motions in Protein Design."
76. Davis et al., "The Backrub Motion."
77. Davis et al., "The Backrub Motion," 264, emphasis added.
78. Davis et al., "The Backrub Motion," 270.
79. Goh et al., "Conformational Changes Associated with Protein-Protein Interactions," 108.
80. See chapters 6 and 7 for a fuller discussion of the implications of such insights for evolutionary theory.
81. On "ontological choreography," see Thompson, *Making Parents*.
82. On the dance between meaning and matter, see Barad, "Meeting the Universe Halfway."

SIX | MACHINIC LIFE

1. On molecular machines, see Goodsell, *The Machinery of Life*. See also Tanford and Reynolds, *Nature's Robots*.
2. See, for example, Hill and Rich, "A Physical Interpretation for the Natural Photosynthetic Process"; Bourne, "One Molecular Machine Can Transduce Diverse Signals"; Hoffman, "An RNA First"; Kreisberg et al., "The Interdigitated Beta-Helix Domain of the P22 Tailspike Protein Acts as a Molecular Clamp in Trimer Stabilization"; Harrison, "Whither Structural Biology?"; Chiu et al., "Electron Cryomicroscopy of Biological Machines at Subnanometer Resolution."
3. See also Calvert, "The Commodification of Emergence"; Roosth, "Crafting Life."
4. The time-lapse movie of the pullulating cell was generated in Dan's laboratory, and in other contexts Dan can be seen animating the cell as if it were like a rock climber inching its way along a sheer surface. See Myers and Dumit, "Haptic Creativity."
5. Morgan and Morrison, *Models as Mediators*.
6. Taylor and Blum, "Ecosystems as Circuits," 276.
7. Lynch, "The Externalized Retina."
8. See Haraway, *Modest_Witness*.
9. On haptic aesthetics, see Fisher, "Relational Sense"; and Paterson, *The Senses of Touch*.
10. See Lakoff and Johnson, *Philosophy in the Flesh*, 124.
11. Martin, "The Egg and the Sperm," 501.
12. Fleck, *Genesis and Development of a Scientific Fact*.
13. See Geison, "The Protoplasmic Theory of Life"; Huxley, *A Manual of the Anatomy of Invertebrated Animals*.
14. Cited in Beale, "The Address of the President of the Royal Microscopial Society," 297. See also Huxley, *A Manual of the Anatomy of Invertebrated Animals*, 15.
15. Georges Canguilhem documents the history of machinic figurations that have shaped

concepts of the organism. These tropes and logics date as far back as Aristotle, who likened animal movements to the parts of "war machines," specifically the "arms of catapults" (Canguilhem, *Knowledge of Life*, 78–79).

16. See Geison, "The Protoplasmic Theory of Life."

17. Beale, "The Address of the President of the Royal Microscopial Society," 297, emphasis added.

18. Beale, "The Address of the President of the Royal Microscopial Society," 297.

19. See, for example, Hacking, *Representing and Intervening*; Dumit, *Picturing Personhood*; Lynch and Woolgar, *Representation in Scientific Practice*; Coopmans et al., *Representation in Scientific Practice Revisited*.

20. Gilbert and Mulkay, *Opening Pandora's Box*, quoted in Lynch, "Science in the Age of Mechanical Reproduction," 209.

21. This is a phenomenon that Alberto Cambrosio and his colleagues have examined in their historical study of the cartoon diagrams that early immunologist Paul Erlich drew in the process of his inquiry into the invisible chemical interactions among antibodies and antigens (see "Erlich's 'Beautiful Pictures'"). In this case, Erlich's "inventive" diagrams conjured antibodies in ways that allowed him to hypothesize their interactions with other molecules. While these diagrams were not grounded in empirical evidence, they did help him to organize his experimental protocols and to establish a successful research program.

22. See "Situated Knowledges" in Haraway, *Simians, Cyborgs, and Women*, 190.

23. On the visible and the speakable, see Foucault, *The Birth of the Clinic*.

24. For more on molecular storytelling, see chapter 7.

25. See, for example, Geison, "The Protoplasmic Theory of Life"; Hopwood, "'Giving Body' to Embryos"; Keller, *Refiguring Life*, *Making Sense of Life*; and Pauly, *Controlling Life*.

26. Canguilhem, *Knowledge of Life*, 77.

27. See also Pauling et al., "The Structure of Proteins."

28. Meinel, "Molecules and Croquet Balls."

29. Jacobus H. van't Hoff published the first theory of three-dimensional molecular structure in his *Arrangement of Atoms in Space*, in 1874. However, because molecular structures could not be visualized directly, this concept did not gain traction for some time.

30. Hofmann cited in Meinel, "Molecules and Croquet Balls," 250.

31. Meinel, "Molecules and Croquet Balls," 252.

32. Meinel, "Molecules and Croquet Balls," 267.

33. Meinel, "Molecules and Croquet Balls," 266.

34. Meinel, "Molecules and Croquet Balls," 270.

35. Francoeur, "The Forgotten Tool."

36. Harrison, "Whither Structural Biology?" 15.

37. See Fujimura, "Postgenomic Futures."

38. Inside the journal, the caption for the cover reads: "The mechanism by which receptors transduce signals across membranes remains an important open question in molecular biology. Various models have been proposed. . . . Based on the interdigi-

tation of the side chains in this structure, the authors propose a cogwheel model for signal transduction, which involves the concerted rotation of the helices in a plane perpendicular to the membrane. The model is illustrated conceptually on the cover by a gear box with four cogwheels." The image is by Martin Voetsch, Max Planck Institute for Developmental Biology. See inside the cover of the journal *Cell* 126, no. 5 (September 8, 2006).

39. For more on body experiments, see chapter 8.

40. Fisher, "Relational Sense," 4.

41. See Townley and Shapiro, "Crystal Structure of the Adenylate Sensor."

42. Haraway, *Modest_Witness*, 161, emphasis in the original.

43. Haraway, *Modest_Witness*, 135.

44. Haraway, *Modest_Witness*, 64; Haraway, *Simians, Cyborgs, and Women*, 200. See also Latour, "Drawing Things Together."

45. Haraway, *Modest_Witness*, 97.

46. Haraway, *Modest_Witness*, 141.

47. Haraway, *Modest_Witness*, 97.

48. Geoff eventually joined Jim as an instructor in a course on protein folding. For more on that course and on protein folding, see chapter 7, and Myers, "Performing the Protein Fold."

49. For an ethnography of engineers working with CAD/CAM technologies, see Downey, *The Machine in Me*.

50. See, for example, Gilman and Arkin, "Genetic 'Code.'"

51. On standardized biological parts, see also Roosth, "Crafting Life."

52. Bourne, "One Molecular Machine Can Transduce Diverse Signals," 814.

53. Bourne, "One Molecular Machine Can Transduce Diverse Signals," 814.

54. Haraway, *Modest_Witness*, 133–37.

55. For insight into how molecular machines are used by proponents of intelligent design, see the remarkable animations featured on the page "An Introduction to Molecular Machines and Irreducible Complexity" on the Access Research Network website: http://www.arn.org/mm/mm.htm.

56. Haraway, *Modest_Witness*, 137.

57. Haraway, "A Cyborg Manifesto," 180.

58. Haraway, "A Cyborg Manifesto," 180.

SEVEN | LIVELY MACHINES

1. *The Inner Life of the Cell* is available to view online at http://www.studiodaily.com /main/technique/tprojects/6850.html. This animation is the product of a collaboration between Harvard University and the Howard Hughes Medical Institute Biological Sciences Multimedia Project. Harvard scientists Alain Viel and Robert A. Lue directed the conception and scientific content, and the animation was produced by John Liebler and directed by David Bolinsky at XVIVO Studios. See the "Multimedia Production Site" for descriptions of other projects, guided tours through contemporary innovations in the life sciences, interviews with scientists, and clips of other

animations, at http://multimedia.mcb.harvard.edu/. Currently clips of this video are widely circulated on YouTube.

2. See David Bolinsky's TED Talk, "Visualizing the Wonder of a Living Cell," at http://www.ted.com/talks/david_bolinsky_animates_a_cell. As of January 1, 2015, this TED Talk has been viewed 1,236,266 times.

3. Marchant, "Cellular Visions."

4. On neoliberal subjectivities, see, for example, Ong, *Neoliberalism as Exception*; Ong and Collier, *Global Assemblages*; Comaroff and Comaroff, "Millennial Capitalism"; Cooper, *Life as Surplus*.

5. On animation more generally, see, for example, Cartwright, "The Hands of the Animator"; Vivian Sobchack, "Animation and Automation, or, the Incredible Effortfulness of Being."

6. Some modelers used physical materials to animate biological processes. For example, in the late nineteenth century, embryologist Wilhelm Roux (1859–1924) incubated balls of dough containing varying quantities of yeast, joined them together in cellular formations, and observed the patterns they formed as they rose. In so doing Roux effectively produced an "animation" of the physical forces that affected the differential growth of cells during embryogenesis. See Hopwood, "'Giving Body' to Embryos." See also Keller, *Making Sense of Life*, for a vivid description of Stefan Leduc's "artificial organisms," which animated living process using the medium of organic salts.

7. Compare *Fantastic Voyage* to the body politic narrated in the Hollywood animation *Osmosis Jones* (2001, directed by Peter and Bobby Farrelly). Bill Murray stars as "Frank," an aging, uncouth zookeeper who picks up a pathogen from a caged monkey. His internal body is animated as a city-state, and the "City of Frank" is under invasion by lethal germs. "Good guy" white blood cell "cops" patrol the city to defend Frank's organs and tissues from "bad guy" germ and virus gangsters and Mafioso. This vision of the body as a fortress to be defended by heroic acts against evildoers animates a cellular and molecular life that is thoroughly racialized and sexualized. For other renderings of bodies as cities and states, see Emily Martin, *Flexible Bodies*.

8. Cartwright, *Screening the Body*; Landecker, "Cellular Features," "Creeping, Drinking, Dying," and *Culturing Life*; Gaycken, "'The Swarming of Life.'"

9. See also Doane, *The Emergence of Cinematic Time*; Deleuze, *Cinema 1*.

10. Hannah Landecker documents the incredible effort and energy expended in the work of setting up microcinematographic experiments with cells and tissue culture. See *Culturing Life*.

11. Kelty and Landecker, "A Theory of Animation."

12. Kelty and Landecker, "A Theory of Animation," 45.

13. Kelty and Landecker, "A Theory of Animation," 32.

14. Hartl and Hayer-Hartl, "Converging Concepts of Protein Folding in Vitro and in Vivo," 574.

15. Bartlett and Radford, "An Expanding Arsenal of Experimental Methods Yields an Explosion of Insights into Protein Folding Mechanisms," 582.

16. Creutzfeldt-Jakob disease (CJD), a human prion disease, is also grouped within the protein folding diseases. It is the human form of bovine spongiform encephalopathy (BSE), otherwise known as mad cow disease.

17. Dobson, "Protein Folding and Misfolding," 884.

18. Today, increasingly powerful computers and distributed computing are being applied to the as yet intractable problems of protein structure prediction and protein folding. Computer scientists, mathematicians, and protein modelers participate in competitions to test their algorithms for protein structure prediction. "Critical Assessment of Techniques for Protein Structure Prediction," or CASP, is a competition funded by the National Institutes of Health and the National Library of Medicine. This competition has now hosted a series of experiments in which protein modelers apply their algorithms to see which methods best predict the structures of proteins (see http://predictioncenter.org). One of the biggest challenges these researchers face turns out to be predicting where and when water molecules interact with the polypeptide chain as it folds up in its wet, cellular environment. Predictive modeling also requires extensive computer power to run protein folding simulations. Folding@ Home is a protein prediction project that has been modeled on SETI's (the Search for Extra-terrestrial Intelligence) use of volunteers' home computers to run predictive algorithms. These projects continue, with uneven success, to devise algorithms in attempts to predict structure from protein sequence. Researchers have recently realized that if, rather than depending entirely on algorithms, they integrated human judgment into the process, they would have better results. They developed a game that makes use of players' intuitive skills, trained judgment, and haptic creativity to help predict folding pathways. This "human-computer" symbiosis produced significantly better results. See Cooper et al., "Predicting Protein Structures with a Multiplayer Online Game."

19. For another example of protein folding animations, see Jones, "Successful Ab Initio Prediction," 185–91: "This animated GIF image shows a synthetic folding trajectory for a small alpha-helical protein (porcine NK-lysin) which was predicted using our FRAGFOLD software as part of the 2nd CASP experiment carried out in 1996. This was the first successful prediction of a novel protein fold in CASP. . . . Each frame of the animation was generated by linearly interpolating the coordinates from 'snapshots' taken during the protein folding simulation, and so is not intended to be physically realistic." The animation is available online. See David Jones's website at http://wwwo.cs.ucl.ac.uk/staff/d.jones/t42morph.html.

20. Transfer RNA, or tRNA, is a form of ribonucleic acid that is involved in protein synthesis. For more on protein synthesis, see chapter 8.

21. While neo-Darwinism remains the grounding logic of the life sciences, there are many disparate views of evolution. Several prominent evolutionary theorists challenge the constraints of conventional neo-Darwinisms. See, for example, Margulis, *Symbiosis in Cell Evolution*; Margulis and Sagan, *Microcosmos*; Gould, "Darwinian Fundamentalism"; Gould and Lewontin, "The Spandrels of San Marco"; Shapiro, *Evolution*. Many critics of neo-Darwinism embrace Darwin's less dogmatic views, which did not entirely banish Lamarckian modes of inheritance.

22. Kirschner et al., "Molecular 'Vitalism.'"
23. "Visualizing the Wonder of a Living Cell" TED Talks discussion thread, accessed August 18, 2011, http://www.ted.com/talks/david_bolinsky_animates_a_cell.html.
24. Latour, *On the Modern Cult of the Factish Gods*. On fetishism in molecular genetics, see Haraway, *Modest_Witness*.
25. Latour, *On the Modern Cult of the Factish Gods*, 3.
26. Latour, *On the Modern Cult of the Factish Gods*, 21.
27. On the relationship between facts, fetishes, and factishes, see Latour, *On the Modern Cult of the Factish Gods*. See also Stengers, *Cosmopolitics I*.
28. Latour, *On the Modern Cult of the Factish Gods*, 22.
29. See Elizabeth Grosz, *The Nick of Time*, for a feminist reading of random and directionless change in Darwin's theory of evolution. For a feminist critique of neo-Darwinian models of randomness, see Hustak and Myers, "Involutionary Momentum."
30. Riskin, "Eighteenth-Century Wetware," 99; see also in Stacey and Suchman, "Animation and Automation," 10.
31. See, for example, Samuel Butler's lively treatment of machines in "The Book of the Machines" (*Erewhon*), and Deleuze and Guattari's reading of Butler in *Anti-Oedipus*. See also Sagan, "Samuel Butler's Willful Machines."
32. Animism and mechanism have long been intertwined. See Lenoir, *The Strategy of Life*; Tresch, *The Romantic Machine*; Keller, *Refiguring Life*, and *Making Sense of Life*.
33. Thanks to Dorion Sagan for this concept.
34. On "master molecules" governing cells, see Kay, *The Molecular Vision of Life*; Keller, *Refiguring Life*.
35. See Keller, *Reflections on Gender and Science*, and "Secrets of God, Nature, and Life" in *Secrets of Life, Secrets of Death*.
36. On the "crossed-out" God of the moderns, see Latour, *We Have Never Been Modern*.

EIGHT | MOLECULAR CALISTHENICS

1. Despret, "From Secret Agents to Interagency," 38.
2. Myers, "Animating Mechanism."
3. For more on the use of metonymic models and analogies in the protein folding course, see Myers, "Performing the Protein Fold."
4. Alač, *Handling Digital Brains*, "Moving Android"; Vertesi, "Seeing Like a Rover"; Prentice, "The Anatomy of a Surgical Simulation"; Ochs et al., "Interpretive Journeys"; Goodwin, "Action and Embodiment within Situated Human Interaction."
5. Ochs et al., "Interpretive Journeys."
6. Ochs et al., "Interpretive Journeys," 158.
7. Ochs et al., "Interpretive Journeys," 161.
8. Bourdieu, *Outline of a Theory of Practice*, 1.
9. Bourdieu, *Outline of a Theory of Practice*, 1.
10. Rotman, *Becoming beside Ourselves*, and "Gesture, or the Body without Organs of Speech."
11. Foucault, *The Order of Things*, 22.

12. Foucault, *The Order of Things*, 22.
13. Taussig, *Mimesis and Alterity*. See also Suchman, *Human-Machine Reconfigurations*; Myers, "Animating Mechanism."
14. Taussig, *Mimesis and Alterity*, xiii.
15. Taussig, *Mimesis and Alterity*, 21.
16. Taussig, *Mimesis and Alterity*, 21.
17. Taussig, *Mimesis and Alterity*, 34–35.
18. Taussig, *Mimesis and Alterity*, 45.
19. See Prentice, "The Anatomy of a Surgical Simulation," on mutual articulation.
20. Taussig, *Mimesis and Alterity*, 40.
21. Taussig, *Mimesis and Alterity*, 34.
22. Stacey and Suchman, "Animation and Automation," 4.
23. Stacey and Suchman, "Animation and Automation," 4–5.
24. Stacey and Suchman, "Animation and Automation," 6.
25. Stacey and Suchman, "Animation and Automation," 19.
26. Stacey and Suchman, "Animation and Automation," 23.
27. Despret, "From Secret Agents to Interagency," 29.
28. Despret, "From Secret Agents to Interagency," 38.
29. See, for example, Bourne, "One Molecular Machine Can Transduce Diverse Signals"; Harrison, "Whither Structural Biology?"
30. The term "transduction" has at least two other lineages in the history of science: one in acoustics and the other in biology. It refers both to the "action or process of transducing a signal," such as sound, through one medium to another, and "the transfer of genetic material from one cell to another by a virus or virus-like particle" (*Oxford English Dictionary*, 3rd ed. [December 2009]). For ways of thinking through transduction in both ethnographic research and technologically mediated communication, see Helmreich, "An Anthropologist Underwater." See also Mackenzie, *Transductions*; Simondon, "The Position of the Problem of Ontogenesis"; Myers, "Animating Mechanism."
31. Contact improvisation is an "intra-active" movement practice in which two or more dancers engage in an improvised conversation. Dancers keep close physical contact while exploring the play between their bodies and experimenting with balance, gravity, weight tension, gesture, and tacit modes of communication. Unlike other social dance techniques, there is no leader and no follower. The dance is an *agencement* that is collaboratively articulated. On contact improvisation, see Myers and Dumit, "Haptic Creativity"; Dumit, "Don't Know Where You Are Going."
32. Bergson, *Matter and Memory*, 28.
33. Bergson, *Matter and Memory*, 55–56.
34. See also Myers and Dumit, "Haptic Creativity."
35. See, for example, Fischer, *Anthropological Futures*.
36. Paterson, *The Senses of Touch*, 66.
37. Paterson, *The Senses of Touch*, 72.
38. Rheinberger, *Toward a History of Epistemic Things*, 28.
39. While at a weeklong meeting on protein folding that gathered protein crystallog-

raphers, physicists, mathematicians, mechanical engineers, and others together in Banff, Alberta, I was told about one U.S.-based crystallographer who actually sought out training in expressive dance as a means to better communicate the forms and folds of his proteins. At another conference I met a Harvard-based crystallographer who had commissioned a Boston dance company to choreograph and perform the molecular phenomena he was investigating to accompany his acceptance of a prestigious award. For more on dance and science, see Myers, "Dance Your PhD."

40. As of December 28, 2014, one online version of *Protein Synthesis: An Epic on the Cellular Level* had received 956,445 views on YouTube, http://www.youtube.com/watch?v=u9dhooiCLww.

Protein Synthesis: An Epic on the Cellular Level was remounted in 2006 at Kenyon College in Ohio by students and faculty of the Department of Dance and Drama and the Department of Biology. In place of Paul Berg, this reenactment was narrated by MIT/Whitehead Institute biochemist Harvey Lodish; see "The Protein Synthesis Dance at Kenyon College," http://biology.kenyon.edu/courses/bio1114/proteindance/.

And in 2012, MIT hosted a "bio flash mob" that emulated the large scale format of the epic, but this time to stage a kinetic diagram of cancer cell biology. See the MIT/Koch Institute YouTube channel for the video, http://www.youtube.com/watch?v=jis2mXXY90Y&feature=youtube_gdata_player.

41. Tierney, "Dancing Dissertations."

42. For videos of the 2012 competition, see John Bohannon's Gonzo Labs website, http://gonzolabs.org/2012-videos/.

43. For more on the contests, see Myers, "Dance Your PhD."

44. To view LiCata's YouTube submission and his explanation of the choreography *A Molecular Dance in the Blood, Observed*, see http://www.youtube.com/watch?v=2L1UJgYH6bU&feature=channel_page.

45. On the carnivalesque, see Bakhtin, *Rabelais and His World*.

46. Deleuze, *Cinema 1*. See also Canales, *Tenth of a Second*; Mary Anne Doane, *Emergence of Cinematic Time*.

47. Deleuze, *Cinema 1*, 10. This is one reason that video ethnography poses a challenge to my research. To make movies, or isolate snapshots of modelers mid-gesture, is to cut into what I see as a larger social and semiotic context for expression and meaning making.

48. Deleuze, *Cinema 1*, 3–11.

49. Marcin Khedzior, an architect, experimental musician, and break-dancer based in Toronto, has helped me to think through ways that temporality can be made elastic. He invokes the musical term "tempo rubato." Translated directly as "stolen time," it indicates where a musician can play with and alter the tempo of a musical score. Rather than thinking of it as "robbed time," Khedzior treats rubato as a way of "rubbing" or "massaging time." This practice of pulling at the tempo allows a performer to rework and transform the flow of time, and so inflect the performance.

50. See "Sydney PhD student turns thesis into dance," ABC News, December 1, 2008, http://www.abc.net.au/news/stories/2008/12/01/2434873.htm.

51. Bohannon quoted in Robert Mitchum, "In Chicago, Researchers Hold Finale to

'Dance Your PhD,'" *Chicago Tribune*, February 19, 2009, http://articles.chicagotribune
.com/2009–02–16/news/0902150241_1_dance-researchers-escherichia.

52. John Bohannon interviewed on NPR's *It's Only a Game*, December 6, 2008, http://
www.bu.edu/wbur/storage/2008/12/onlyagame_1206_4.mp3.

53. John Tierney, "Dancing Dissertations," *New York Times*, February 14, 2008, http://
tierneylab.blogs.nytimes.com/2008/02/14/dancing-dissertations/.

54. See also Traweek, *Beamtimes and Lifetimes*; Kaiser, "The Postwar Suburbanization of
American Physics"; Gusterson, "A Pedagogy of Diminishing Returns."

CONCLUSION | WHAT IS LIFE BECOMING?

1. Foucault, *The Order of Things*, 139.
2. Foucault, *The Order of Things*, 303.
3. Foucault, *The Order of Things*, 303.
4. On secrets of life, see Keller, *Secrets of Life, Secrets of Death.*
5. See, for example, Crick, *Life Itself.*
6. François Jacob, *Logic of Life*, 331. See also Doyle, *On Beyond Living*, 13.
7. Doyle, *On Beyond Living.*
8. See also Haraway, *Modest_Witness*; Kay, *Who Wrote the Book of Life?*
9. Doyle, *On Beyond Living*, 17.
10. Doyle, *On Beyond Living*, 17.
11. Doyle, *On Beyond Living*, 10.
12. See Keller, *Refiguring Life*, and *Making Sense of Life.*
13. Keller, *Making Sense of Life*, 217.
14. Keller, *Making Sense of Life*, 217.
15. Haraway, *Modest_Witness*, 137.
16. For critical histories of the central dogma, see Keller, *Refiguring Life*; Kay, *Who Wrote
the Book of Life*; and Doyle, *On Beyond Living*. For examples of ways life scientists are
reworking the central dogma, see Coen, *The Art of Genes*; Shapiro, *Evolution*; Lewontin, *Biology as Ideology*; Holdrege, *Genetics and the Manipulation of Life.*
17. Shapiro, "Revisiting the Central Dogma in the 21st Century," 7.
18. Shapiro, "Revisiting the Central Dogma in the 21st Century," 14.
19. Shapiro, "Revisiting the Central Dogma in the 21st Century," 7. On epigenetics, see
Landecker, "Food as Exposure."
20. Robinson et al., "The Molecular Sociology of the Cell."
21. Shapiro, "Revisiting the Central Dogma," 15.
22. Shapiro, "Revisiting the Central Dogma," 23.
23. Mae-Wan Ho offers this insight: "In a very real sense, we never cease to write and
overwrite our evolutionary history. DNA may be seen as a specific text in which organisms record their evolutionary experiences." See Ho, "DNA and the New Organicism," 91.
24. See Margulis et al., *Chimeras and Consciousness*; Ben-Jacob et al., "Smart Bacteria";
Thompson, "Living Ways of Sense Making." See also Melissa Atkinson-Graham, "Re-
scaling Sentience: An Anthropology of Cellular Life."
25. See Merleau-Ponty, *The Visible and Invisible*, 142.

26. Heller-Roazen, *The Inner Touch*.

27. This is an argument well supported by the endosymbiotic view of life put forward by evolutionary ecologist Lynn Margulis and philosopher Dorion Sagan. See Margulis and Sagan, *What Is Life?*, and *Microcosmos*.

28. On the relation between organisms and their milieu, see Canguilhem, "The Living and Its Milieu."

29. For a fuller account of the economizing logics of evolution, see Hustak and Myers, "Involutionary Momentum."

30. On such an "ecology of practices," see Stengers, *Cosmopolitics I*.

31. See Helmreich, "An Anthropologist Underwater."

32. On tactics, see de Certeau, *The Practice of Everyday Life*; Da Costa and Philip, *Tactical Biopolitics*.

APPENDIX | A PROTEIN PRIMER

1. There are many ways to look at an egg and ask, "What is it made of?" We could start by asking how it is made; that is, what are the conditions of its production? How are the reproductive labors of chickens exploited on factory farms to produce food for human consumption? (See, for example, Squire, *Poultry Science, Chicken Culture*.) Alternatively, it is possible to be on intimate terms with chickens and raise them in our backyards (see, for example, Shulman, *Eat the City*). In all cases eggs raise crucial questions about capitalism, consumption, production, and reproduction (for insights into the ways capital is rendered from animal bodies, see Shukin, *Animal Capital*). These are salient analyses when thinking about other foods as well, including plant-based agricultural products. If I had picked up a piece of tofu and inquired into what it is made of, I would soon discover that soybeans and their farmers are not immune to the violence of capital (see, for example, Hetherington, "Beans before the Law").

2. For a lively tutorial on this process, see the YouTube video *Protein Synthesis: An Epic on the Cellular Level*, http://www.youtube.com/watch?v=u9dhooiCLww. For more on this remarkable rendering of protein synthesis, see chapter 8. See the conclusion of this book for a critique of the central dogma.

3. Voisine et al., "Chaperone Networks."

Abir-Am, Pnina G. "The Politics of Macromolecules." *Osiris*, 2nd ser., 7 (1992): 164–91.

Abir-Am, Pnina G., and Dorinda Outram. *Uneasy Careers and Intimate Lives: Women in Science, 1789–1979*. New Brunswick, NJ: Rutgers University Press, 1987.

Adam, Frédéric, Marie-Christine Bouton, Marie-Geneviève Huisse, and Martine Jandrot-Perrus. "Thrombin Interaction with Platelet Membrane Glycoprotein Ib Alpha." *Trends in Molecular Medicine* 9, no. 11 (November 2003): 461–64.

Ahmed, Sara. "Affective Economies." *Social Text* 79, 22, no. 2 (June 1, 2004): 117–39.

Ahmed, Sara. *The Cultural Politics of Emotion*. Edinburgh: Edinburgh University Press, 2004.

Ajees, A. Abdul, Krishnasamy Gunasekaran, Sthanam V. L. Narayana, and H. M. Krishna Murthy. "Crystallography: Crystallographic Evidence for Deviating C3b Structure (Reply)." *Nature* 448, no. 7154 (August 9, 2007): E2–E3.

Ajees, A. Abdul, K. Gunasekaran, John E. Volanakis, Sthanam V. L. Narayana, Girish J. Kotwal, and H. M. Krishna Murthy. "The Structure of Complement C3b Provides Insights into Complement Activation and Regulation." *Nature* 444, no. 7116 (October 2006): 221–25.

Alač, Morana. *Handling Digital Brains A Laboratory Study of Multimodal Semiotic Interaction in the Age of Computers*. Cambridge, MA: MIT Press, 2011.

Alač, Morana. "Moving Android: On Social Robots and Body-in-Interaction." *Social Studies of Science* 39, no. 4 (August 2009): 491–528.

Alač, Morana, and Edwin Hutchins. "I See What You Are Saying: Action as Cognition in fMRI Brain Mapping Practice." *Journal of Cognition and Culture* 4, no. 3 (2004): 629–61.

Alaimo, Stacy, and Susan Heckman. *Material Feminisms*. Bloomington: Indiana University Press, 2008.

Atkinson-Graham, Melissa. "Re-scaling Sentience: An Anthropology of Cellular Life." Dissertation research proposal, Graduate Program in Social Anthropology, York University, 2012.

Austin, J. L. *How to Do Things with Words*. Cambridge, MA: Harvard University Press, 1975.

Bakhtin, Mikhail. *Rabelais and His World*. Bloomington: Indiana University Press, 2009.

Barad, Karen. *Meeting the Universe Halfway: Quantum Physics and the Entanglement of Matter and Meaning.* Durham, NC: Duke University Press, 2007.

Barad, Karen. "Meeting the Universe Halfway: Realism and Social Constructivism without Contradiction." In *Feminism, Science, and Philosophy of Science*, edited by Lynn Hankinson Nelson and Jack Nelson, 161–94. Boston: Kluwer, 1996.

Barad, Karen. "Posthumanist Performativity: Toward an Understanding of How Matter Comes to Matter." *Signs* 28, no. 3 (2003): 801–31.

Barley, Stephen R. "The Social Construction of a Machine: Ritual, Superstition, Magical Thinking and Other Pragmatic Responses to Running a CT Scanner." In *Biomedicine Examined*, edited by Margaret Lock and Deborah Gordon, 497–539. Boston: Kluwer Academic Publishers, 1988.

Barry, C. D., H. E. Bosshard, R. A. Ellis, and G. R. Marshall. "Evolving Macromodular Molecular Modeling System." *Federation Proceedings* 33, no. 12 (1974): 2368–72.

Bartlett, Alice I., and Sheena E. Radford. "An Expanding Arsenal of Experimental Methods Yields an Explosion of Insights into Protein Folding Mechanisms." *Nature Structural and Molecular Biology* 16, no. 6 (June 2009): 582–88.

Beale, Lionel S. "The Address of the President of the Royal Microscopial Society." *Science* 2, no. 52 (1881): 294–97.

Ben-Jacob, Eshel, Yoash Shapira, and Alfred I. Tauber. "Smart Bacteria." In *Chimeras and Consciousness: Evolution of the Sensory Self*, edited by Lynn Margulis, Celeste A. Asikainen, and Wolfgang E. Krumbein, 55–62. Cambridge, MA: MIT Press, 2011.

Bennett, Jane. *Vibrant Matter: A Political Ecology of Things.* Durham, NC: Duke University Press, 2010.

Berman, Helen M. "The Protein Data Bank: A Historical Perspective." *Acta Crystallographica Section A: Foundations of Crystallography* 64, no. 1 (2008): 88–95.

Berman, Helen M., John Westbrook, Zukang Feng, Gary Gilliland, T. N. Bhat, Helge Weissig, Ilya N. Shindyalov, and Philip E. Bourne. "The Protein Data Bank." *Nucleic Acids Research* 28, no. 1 (January 1, 2000): 235–42.

Bergson, Henri. *Matter and Memory.* New York: Zone Books, 1991.

Bernstein, Robin. "Dances with Things: Material Culture and the Performance of Race." *Social Text* 27, no. 4 (2009): 67–94.

Bhardwaj, Nitin, and Mark Gerstein. "Relating Protein Conformational Changes to Packing Efficiency and Disorder." *Protein Science* 18, no. 6 (June 1, 2009): 1230–40.

Block, Jeremy N., David J. Zielinski, Vincent B. Chen, Ian W. Davis, E. Claire Vinson, Rachael Brady, Jane S. Richardson, and David C. Richardson. "KinImmerse: Macromolecular VR for NMR Ensembles." *Source Code for Biology and Medicine* 4 (2009): 3.

Borrell, Brendan. "Fraud Rocks Protein Community." *Nature News* 462, no. 7276 (December 22, 2009): 970–70.

Bourdieu, Pierre. *The Logic of Practice.* Stanford, CA: Stanford University Press, 1990.

Bourdieu, Pierre. *Outline of a Theory of Practice.* Cambridge: Cambridge University Press, 1977.

Bourdieu, Pierre. *Practical Reason: On the Theory of Action.* Stanford, CA: Stanford University Press, 1998.

Bourdieu, Pierre, and Richard Nice. *Science of Science and Reflexivity*. Chicago: University of Chicago Press, 2004.

Bourne, Henry. "One Molecular Machine Can Transduce Diverse Signals." *Nature* 321 (1986): 814–16.

Buck-Morss, Susan. "Aesthetics and Anaesthetics: Walter Benjamin's Artwork Essay Reconsidered." *October* 62 (October 1, 1992): 3–41.

Burri, Regula Valérie, Cornelius Schubert, and Jörg Strübing. "The Five Senses of Science." *Science, Technology and Innovation Studies* 7, no. 1 (2011): 3–7.

Butler, Judith P. *Bodies That Matter: On the Discursive Limits of "Sex."* New York: Routledge, 1993.

Butler, Samuel. *Erewhon* [1872]. London: Penguin, 1985.

Caldwell, John W. "Some Analogies between Molecules and Crystals." *Science* 20, no. 497 (1892): 88–89.

Calvert, Jane. "The Commodification of Emergence: Systems Biology, Synthetic Biology and Intellectual Property." *BioSocieties* 3, no. 4 (2008): 383–98.

Cambrosio, Alberto, Daniel Jacobi, and Peter Keating. "Erlich's 'Beautiful Pictures' and the Controversial Beginnings of Immunological Imagery." *Isis* 84 (1993): 662–99.

Canales, Jimena. *A Tenth of a Second: A History*. Chicago: University of Chicago Press, 2009.

Canguilhem, Georges. *Knowledge of Life*. 3rd ed. New York: Fordham University Press, 2008.

Canguilhem, Georges. "The Living and Its Milieu." *Grey Room* 3 (spring 2001): 7–31.

Cartwright, Lisa. *Screening the Body: Tracing Medicine's Visual Culture*. Minneapolis: University of Minnesota Press, 1995.

Cartwright, Lisa. "The Hands of the Animator: Rotoscopic Projection, Condensation, and Repetition Automatism in the Fleischer Apparatus." *Body and Society* 18, no. 1 (March 1, 2012): 47–78.

Cartwright, Nancy. "Models: The Blueprints for Laws." In "Proceedings of the 1996 Biennial Meetings of the Philosophy of Science Association," supplement, *Philosophy of Science* 64 (1997): S292–S303.

Cartwright, Nancy, Towfic Shomar, and Mauricio Suárez. "The Tool Box of Science." *Poznian Studies in the Philosophy of the Sciences and the Humanities* 44 (1995): 137–49.

Çelik, Zeynep. "Kinaesthesia." In *Sensorium: Embodied Experience, Technology, and Contemporary Art*, edited by Caroline A. Jones, 159–62. Cambridge, MA: MIT Press, 2006.

Celikel, Reha, Richard A. McClintock, James R. Roberts, G. Loredana Mendolicchio, Jerry Ware, Kottayil I. Varughese, and Zaverio M. Ruggeri. "Modulation of α-Thrombin Function by Distinct Interactions with Platelet Glycoprotein Ibα." *Science* 301, no. 5630 (July 11, 2003): 218–21.

Chang, Geoffrey, Christopher Roth, Christopher Reyes, Owen Pornillos, Yen-Ju Chen, and Andy Chen. "Retraction." *Science* 314, no. 5807 (2006): 1875.

Chiu, Wah, Matthew Baker, Wen Jiang, Matthew Dougherty, and Michael Schmid. "Electron Cryomicroscopy of Biological Machines at Subnanometer Resolution." *Structure* 13 (2005): 363–72.

Clardy, Jon. "Borrowing to Make Ends Meet." *Proceedings of the National Academy of Sciences* 96, no. 5 (1999): 1836–999.

Clarke, Adele, and Joan H. Fujimura. *The Right Tools for the Job: At Work in the Twentieth-Century Life Sciences*. Princeton, NJ: Princeton University Press, 1992.

Clough, Patricia Ticineto, and Jean Halley. *The Affective Turn: Theorizing the Social*. Durham, NC: Duke University Press, 2007.

Coen, Enrico S. *The Art of Genes: How Organisms Make Themselves*. Oxford: Oxford University Press, 1999.

Collins, D. M., F. A. Cotton, E. E. Hazen, E. F. Meyer, and C. N. Morimoto. "Protein Crystal-Structures—Quicker, Cheaper Approaches." *Science* 190, no. 4219 (1975): 1047–53.

Collins, H. M. *Changing Order: Replication and Induction in Scientific Practice*. London: Sage, 1985.

Comaroff, Jean, and John Comaroff. "Millennial Capitalism: First Thoughts on a Second Coming." *Public Culture* 12, no. 2 (2000): 291–343.

Cooper, Melinda. *Life as Surplus: Biotechnology and Capitalism in the Neoliberal Era*. Seattle: University of Washington Press, 2008.

Cooper, Melinda. "Pre-empting Emergence: The Biological Turn in the War on Terror." *Theory, Culture and Society* 23, no. 4 (2006): 113–35.

Cooper, Seth, Firas Khatib, Adrien Treuille, Janos Barbero, Jeehyung Lee, Michael Beenen, Andrew Leaver-Fay, David Baker, Zoran Popovic, and Foldit Players. "Predicting Protein Structures with a Multiplayer Online Game." *Nature* 466, no. 7307 (2010): 756–60.

Coopmans, Catelijne, Janet Vertesi, Michael E. Lynch, and Steve Woolgar. *Representation in Scientific Practice Revisited*. Cambridge, MA: MIT Press, 2013.

Cowen, Leah E. "The Evolution of Fungal Drug Resistance: Modulating the Trajectory from Genotype to Phenotype." *Nature Reviews Microbiology* 6, no. 3 (2008): 187–98.

Cowen, Leah E., and Susan Lindquist. "Hsp90 Potentiates the Rapid Evolution of New Traits: Drug Resistance in Diverse Fungi." *Science* 309, no. 5744 (September 30, 2005): 2185–89.

Creager, Angela N. H. *The Life of a Virus: Tobacco Mosaic Virus as an Experimental Model, 1930–1965*. Chicago: University of Chicago Press, 2002.

Crick, Francis. *Life Itself: Its Origin and Nature*. New York: Simon and Schuster, 1981.

Csordas, Thomas J. "Somatic Modes of Attention." *Cultural Anthropology* 8, no. 2 (1993): 135–56.

da Costa, Beatriz, and Kavita Philip. *Tactical Biopolitics: Art, Activism, and Technoscience*. Cambridge, MA: MIT Press, 2008.

Daston, Lorraine. "The Glass Flowers." *Things That Talk, Max Planck Institute for the History of Science* Preprint 233 (2003): 5–32.

Daston, Lorraine. "The Moral Economy of Science." *Osiris*, 2nd ser., 10 (January 1, 1995): 3–24.

Daston, Lorraine, and Peter Galison. "The Image of Objectivity." *Representations* 40 (1992): 81–128.

Daston, Lorraine, and Peter Galison. *Objectivity*. New York: Zone Books, 2007.

Davis, Ian W., W. Bryan Arendall III, David C. Richardson, and Jane S. Richardson. "The Backrub Motion: How Protein Backbone Shrugs When a Sidechain Dances." *Structure* 14, no. 2 (February 2, 2006): 265–74.

Dawson, J. W. "American Association for the Advancement of Science: Address of the Retiring President, Dr. J. W. Dawson, at Minneapolis, August 15, 1883." *Science* 2 (1883): 190–99.

Dawson, Roger J. P., and Kaspar P. Locher. "Structure of a Bacterial Multidrug ABC Transporter." *Nature* 443, no. 7108 (September 14, 2006): 180–85.

de Certeau, Michel. *The Practice of Everyday Life.* Berkeley: University of California Press, 1984.

de Chadarevian, Soraya. *Designs for Life: Molecular Biology after World War II.* Cambridge: Cambridge University Press, 2002.

de Chadarevian, Soraya, and Nick Hopwood. *Models: The Third Dimension of Science.* Stanford, CA: Stanford University Press, 2004.

Deleuze, Gilles. *Cinema 1: The Movement Image.* Minneapolis: University of Minnesota Press, 1986.

Deleuze, Gilles, and Félix Guattari. *Anti-Oedipus: Capitalism and Schizophrenia.* Minneapolis: University of Minnesota Press, 1983.

Deleuze, Gilles, and Félix Guattari. *A Thousand Plateaus: Capitalism and Schizophrenia.* Minneapolis: University of Minnesota Press, 1980.

Despret, Vinciane. "The Body We Care For: Figures of Anthropo-Zoo-Genesis." *Body and Society* 10, nos. 2–3 (2004): 111–34.

Despret, Vinciane. "From Secret Agents to Interagency." *History and Theory* 52, no. 4 (2013): 29–44.

Diamond, R., A. Wynn, K. Thomsen, and J. Turner. "Three-Dimensional Perception for One-Eyed Guys, or The Use of Dynamic Paralax." In *Computational Crystallography: Papers Presented at the International Summer School on Crystallographic Computing Held at Carleton University, Ottawa, Canada, August 7–15, 1981,* edited by David Sayre, 286–93. Oxford: Clarendon Press, 1982.

Doane, Mary Ann. *The Emergence of Cinematic Time: Modernity, Contingency; The Archive.* Cambridge, MA: Harvard University Press, 2002.

Dobson, Christopher M. "Protein Folding and Misfolding." *Nature* 426, no. 6968 (December 18, 2003): 884–90.

Doing, Park. *Velvet Revolution at the Synchrotron: Biology, Physics, and Change in Science.* Cambridge, MA: MIT Press, 2009.

Downes, Nick. *Big Science.* Washington, DC: American Academy for the Advancement of Science (AAAS) Press, 1992.

Downey, Gary Lee. *The Machine in Me: An Anthropologist Sits among Computer Engineers.* New York: Routledge, 1998.

Downey, Gary Lee, and Joseph Dumit. *Cyborgs and Citadels: Anthropological Interventions in Emerging Sciences and Technologies.* Santa Fe, NM: School of American Research Press, 1997.

Doyle, Richard. *On Beyond Living: Rhetorical Transformations of the Life Sciences.* Writing Science. Stanford, CA: Stanford University Press, 1997.

Doyle, Richard. *Wetwares: Experiments in Postvital Living*. Minneapolis: University of Minnesota Press, 2003.

Dumas, John J., Ravindra Kumar, Jasbir Seehra, William S. Somers, and Lidia Mosyak. "Crystal Structure of the GpIbα-Thrombin Complex Essential for Platelet Aggregation." *Science* 301, no. 5630 (July 11, 2003): 222–26.

Dumit, Joseph. "Don't Know Where You Are Going: Embodying Improvisation in Dance, Theatre, Neuroscience, and Anthropology." Paper presented at the American Anthropological Association Meetings, Washington, DC, December 3–7, 2014.

Dumit, Joseph. *Drugs for Life: How Pharmaceutical Companies Define Our Health*. Durham, NC: Duke University Press, 2012.

Dumit, Joseph. *Picturing Personhood: Brain Scans and Biomedical Identity*. Princeton, NJ: Princeton University Press, 2004.

Durkheim, Émile. *The Rules of Sociological Method: And Selected Texts on Sociology and Its Method*. London: Palgrave Macmillan, 1895.

Echols, Nathaniel, Duncan Milburn, and Mark Gerstein. "MolMovDB: Analysis and Visualization of Conformational Change and Structural Flexibility." *Nucleic Acids Research* 31, no. 1 (January 1, 2003): 478–82.

Edwards, P. N., M. S. Mayernik, A. L. Batcheller, G. C. Bowker, and C. L. Borgman. "Science Friction: Data, Metadata, and Collaboration." *Social Studies of Science* 41, no. 5 (August 15, 2011): 667–90.

Esposito, Roberto. Translated by Zakiya Hanafi. "Community, Immunity, Biopolitics." *Angelaki* 18, no. 3 (2013): 83–90.

Fersht, Alan R. "From the First Protein Structures to Our Current Knowledge of Protein Folding: Delights and Scepticisms." *Nature Reviews Molecular Cell Biology* 9, no. 8 (2008): 650–54.

Fischer, Michael M. J. *Anthropological Futures*. Durham, NC: Duke University Press, 2009.

Fischer, Michael M. J. *Emergent Forms of Life and the Anthropological Voice*. Durham, NC: Duke University Press, 2003.

Fisher, Jennifer. "Relational Sense: Towards a Haptic Aesthetics." *Parachute: Contemporary Art Magazine* no. 87 (1997): 4–11.

Fleck, Ludwik. *Genesis and Development of a Scientific Fact*. Chicago: University of Chicago Press, 1979.

Fleischer, Richard, dir. *Fantastic Voyage*, 1966.

Fortun, Michael A. *Promising Genomics: Iceland and deCODE Genetics in a World of Speculation*. Berkeley: University of California Press, 2008.

Foucault, Michel. *The Birth of the Clinic: An Archaeology of Medical Perception*. New York: Pantheon Books, 1973.

Foucault, Michel. *The Care of the Self*. Vol. 3 of *The History of Sexuality*. New York: Vintage Books, 1986.

Foucault, Michel. *Discipline and Punish: The Birth of the Prison*. New York: Vintage Books, 1995.

Foucault, Michel. *The History of Sexuality*. Vol. 1, *An Introduction*. New York: Pantheon Books, 1978.

Foucault, Michel. *The Order of Things: An Archaeology of the Human Sciences*. New York: Pantheon Books, 1971.

Francoeur, Eric. "Beyond Dematerialization and Inscription: Does the Materiality of Molecular Models Really Matter?" *International Journal for Philosophy of Chemistry* 6, no. 1 (2000): 63–84.

Francoeur, Eric. "Cyrus Levinthal, the Kluge and the Origins of Interactive Molecular Graphics." *Endeavour* 26, no. 4 (2002): 127–31.

Francoeur, Eric. "The Forgotten Tool: The Design and Use of Molecular Models." *Social Studies of Science* 27 (1997): 7–40.

Francoeur, Eric. "Molecular Models and the Articulation of Structural Constraints in Chemistry." In *Tools and Modes of Representation in the Laboratory Sciences*, edited by Ursula Klein, 95–115. Boston: Kluwer Academic Publishers, 2001.

Francoeur, Eric, and Jerome Segal. "From Model Kits to Interactive Computer Graphics." In *Models: The Third Dimension of Science*, edited by Soraya de Chadarevian and Nick Hopwood, 402–29. Stanford, CA: Stanford University Press, 2004.

Franklin, Sarah. *Dolly Mixtures: The Remaking of Genealogy*. Durham, NC: Duke University Press, 2007.

Franklin, Sarah. "Life Itself." In *Global Nature, Global Culture*, edited by Sarah Franklin, Celia Lury, and Jackie Stacey, 188–227. London: Sage, 2000.

Franklin, Sarah. "Science as Culture, Cultures of Science." *Annual Review of Anthropology* (1995): 163–84.

Franklin, Sarah, and Margaret M. Lock. *Remaking Life and Death: Toward an Anthropology of the Biosciences*. Santa Fe, NM: School of American Research Press, 2003.

Fujimura, Joan H. "Postgenomic Futures: Translations across the Machine-Nature Border in Systems Biology." *New Genetics and Society* 24, no. 2 (2005): 195–225.

Galison, Peter. "Einstein's Clocks: The Place of Time." *Critical Inquiry* 26, no. 2 (winter 2000): 355–89.

Galison, Peter. *Image and Logic: A Material Culture of Microphysics*. Chicago: University of Chicago Press, 1997.

Galison, Peter. "Judgment against Objectivity." In *Picturing Science, Producing Art*, edited by Caroline A. Jones and Peter Galison, 327–59. New York: Routledge, 1998.

Ganguli, Ishani. "The Profiles: Ben Ortiz." *Scientist*, November 7, 2005. http://www.the -scientist.com/?articles.view/articleNo/16824/title/Ben-Ortiz/.

Garfinkel, Harold, Eric Livingston, and Michael Lynch. "The Work of a Discovering Science Constructed with Materials from the Optically Discovered Pulsar." *Philosophy of the Social Sciences*, no. 11 (1981): 131–58.

Gawrylewski, Andrea. "Retractions Unsettle Structural Bio: Recent Findings Upend Conclusions from Five Highly Cited Papers." *Scientist*, January 4, 2007. http://www .the-scientist.com/?articles.view/articleNo/24647/title/Retractions-unsettle-structural -bio/.

Gaycken, Oliver. "'The Swarming of Life': Moving Images, Education, and Views through the Microscope." *Science in Context* 24, no. 3 (September 2011): 361–80.

Geertz, Clifford. *The Interpretation of Cultures*. New York: Basic Books, 1973.

Geison, Gerald. "The Protoplasmic Theory of Life and the Vitalist-Mechanist Debate." *Isis* 60, no. 30 (1969): 272–92.

Georgiev, Ivelin, Daniel Keedy, Jane S. Richardson, David C. Richardson, and Bruce R. Donald. "Algorithm for Backrub Motions in Protein Design." *Bioinformatics* 24, no. 13 (July 1, 2008): 196–204.

Gerstein, Mark, and Cyrus Chothia. "Proteins in Motion." *Science* 285, no. 5434 (1999): 1682–83.

Geurts, Kathryn Linn. "On Rocks, Walks, and Talks in West Africa: Cultural Categories and an Anthropology of the Senses." *Ethos* 30, no. 3 (2002): 178–98.

Gilbert, Nigel, and M. J. Mulkay. *Opening Pandora's Box: A Sociological Analysis of Scientists Discourse.* Cambridge: Cambridge University Press, 1984.

Gilman, Alex, and Adam Arkin. "Genetic 'Code': Representations and Dynamical Models of Genetic Components and Networks." *Annual Review of Genomics and Human Genetics* 3 (2002): 341–69.

Glusker, Jenny Pickworth. *Structural Crystallography in Chemistry and Biology.* Stroudsburg, PA: Hutchinson Ross, 1981.

Glusker, Jenny Pickworth, and Kenneth Trueblood. *Crystal Structure Analysis: A Primer.* Oxford: Oxford University Press, 1985.

Goffman, Erving. *The Presentation of Self in Everyday Life.* Garden City, NY: Doubleday, 1959.

Goh, Chern-Sing, Duncan Milburn, and Mark Gerstein. "Conformational Changes Associated with Protein-Protein Interactions." *Current Opinion in Structural Biology* 14, no. 1 (February 2004): 104–9.

Goodsell, David. *The Machinery of Life.* New York: Springer-Verlag, 1993.

Goodwin, Charles. "Action and Embodiment within Situated Human Interaction." *Journal of Pragmatics* 32 (2000): 1489–522.

Goodwin, Charles. "Practices of Seeing: Visual Analysis; An Ethnomethodological Approach." In *Handbook of Visual Analysis*, edited by Theo van Leewen and Carey Jewitt, 157–82. London: Sage, 2000.

Gould, S. J., and R. C. Lewontin. "The Spandrels of San Marco and the Panglossian Paradigm: A Critique of the Adaptationist Programme." *Proceedings of the Royal Society of London*, ser. B, *Biological Sciences* 205, no. 1161 (1979): 581–98.

Gould, Stephen Jay. "Darwinian Fundamentalism." *New York Review of Books*, June 12, 1997. http://www.nybooks.com/articles/archives/1997/jun/12/darwinian-fundamentalism/.

Gregg, Melissa, and Gregory J. Seigworth. *The Affect Theory Reader.* Durham, NC: Duke University Press, 2010.

Griesemer, James R. "Material Models in Biology." *PSA: Proceedings of the Biennial Meeting of the Philosophy of Science Association, Vol. 2: Symposia and Invited Papers* (1990): 79–93.

Griesemer, James R. "Three-Dimensional Models in Philosophical Perspective." In *Models: The Third Dimension of Science*, edited by Soraya de Chadarevian and Nick Hopwood, 433–42. Stanford, CA: Stanford University Press, 2004.

Grosz, Elizabeth. *The Nick of Time: Politics, Evolution, and the Untimely.* Durham, NC: Duke University Press, 2004.

Gusterson, Hugh. *Nuclear Rites: A Weapons Laboratory at the End of the Cold War.* Berkeley: University of California Press, 1996.

Gusterson, Hugh. "A Pedagogy of Diminishing Returns: Scientific Involution across Three Generations of Nuclear Weapons Scientists." In *Pedagogy and the Practice of Science,* edited by David Kaiser, 75–107. Cambridge, MA: MIT Press, 2005.

Hacking, Ian. "Making Up People." In *The Science Studies Reader,* edited by Mario Biagioli, 161–71. New York: Routledge, 1999.

Hacking, Ian. *Representing and Intervening: Introductory Topics in the Philosophy of Natural Science.* Cambridge: Cambridge University Press, 1983.

Haraway, Donna J. "The Biopolitics of Postmodern Bodies: Constitutions of Self in Immune System Discourse." In *Simians, Cyborgs, and Women: The Reinvention of Nature.* New York: Routledge, 1991.

Haraway, Donna J. "Crittercam: Compounding Eyes in NatureCultures." In *Postphenomenology: A Critical Companion to Ihde,* edited by Evan Selinger, 175–88. Albany: State University of New York Press, 2006.

Haraway, Donna J. *Crystals, Fabrics, and Fields: Metaphors of Organicism in Twentieth-Century Developmental Biology.* New Haven, CT: Yale University Press, 1976.

Haraway, Donna J. "A Cyborg Manifesto: Science, Technology, and Socialist-Feminism in the Late Twentieth Century." In *Simians, Cyborgs, and Women: The Reinvention of Nature,* 149–81. New York: Routledge, 1991.

Haraway, Donna J. "From Cyborgs to Companion Species: Kinship in Technoscience." Keynote Lecture, *Taking Nature Seriously: Citizens, Science and Environment,* Conference, Eugene, Oregon, February 2001.

Haraway, Donna J. "A Game of Cat's Cradle: Science Studies, Feminist Theory, Cultural Studies." *Configurations* 2, no. 1 (1994): 59–71.

Haraway, Donna J. *Modest_Witness@Second_Millennium.FemaleMan_Meets_OncoMouse: Feminism and Technoscience.* New York: Routledge, 1997.

Haraway, Donna J. *Primate Visions: Gender, Race, and Nature in the World of Modern Science.* New York: Routledge, 1989.

Haraway, Donna J. "The Promises of Monsters: A Regenerative Politics for Inappropriate/d Others." In *Cultural Studies,* edited by Lawrence Grossberg, Cary Nelson, and Paula A. Treichler, 295–337. New York: Routledge, 1992.

Haraway, Donna J. *Simians, Cyborgs, and Women: The Reinvention of Nature.* New York: Routledge, 1991.

Haraway, Donna J. "Situated Knowledges: The Science Question in Feminism and the Privilege of Partial Perspectives." *Feminist Studies* 14, no. 3 (1988): 575–99.

Haraway, Donna J. *When Species Meet.* Minneapolis: University of Minnesota Press, 2008.

Harding, Sandra G. *The Science Question in Feminism.* Ithaca, NY: Cornell University Press, 1986.

Hardt, Michael. "Affective Labor." *boundary 2* 26, no. 2 (1999): 89–100.

Hardt, Michael, and Antonio Negri. *Empire.* Cambridge, MA: Harvard University Press, 2000.

Harrison, Paul M., and Mark Gerstein. "Studying Genomes through the Aeons: Protein

Families, Pseudogenes and Proteome Evolution." *Journal of Molecular Biology* 318, no. 5 (May 17, 2002): 1155–74.

Harrison, Stephen. "Whither Structural Biology?" *Nature Structural and Molecular Biology* 11 (2004): 12–15.

Hartl, F. Ulrich, and Manajit Hayer-Hartl. "Converging Concepts of Protein Folding in Vitro and in Vivo." *Nature Structural and Molecular Biology* 16, no. 6 (June 2009): 574–81.

Hartmann, Betsy, Banu Subramaniam, and Charles Zerner. *Making Threats: Biofears and Environmental Anxieties.* Lanham, MD: Rowman and Littlefield, 2005.

Hayden, Cori. *When Nature Goes Public: The Making and Unmaking of Bioprospecting in Mexico.* Princeton, NJ: Princeton University Press, 2003.

Hayles, N. Katherine. "Constrained Constructivism: Locating Scientific Inquiry in the Theater of Representation." *New Orleans Review* 18 (1991).

Hayles, N. Katherine. *How We Became Posthuman: Virtual Bodies in Cybernetics, Literature, and Informatics.* Chicago: University of Chicago Press, 1999.

Hayward, Eva. "Enfolded Vision: Refracting the Love Life of the Octopus." *Octopus: A Visual Studies Journal* 1 (2005): 29–44.

Hayward, Eva. "Fingery Eyes: Impressions of Cup Corals." *Cultural Anthropology* 25, no. 4 (2010): 577–99.

Heath, Deborah. "Bodies, Antibodies, and Modest Interventions." In *Cyborgs and Citadels: Anthropological Interventions in Emerging Sciences and Technologies*, edited by Gary Downey and Joseph Dumit, 67–82. Santa Fe, NM: School of American Research Press, 1997.

Heller-Roazen, Daniel. *The Inner Touch: Archaeology of a Sensation.* Cambridge, MA: Zone Books, 2007.

Helmreich, Stefan. *Alien Ocean: Anthropological Voyages in Microbial Seas.* Berkeley: University of California Press, 2008.

Helmreich, Stefan. "An Anthropologist Underwater: Immersive Soundscapes, Submarine Cyborgs, and Transductive Ethnography." *American Anthropologist* 34, no. 4 (2007): 621–41.

Helmreich, Stefan. *Silicon Second Nature: Culturing Artificial Life in a Digital World.* Berkeley: University of California Press, 1998.

Helmreich, Stefan. "Species of Biocapital." *Science as Culture* 17, no. 4 (2008): 463–78.

Herzig, Rebecca. "On Performance, Productivity, and Vocabularies of Motive in Recent Studies of Science." *Feminist Theory* 5, no. 2 (2004): 127–47.

Hetherington, Kregg. "Beans before the Law: Knowledge Practices, Responsibility, and the Paraguayan Soy Boom." *Cultural Anthropology* 28, no. 1 (2013): 65–85.

Hill, Robert, and Peter Rich. "A Physical Interpretation for the Natural Photosynthetic Process." *Proceedings of the National Academy of Sciences* 80 (February 1983): 978–82.

Ho, Mae-Wan. "DNA and the New Organicism." In *The Future of DNA*, edited by Johannes Wirz and E. L. van Bueren, 78–93. Dordrecht: Kluwer Academic Publishers, 1997.

Hodgkin, Dorothy Crowfoot. "The X-ray Analysis of Complicated Molecules, Nobel

Lecture, December 11, 1964." In *Nobel Lectures, Chemistry 1963–1970*, 71–86. Amsterdam: Elsevier, 1972.

Hoffman, Michelle. "An RNA First: It's Part of the Gene-Copying Machinery." *Science* 252, no. 5005 (1991): 506–7.

Holdrege, Craig. *Genetics and the Manipulation of Life: The Forgotten Factor of Context.* Hudson, NY: Lindisfarne Press, 1996.

Hopwood, Nick. "'Giving Body' to Embryos: Modeling, Mechanism, and the Microtome in Late Nineteenth-Century Anatomy." *Isis* 90 (1999): 462–96.

Howes, David. *Empire of the Senses: The Sensual Culture Reader.* Oxford: Berg, 2005.

Hubbard, Ruth, and Elijah Wald. *Exploding the Gene Myth: How Genetic Information Is Produced and Manipulated by Scientists, Physicians, Employers, Insurance Companies, Educators, and Law Enforcers.* Boston: Beacon Press, 1993.

Hustak, Carla, and Natasha Myers. "Involutionary Momentum: Affective Ecologies and the Sciences of Plant/Insect Encounters." *Differences* 23, no. 3 (2012): 74–118.

Huxley, Thomas Henry. *A Manual of the Anatomy of Invertebrated Animals.* New York: D. Appleton, 1878.

Ihde, Don. *Bodies in Technology.* Minneapolis: University of Minnesota Press, 2001.

Jacob, François. *The Logic of Life: A History of Heredity.* Princeton, NJ: Princeton University Press, 1973.

Janssen, Bert J. C., Agni Christodoulidou, Andrew McCarthy, John D. Lambris, and Piet Gros. "Structure of C3b Reveals Conformational Changes That Underlie Complement Activity." *Nature* 444, no. 7116 (October 2006): 213–16.

Janssen, Bert J. C., Randy J. Read, Axel T. Brunger, and Piet Gros. "Crystallography: Crystallographic Evidence for Deviating C3b Structure." *Nature* 448, no. 7154 (2007): E1–E2.

Jones, Caroline A. "The Mediated Sensorium." In *Sensorium: Embodied Experience, Technology, and Contemporary Art*, edited by Caroline A. Jones, 5–49. Cambridge, MA: MIT Press, 2006.

Jones, David T. "Successful Ab Initio Prediction of the Tertiary Structure of NK-Lysin Using Multiple Sequences and Recognized Supersecondary Structural Motifs." *Proteins*, supplement 1 (1997): 185–91.

Kaiser, David. *Drawing Theories Apart: The Dispersion of Feynman Diagrams in Postwar Physics.* Chicago: Chicago University Press, 2005.

Kaiser, David, ed. *Pedagogy and the Practice of Science: Historical and Contemporary Perspectives.* Cambridge, MA: MIT Press, 2005.

Kaiser, David. "The Postwar Suburbanization of American Physics." *American Quarterly* 56, no. 4 (2004): 851–88.

Kay, Lily E. *The Molecular Vision of Life: Caltech, the Rockefeller Foundation, and the Rise of the New Biology.* New York: Oxford University Press, 1993.

Kay, Lily E. *Who Wrote the Book of Life? A History of the Genetic Code.* Stanford, CA: Stanford University Press, 2000.

Keating, Peter, Alberto Cambrosio, and Michael Mackenzie. "The Tools of the Discipline: Standards, Models, and Measures in the Affinity/Avidity Controversy in Immunology." In *The Right Tools for the Job: At Work in the Twentieth-Century Life Sciences,* edited by Adele Clarke and Joan H. Fujimura, 312–49. Princeton, NJ: Princeton University Press, 1992.

Keller, Evelyn Fox. *The Century of the Gene*. Cambridge, MA: Harvard University Press, 2000.

Keller, Evelyn Fox. *A Feeling for the Organism: The Life and Work of Barbara McClintock*. New York: W. H. Freeman, 1983.

Keller, Evelyn Fox. *Making Sense of Life: Explaining Biological Development with Models, Metaphors, and Machines*. Cambridge, MA: Harvard University Press, 2002.

Keller, Evelyn Fox. "Models of and Models for: Theory and Practice in Contemporary Biology." In "Proceedings of the 1998 Biennial Meetings of the Philosophy of Science Association. Part II: Symposia Papers," supplement, *Philosophy of Science* 67 (2000): S72–S86.

Keller, Evelyn Fox. *Refiguring Life: Metaphors of Twentieth-Century Biology*. New York: Columbia University Press, 1995.

Keller, Evelyn Fox. *Reflections on Gender and Science*. New Haven, CT: Yale University Press, 1985.

Keller, Evelyn Fox. *Secrets of Life, Secrets of Death: Essays on Language, Gender, and Science*. New York: Routledge, 1992.

Kelty, Christopher, and Hannah Landecker. "A Theory of Animation: Cells, L-systems, and Film." *Grey Room* 1, no. 17 (2004): 30–63.

Kendrew, John C. "Myoglobin and the Structure of Proteins. Nobel Lecture, December 11, 1962." In *Nobel Lectures, Chemistry 1942–1962*, 676–98. Amsterdam: Elsevier, 1964.

Kendrew, John C., G. Bodo, Howard M. Dintzis, et al. "A Three-Dimensional Model of the Myoglobin Molecule Obtained by X-ray Analysis." *Nature* 181, no. 4610 (1958): 662–66.

Kirschner, Marc, John Gerhart, and Tim Mitchison. "Molecular 'Vitalism.'" *Cell* 100 (2000): 79–88.

Knorr-Cetina, Karin. *Epistemic Cultures: How the Sciences Make Knowledge*. Cambridge, MA: Harvard University Press, 1999.

Kobe, B., G. Guncar, R. Buchholz, T. Huber, and B. Maco. "The Many Faces of Platelet Glycoprotein Ibα-Thrombin Interaction." *Current Protein and Peptide Science* 10, no. 6 (December 2009): 551–58.

Kohler, Robert. *Lords of the Fly: Drosophila Genetics and the Experimental Life*. Chicago: University of Chicago Press, 1994.

Kohler, Robert. "Moral Economy, Material Culture, and Community in Drosophila Genetics." In *The Science Studies Reader*, edited by Mario Biagioli, 243–57. New York: Routledge, 1999.

Kosek, Jake. "Ecologies of Empire: On the New Uses of the Honeybee." *Cultural Anthropology* 25, no. 4 (November 1, 2010): 650–78.

Kreisberg, Jason, Scott Betts, Cameron Hasse-Pettingell, and Jonathan King. "The Interdigitated Beta-Helix Domain of the P22 Tailspike Protein Acts as a Molecular Clamp in Trimer Stabilization." *Protein Science* 11 (2002): 820–30.

Kuhn, Thomas S. *The Structure of Scientific Revolutions*. 3rd ed. Chicago: University of Chicago Press, 1996.

Lakoff, George, and Mark Johnson. *Philosophy in the Flesh: The Embodied Mind and Its Challenge to Western Thought*. New York: Basic Books, 1999.

Lamond, Angus I., Mathias Uhlen, Stevan Horning, Alexander Makarov, Carol V. Robinson, Luis Serrano, F. Ulrich Hartl, et al. "Advancing Cell Biology through Proteomics in Space and Time (PROSPECTS)." *Molecular and Cellular Proteomics* 11, no. 3 (March 1, 2012). http://www.mcponline.org/content/11/3/O112.017731.

Landecker, Hannah. "Cellular Features: Microcinematography and Film Theory." *Critical Inquiry* 31, no. 4 (2005): 903-37.

Landecker, Hannah. "Creeping, Drinking, Dying: The Cinematic Portal and the Microscopic World of the Twentieth-Century Cell." *Science in Context* 24, no. 3 (2011): 381-416.

Landecker, Hannah. *Culturing Life: How Cells Became Technologies*. Cambridge, MA: Harvard University Press, 2006.

Landecker, Hannah. "Food as Exposure: Nutritional Epigenetics and the New Metabolism." *BioSocieties* 6, no. 2 (2011): 167-94.

Landecker, Hannah. "Living Differently in Time: Plasticity, Temporality and Cellular Biotechnologies." In *Technologized Images: Technologized Bodies*, edited by Jeanette Edwards, Penny Harvey, and Peter Wade, 211-36. New York: Berghahn Books, 2010.

Langridge, R., T. E. Ferrin, I. D. Kuntz, and M. L. Connolly. "Real-Time Color Graphics in Studies of Molecular-Interactions." *Science* 211, no. 4483 (1981): 661-66.

Langridge, Robert. "Interactive Three-Dimensional Computer Graphics in Molecular Biology." *Federation Proceedings* 33, no. 12 (1974): 2332-35.

Latour, Bruno. "Drawing Things Together." In *Representation in Scientific Practice*, edited by Michael Lynch and Steve Woolgar, 19-68. Cambridge, MA: MIT Press, 1990.

Latour, Bruno. "How to Talk about the Body? The Normative Dimensions of Science Studies." *Body and Society* 10, nos. 2-3 (2004): 205-29.

Latour, Bruno. *On the Modern Cult of the Factish Gods*. Durham, NC: Duke University Press, 2010.

Latour, Bruno. *Science in Action: How to Follow Scientists and Engineers through Society*. Cambridge, MA: Harvard University Press, 1987.

Latour, Bruno. *We Have Never Been Modern*. Cambridge, MA: Harvard University Press, 1991.

Latour, Bruno, and Steve Woolgar. *Laboratory Life: The Construction of Scientific Facts*. Princeton, NJ: Princeton University Press, 1979.

Lave, Jean, and Etienne Wenger. *Situated Learning: Legitimate Peripheral Participation*. Cambridge: Cambridge University Press, 1991.

Lave, Rebecca, Philip Mirowski, and Samuel Randalls. "Introduction: STS and Neoliberal Science." *Social Studies of Science* 40, no. 5 (2010): 659-75.

Law, John. "The Development of Specialties in Science: The Case of X-ray Protein Crystallography." *Science Studies* 3, no. 3 (1973): 275-303.

Lenoir, Timothy. *The Strategy of Life: Teleology and Mechanics in Nineteenth Century German Biology*. Boston: D. Reidel, 1982.

Leonelli, Sabina. "When Humans Are the Exception: Cross-Species Databases at the

Interface of Biological and Clinical Research." *Social Studies of Science* 42, no. 2 (April 1, 2012): 214–36.

Levin, Nadine. "Multivariate Statistics and the Enactment of Metabolic Complexity." *Social Studies of Science* 44, no. 4 (August 1, 2014): 555–78.

Levinthal, Cyrus. "Molecular Model-Building by Computer." *Scientific American* 214 (1966): 42–52.

Lewontin, Richard C. *Biology as Ideology: The Doctrine of DNA*. New York: Harper Perennial, 1992.

Licklider, J. C. R. "Man-Computer Symbiosis." *IRE Transactions on Human Factors in Electronics* 1 (March 1960): 4–11.

Liebler, John. *The Inner Life of the Cell*. XVIVO Studios. http://multimedia.mcb.harvard .edu.

Lock, Margaret M., and Judith Farquhar. *Beyond the Body Proper: Reading the Anthropology of Material Life*. Durham, NC: Duke University Press, 2007.

Lynch, Michael. "Discipline and the Material Form of Images: An Analysis of Scientific Visibility." *Social Studies of Science* 15, no. 1 (1985): 37–66.

Lynch, Michael. "The Externalized Retina: Selection and Mathematization in the Visual Documentation of Objects in the Life Sciences." In *Representation in Scientific Practice*, edited by Michael Lynch and Steve Woolgar, 153–86. Cambridge, MA: MIT Press, 1990.

Lynch, Michael. "Sacrifice and the Transformation of the Animal Body into a Scientific Object: Laboratory Culture and Ritual Practice in the Neurosciences." *Social Studies of Science* 18, no. 2 (1988): 265–89.

Lynch, Michael. "Science in the Age of Mechanical Reproduction: Moral and Epistemic Relations between Diagrams and Photographs." *Biology and Philosophy* 6 (1991): 205–26.

Lynch, Michael. *Scientific Practice and Ordinary Action: Ethnomethodology and Social Studies of Science*. Cambridge: Cambridge University Press, 1997.

Lynch, Michael E., and Steve Woolgar, eds. *Representation in Scientific Practice*. Cambridge, MA: MIT Press, 1990.

Mackenzie, Adrian. *Transductions: Bodies and Machines at Speed*. London: Continuum, 2002.

Maienschein, Jane. *Transforming Traditions in American Biology, 1880–1915*. Baltimore: Johns Hopkins University Press, 1991.

Mahmood, Saba. "Feminist Theory, Embodiment, and the Docile Agent: Some Reflections on the Egyptian Islamic Revival." *Cultural Anthropology* 16, no. 2 (May 2001): 202–36.

Marchant, Beth. "Cellular Visions: The Inner Life of a Cell." July 20, 2006. http://www .studiodaily.com/main/technique/tprojects/6850.html.

Marcus, George E. *Para-Sites: A Casebook against Cynical Reason*. Chicago: University of Chicago Press, 2000.

Marcus, George E., and Michael M. J. Fischer. *Anthropology as Cultural Critique: An Experimental Moment in the Human Sciences*. Chicago: University of Chicago Press, 1999.

Margulis, Lynn. *Symbiosis in Cell Evolution*. New York: W. H. Freeman, 1981.

Margulis, Lynn, Celeste A. Asikainen, and Wolfgang E. Krumbein. *Chimeras and Consciousness: Evolution of the Sensory Self.* Cambridge: MA: MIT Press, 2011.

Margulis, Lynn, and Dorion Sagan. *Microcosmos: Four Billion Years of Evolution from Our Microbial Ancestors.* Berkeley: University of California Press, 1997.

Margulis, Lynn, and Dorion Sagan. *What Is Life?* Berkeley: University of California Press, 1995.

Marks, Laura U. *The Skin of the Film: Intercultural Cinema, Embodiment, and the Senses.* Durham, NC: Duke University Press, 2000.

Martin, Emily. "Anthropology and the Cultural Study of Science." In "Anthropological Approaches in Science and Technology Studies," special issue of *Science, Technology and Human Values* 23, no. 1 (1998): 24–44.

Martin, Emily. "The Egg and the Sperm: How Science Has Constructed a Romance Based on Stereotypical Male-Female Roles." *Signs* 16, no. 3 (spring 1991): 485–501.

Martin, Emily. *Flexible Bodies: Tracking Immunity in American Culture from the Days of Polio to the Age of AIDS.* Boston: Beacon Press, 1994.

Marx, Karl. *Capital: A Critique of Political Economy, Vol. 1.* London: Lawrence and Wishart, 1970.

Mascia-Lees, Frances E. *A Companion to the Anthropology of the Body and Embodiment.* Oxford: John Wiley and Sons, 2011.

Masco, Joseph. "Mutant Ecologies: Radioactive Life in Post–Cold War New Mexico." *Cultural Anthropology* 19, no. 4 (November 1, 2004): 517–50.

Massumi, Brian. *Parables for the Virtual: Movement, Affect, Sensation.* Durham, NC: Duke University Press, 2002.

Mauss, Marcel. "Techniques of the Body." *Economy and Society* 2, no. 1 (1935): 70–88.

McCullough, Malcolm. *Abstracting Craft: The Practiced Digital.* Cambridge, MA: MIT Press, 1998.

Meinel, Christoph. "Molecules and Croquet Balls." In *Models: The Third Dimension of Science*, edited by Soraya de Chadarevian and Nick Hopwood, 242–75. Stanford, CA: Stanford University Press, 2004.

Merleau-Ponty, Maurice. *Phenomenology of Perception.* New York: Humanities Press, 1962.

Merleau-Ponty, Maurice. *The Visible and the Invisible.* Evanston, IL: Northwestern University Press, 1968.

Milburn, Colin. "Digital Matters: Video Games and the Cultural Transcoding of Nanotechnology." In *Governing Future Technologies*, edited by Mario Kaiser, Monika Kurath, Sabine Maasen, and Christoph Rehmann-Sutter, 109–27. Dordrecht: Springer Netherlands, 2010.

Miller, Chris. "Pretty Structures, but What about the Data?" *Science* 315, no. 5811 (2007): 459.

Miller, Greg. "Scientific Publishing: A Scientist's Nightmare: Software Problem Leads to Five Retractions." *Science* 314, no. 5807 (2006): 1856–57.

Mitchum, Robert. "In Chicago, Researchers Hold Finale to 'Dance Your PhD.'" *Chicago Tribune*, February 16, 2009. http://articles.chicagotribune.com/2009-02-16/news /0902150241_1_dance-researchers-escherichia.

Mody, Cyrus C. M. "A Little Dirt Never Hurt Anyone: Knowledge-Making and Contamination in Materials Science." *Social Studies of Science* 31, no. 1 (February 1, 2001): 7–36.

Mody, Cyrus C. M. "The Sounds of Science: Listening to Laboratory Practice." *Science, Technology and Human Values* 30, no. 2 (spring 2005): 175–98.

Mol, Annemarie. *The Body Multiple: Ontology in Medical Practice*. Durham, NC: Duke University Press, 2002.

Morgan, Mary S., and Margaret Morrison. *Models as Mediators: Perspectives on Natural and Social Science*. Cambridge: Cambridge University Press, 1999.

Morus, Iwan Rhys. "Placing Performance." *Isis* 101, no. 4 (2010): 775–78.

Murphy, Michelle. "The Economization of Life." Manuscript.

Murphy, Michelle. *Seizing the Means of Reproduction: Entanglements of Feminism, Health, and Technoscience*. Durham, NC: Duke University Press, 2012.

Murphy, Michelle. *Sick Building Syndrome and the Problem of Uncertainty: Environmental Politics, Technoscience, and Women Workers*. Durham, NC: Duke University Press, 2006.

Myers, Natasha. "Animating Mechanism: Animation and the Propagation of Affect in the Lively Arts of Protein Modeling." In "The Future of Feminist Technoscience," special issue of *Science Studies* 19, no. 2 (2006): 6–30.

Myers, Natasha. "Anthropologist as Transducer in a Field of Affects." Paper presented at the *Research-Creation Think Tank*, University of Alberta, March 25, 2014.

Myers, Natasha. "Dance Your PhD: Embodied Animations, Body Experiments, and the Affective Entanglements of Life Science Research." *Body and Society* 18, no. 1 (March 1, 2012): 151–89.

Myers, Natasha. "Interrupting the Order of Things: Experiments in Plant Sentience." Keynote Lecture, *Interruptions: Science, Feminisms, Knowledges*, Berkeley Center for Science, Technology, Medicine and Society, April 18–19, 2014.

Myers, Natasha. "Pedagogy and Performativity: Rendering Laboratory Lives in the Documentary *Naturally Obsessed: The Making of a Scientist*." *Isis* 101, no. 4 (December 1, 2010): 817–28.

Myers, Natasha. "Performing the Protein Fold." In *Simulation and Its Discontents*, edited by Sherry Turkle, 171–88. Cambridge, MA: MIT Press, 2009.

Myers, Natasha. "Visions for Embodiment in Technoscience." In *Teaching as Activism: Equity Meets Environmentalism*, edited by Peggy Tripp and Linda Muzzin, 255–67. Montreal: McGill-Queen's University Press, 2005.

Myers, Natasha, and Joseph Dumit. "Haptic Creativity and the Mid-embodiments of Experimental Life." In *Wiley-Blackwell Companion to the Anthropology of the Body and Embodiment*, edited by Fran Mascia-Lees, 239–61. Oxford: Wiley-Blackwell, 2011.

Negri, Antonio. "Value and Affect." *boundary 2* 26, no. 2 (1999): 77–88.

Nelson, Alondra. "DNA Ethnicity as Black Social Action?" *Cultural Anthropology* 28, no. 3 (2013): 527–36.

Nye, Mary Jo. "Paper Tools and Molecular Architecture in the Chemistry of Linus Pauling." In *Tools and Modes of Representation in the Laboratory Sciences*, edited by Ursula Klein, 117–32. Dordrecht: Kluwer, 2001.

Ochs, Elinor, Sally Jacobi, and Patrick Gonzales. "Interpretive Journeys: How Physicists Talk and Travel through Graphic Space." *Configurations* 2, no. 1 (1994): 151–71.

Olesko, Kathryn. "Tacit Knowledge and School Formation." In "Research Schools: Historical Reappraisals," special issue of *Osiris*, 2nd ser., 8 (1993): 16–29.

Ong, Aihwa. "A Milieu of Mutations: The Pluripotency and Fungibility of Life in Asia." *East Asian Science, Technology and Society* 7, no. 1 (2013): 69–85.

Ong, Aihwa. *Neoliberalism as Exception: Mutations in Citizenship and Sovereignty.* Durham, NC: Duke University Press, 2006.

Ong, Aihwa, and Stephen J. Collier. *Global Assemblages: Technology, Politics, and Ethics as Anthropological Problems.* Malden, MA: Blackwell, 2005.

Ortner, Sherry B. "Theory in Anthropology since the Sixties." *Comparative Studies in Society and History* 26, no. 1 (1984): 126–66.

Paterson, Mark. *The Senses of Touch: Haptics, Affects and Technologies.* New York: Berg, 2007.

Pauling, Linus, Robert Corey, and H. R. Branson. "The Structure of Proteins: Two Hydrogen Bonded Helical Configurations of the Polypeptide Chain." *Proceedings of the National Academy of Sciences* 37 (1951): 205–10.

Pauly, Philip J. *Controlling Life: Jacques Loeb and the Engineering Ideal in Biology.* Berkeley: University of California Press, 1996.

Paxson, Heather. *The Life of Cheese: Crafting Food and Value in America.* Berkeley: University of California Press, 2013.

Perutz, Max F. "The Hemoglobin Molecule (November 1964)." In *The Molecular Basis of Life: An Introduction to Molecular Biology (Articles from 1948–1968)*, edited by Robert Haynes and Philip Hanawalt, 39–51. San Francisco: W. H. Freeman, 1968.

Perutz, Max F. "The Medical Research Council Laboratory of Molecular Biology." http://nobelprize.org/nobel_prizes/medicine/articles/perutz/index.html.

Perutz, Max F. "Obituary: Linus Pauling." *Structural Biology* 1, no. 10 (1994): 667–71.

Pickering, Andrew. *The Mangle of Practice: Time, Agency, and Science.* Chicago: University of Chicago Press, 1995.

Pique, M. E. "Technical Trends in Molecular Graphics." In *Computer Graphics and Molecular Modeling*, edited by Robert J. Fletterick and Mark Zoller, 13–19. Cold Spring Harbor, NY: Cold Spring Harbor Laboratory, 1986.

Polanyi, Michael. "The Logic of Tacit Inference." In *Knowing and Being*, edited by Marjorie Grene, 138–58. Chicago: University of Chicago Press, 1969.

Polanyi, Michael. *Personal Knowledge: Towards a Post-critical Philosophy.* Chicago: University of Chicago Press, 1958.

Pollock, Anne. *Medicating Race: Heart Disease and Durable Preoccupations with Difference.* Durham, NC: Duke University Press, 2012.

Prentice, Rachel. "The Anatomy of a Surgical Simulation: The Mutual Articulation of Bodies in and through the Machine." *Social Studies of Science* 35 (2005): 867–94.

Puig de la Bellacasa, Maria. "Matters of Care in Technoscience: Assembling Neglected Things." *Social Studies of Science* 41, no. 1 (2011): 85–106.

Puig de la Bellacasa, Maria. "Touching Technologies, Touching Vision: The Reclaiming of

Sensorial Experience and the Politics of Speculative Thinking." *Subjectivity* 28 (2009): 297–315.

Rabinow, Paul. *Making PCR: A Story of Biotechnology*. Chicago: University of Chicago Press, 1996.

Rabinow, Paul. *Marking Time: On the Anthropology of the Contemporary*. Princeton, NJ: Princeton University Press, 2007.

Rabinow, Paul, and Nikolas Rose. "Biopower Today." *BioSocieties* 1 (2006): 195–217.

Rancière, Jacques. "The Aesthetic Dimension: Aesthetics, Politics, Knowledge." *Critical Inquiry* 36, no. 1 (January 2009): 1–19.

Rasmussen, Nicolas. *Picture Control: The Electron Microscope and the Transformation of Biology in America, 1940–1960*. Stanford, CA: Stanford University Press, 1997.

Read, Randy J., Paul D. Adams, W. Bryan Arendall, Axel T. Brunger, Paul Emsley, Robbie P. Joosten, Gerard J. Kleywegt, et al. "A New Generation of Crystallographic Validation Tools for the Protein Data Bank." *Structure* 19, no. 10 (October 12, 2011): 1395–412.

Reardon, Jenny. *Race to the Finish: Identity and Governance in an Age of Genomics*. Princeton, NJ: Princeton University Press, 2005.

Rheinberger, Hans-Jörg. *Toward a History of Epistemic Things: Synthesizing Proteins in the Test Tube*. Stanford, CA: Stanford University Press, 1997.

Rhodes, Gale. *Crystallography Made Crystal Clear: A Guide for Users of Macromolecular Models*. 3rd ed. San Diego: Academic Press, 2006.

Richardson, David C., and Jane S. Richardson. "The Kinemage: A Tool for Scientific Communication." *Protein Science* 1, no. 1 (January 1, 1992): 3–9.

Rifkind, Carole, and Richard Rifkind. *Naturally Obsessed: The Making of a Scientist*, 2009. See http://naturallyobsessed.com.

Riskin, Jessica. "Eighteenth-Century Wetware." *Representations* 83, no. 1 (2003): 97–125.

Roach, Joseph R. *Cities of the Dead: Circum-Atlantic Performance*. New York: Columbia University Press, 1996.

Robinson, Carol V., Andrej Sali, and Wolfgang Baumeister. "The Molecular Sociology of the Cell." *Nature* 450, no. 7172 (December 13, 2007): 973–82.

Roosth, Sophia. "Crafting Life: A Sensory Ethnography of Fabricated Biologies." PhD diss., Massachusetts Institute of Technology, 2010.

Roosth, Sophia. "Evolutionary Yarns in Seahorse Valley: Living Tissues, Wooly Textiles, Theoretical Biologies." *Differences* 23, no. 3 (2012): 9–41.

Roosth, Sophia. "Screaming Yeast: Sonocytology, Cytoplasmic Milieus, and Cellular Subjectivities." *Critical Inquiry* 35, no. (winter 2008): 332–50.

Rose, Hilary. *Love, Power, and Knowledge: Towards a Feminist Transformation of the Sciences*. Bloomington: Indiana University Press, 1994.

Rose, Hilary, and Steven P. R. Rose. *The Political Economy of Science: Ideology of/in the Natural Sciences*. London: Macmillan, 1976.

Rose, Nikolas S. *The Politics of Life Itself: Biomedicine, Power and Subjectivity in the Twenty-First Century*. Princeton, NJ: Princeton University Press, 2007.

Rotman, Brian. *Becoming beside Ourselves: The Alphabet, Ghosts, and Distributed Human Being*. Durham, NC: Duke University Press, 2008.

Rotman, Brian. "Gesture, or the Body without Organs of Speech." http://users.wowway .com/~brian_rotman/gesture.html.

Ruggeri, Zaverio M., Alessandro Zarpellon, James R. Roberts, Richard A. McClintock, Hua Jing, and G. Loredana Mendolicchio. "Unraveling the Mechanism and Significance of Thrombin Binding to Platelet Glycoprotein Ib." *Thrombosis and Haemostasis* 104, no. 5 (November 2010): 894–902.

Sacks, Oliver. "The Mind's Eye: What the Blind See." *New Yorker* (July 28, 2003): 48.

Sadler, J. Evan. "A Ménage à Trois in Two Configurations." *Science* 301, no. 5630 (July 11, 2003): 177–79.

Sagan, Dorion. "Samuel Butler's Willful Machines: The Lost Wisdom of Erewhon." *Common Review* (summer 2010): 10–19.

Sagan, Dorion, and Lynn Margulis. *Garden of Microbial Delights: A Practical Guide to the Subvisible World*. Boston: Harcourt Brace Jovanovich, 1988.

Samudra, Jaida Kim. "Memory in Our Body: Thick Participation and the Translation of Kinesthetic Experience." *American Ethnologist* 35, no. 4 (November 1, 2008): 665–81.

Schaffer, Simon. "Self Evidence." *Critical Inquiry* 18, no. 2 (winter 1992): 327–62.

Schrader, A. "Responding to *Pfiesteria Piscicida* (the Fish Killer)." *Social Studies of Science* 40, no. 2 (2010): 275–306.

Schwartz, Hillel. "Torque: The New Kinesthetic of the Twentieth Century." In *Incorporations*, edited by Jonathan Crary and Stanford Kwinter, 70–127. New York: Zone Books, 1992.

Seremetakis, C. Nadia. *The Senses Still*. Chicago: University of Chicago Press, 1996.

Shapin, Steven. *A Social History of Truth: Civility and Science in Seventeenth-Century England*. Chicago: University of Chicago Press, 1994.

Shapin, Steven, and Simon Schaffer. *Leviathan and the Air-Pump: Hobbes, Boyle, and the Experimental Life*. Princeton, NJ: Princeton University Press, 1985.

Shapiro, James A. *Evolution: A View from the 21st Century*. Upper Saddle River, NJ: FT Press Science, 2011.

Shapiro, James A. "Revisiting the Central Dogma in the 21st Century." *Annals of the New York Academy of Sciences* 1178, no. 1 (October 1, 2009): 6–28.

Shukin, Nicole. *Animal Capital: Rendering Life in Biopolitical Times*. Minneapolis: University of Minnesota Press, 2009.

Shulman, Robin. *Eat the City: A Tale of the Fishers, Foragers, Butchers, Farmers, Poultry Minders, Sugar Refiners, Cane Cutters, Beekeepers, Winemakers, and Brewers Who Built New York*. New York: Broadway Books, 2013.

Siler, William, and Donald A. B. Lindberg. *Computers in Life Science Research*. Conference Proceedings, Federation of American Societies for Experimental Biology. New York: Plenum Press, 1975.

Simondon, Gilbert. "The Position of the Problem of Ontogenesis." *Parrhesia* 7 (2009): 4–16.

Sismondo, Sergio. "Models, Simulations, and Their Objects." *Science in Context* 12, no. 2 (1999): 247–60.

Smith, Pamela H. *The Body of the Artisan: Art and Experience in the Scientific Revolution*. Chicago: University of Chicago Press, 2004.

Sobchack, Vivian. "Animation and Automation, or, the Incredible Effortfulness of Being." *Screen* 50, no. 4 (December 21, 2009): 375–91.

Squier, Susan Merrill. *Poultry Science, Chicken Culture: A Partial Alphabet*. New Brunswick, NJ: Rutgers University Press, 2011.

Stacey, Jackie, and Lucy Suchman. "Animation and Automation: The Liveliness and Labours of Bodies and Machines." *Body and Society* 18, no. 1 (March 1, 2012): 1–46.

Star, Susan Leigh. "Craft vs. Commodity, Mess vs. Transcendence: How the Right Tool Became the Wrong One in the Case of Taxidermy and Natural History." In *The Right Tools for the Job: At Work in Twentieth-Century Life Sciences*, edited by Adele Clarke and Joan H. Fujimura, 257–86. Princeton, NJ: Princeton University Press, 1992.

Star, Susan Leigh. "Power, Technologies and the Phenomenology of Conventions: On Being Allergic to Onions." In *A Sociology of Monsters—Essays on Power, Technology and Domination*, edited by John Law, 26–56. London: Routledge, 2001.

Steigerwald, Joan. "Figuring Nature: Ritter's Galvanic Inscriptions." *European Romantic Review* 18, no. 2 (2007): 255–63.

Stengers, Isabelle. *Cosmopolitics I*. Minneapolis: University of Minnesota Press, 2010.

Stengers, Isabelle. "A Constructivist Reading of Process and Reality." *Theory, Culture and Society* 25 (2008): 91–110.

Stent, Gunther. "That Was the Molecular Biology That Was." *Science* 160, no. 3826 (1968): 390–91.

Stewart, Kathleen. *Ordinary Affects*. Durham, NC: Duke University Press, 2007.

Storz, Gisela. "An Expanding Universe of Noncoding RNAs." *Science* 296, no. 5571 (May 17, 2002): 1260–63.

Sturken, Marita, and Lisa Cartwright. *Practices of Looking: An Introduction to Visual Culture*. Oxford: Oxford University Press, 2001.

Suchman, Lucy. *Human-Machine Reconfigurations: Plans and Situated Actions*, 2nd expanded ed. New York: Cambridge University Press, 2007.

Suchman, Lucy. *Plans and Situated Actions: The Problem of Human-Machine Communication*. New York: Cambridge University Press, 1987.

Sunder Rajan, Kaushik. *Biocapital: The Constitution of Postgenomic Life*. Durham, NC: Duke University Press, 2006.

Sunder Rajan, Kaushik. *Lively Capital: Biotechnologies, Ethics, and Governance in Global Markets*. Durham, NC: Duke University Press, 2012.

TallBear, Kimberly. *Native American DNA: Tribal Belonging and the False Promise of Genetic Science*. Minneapolis: University of Minnesota Press, 2013.

Tanford, Charles, and Jacqueline Reynolds. *Nature's Robots: A History of Proteins*. Oxford: Oxford University Press, 2001.

Taussig, Karen-Sue, Rayna Rapp, and Deborah Heath. "Flexible Eugenics: Technologies of the Self in the Age of Genetics." In *Genetic Nature/Culture: Anthropology and Science Beyond the Two-Culture Divide*, edited by Alan Goodman, Deborah Heath, and Susan Lindee, 58–76. Berkeley: University of California Press, 2003.

Taussig, Michael T. *Mimesis and Alterity: A Particular History of the Senses*. New York: Routledge, 1993.

Taylor, Peter J., and Ann S. Blum. "Ecosystems as Circuits: Diagrams and the Limits of Physical Analogies." *Biology and Philosophy* 6 (1991): 275–94.

Thacker, Eugene. *The Global Genome: Biotechnology, Politics, and Culture.* Cambridge, MA: MIT Press, 2005.

Thompson, Charis. *Making Parents: The Ontological Choreography of Reproductive Technologies.* Cambridge, MA: MIT Press, 2005.

Thompson, Evan. "Living Ways of Sense Making." *Philosophy Today*, SPEP Supplement (2001): 114–23.

Tierney, John. "Dancing Dissertations." *New York Times*, February 14, 2008. http://tierneylab.blogs.nytimes.com/2008/02/14/dancing-dissertations/.

Townley, R., and L. Shapiro. "Crystal Structure of the Adenylate Sensor from Fission Yeast AMP-Activated Protein Kinase." *Science* 315 (2007): 1726–29.

Traweek, Sharon. *Beamtimes and Lifetimes: The World of High Energy Physicists.* Cambridge, MA: Harvard University Press, 1988.

Tresch, John. *The Romantic Machine: Utopian Science and Technology after Napoleon.* Chicago: University of Chicago Press, 2012.

Trumpler, Maria. "Converging Images: Techniques of Intervention and Forms of Representation of Sodium-Channel Proteins in Nerve Cell Membranes." *Journal of the History of Biology* 30 (1997): 55–89.

Tsernoglou, D., G. A. Petsko, J. E. McQueen, and J. Hermans. "Molecular Graphics: Application to Structure Determination of a Snake-Venom Neurotoxin." *Science* 197, no. 4311 (1977): 1378–81.

Turkle, Sherry, with additional essays by William Clancey, Yanni Loukissas, Stefan Helmreich, and Natasha Myers. *Simulation and Its Discontents.* Cambridge, MA: MIT Press, 2009.

Turkle, Sherry, Joseph Dumit, Hugh Gusterson, David Mindell, Susan Silbey, Yanni Loukissas, and Natasha Myers. "Information Technologies and Professional Identity: A Comparative Study of the Effects of Virtuality." Report to the National Science Foundation, 2005.

Vanhoorelbeke, Karen, Hans Ulrichts, Roland A. Romijn, Eric G. Huizinga, and Hans Deckmyn. "The GPIbalpha-Thrombin Interaction: Far from Crystal Clear." *Trends in Molecular Medicine* 10, no. 1 (January 2004): 33–39.

Vertesi, Janet. "Seeing Like a Rover: Visualization, Embodiment, and Interaction on the Mars Exploration Rover Mission." *Social Studies of Science* 42, no. 3 (June 1, 2012): 393–414.

Voisine, C., J. S Pedersen, and R. I Morimoto. "Chaperone Networks: Tipping the Balance in Protein Folding Diseases." *Neurobiology of Disease* 40, no. 1 (2010): 12–20.

Waldby, Cathy. *The Visible Human Project: Informatic Bodies and Posthuman Medicine.* New York: Routledge, 2000.

Waldby, Cathy, and Robert Mitchell. *Tissue Economies: Blood, Organs, and Cell Lines in Late Capitalism.* Durham, NC: Duke University Press, 2006.

Wang, Xiangting, Xiaoyuan Song, Christopher K. Glass, and Michael G. Rosenfeld. "The Long Arm of Long Noncoding RNAs: Roles as Sensors Regulating Gene Transcriptional Programs." *Cold Spring Harbor Perspectives in Biology* 3, no. 1 (January 2011): a003756. http://www.ncbi.nlm.nih.gov/pmc/articles/PMC3003465/.

Warwick, Andrew. *Masters of Theory: Cambridge and the Rise of Mathematical Physics.* Chicago: University of Chicago Press, 2003.

Warwick, Andrew, and David Kaiser. "Conclusion: Kuhn, Foucault, and the Power of Pedagogy." In *Pedagogy and the Practice of Science*, edited by David Kaiser, 393–409. Cambridge, MA: MIT Press, 2005.

Watson, James D. *The Double Helix: A Personal Account of the Discovery of the Structure of DNA*. London: Weidenfeld and Nicolson, 1968.

Weber, Max. "Science as a Vocation." In *Max Weber: Essays in Sociology*, edited by H. H. Gerth and C. Wright Mills, 129–58. New York: Oxford University Press, 1946.

Weiss, Robert Alan, dir. *Protein Synthesis: An Epic on the Cellular Level*, 1971.

Wiesmann, Christian, Kenneth J. Katschke, Jianping Yin, Karim Y. Helmy, Micah Steffek, Wayne J. Fairbrother, Scott A. McCallum, et al. "Structure of C3b in Complex with CRIg Gives Insights into Regulation of Complement Activation." *Nature* 444, no. 7116 (October 2006): 217–20.

Wright, Peter E., and H. Jane Dyson. "Intrinsically Unstructured Proteins: Re-assessing the Protein Structure-Function Paradigm." *Journal of Molecular Biology* 293, no. 2 (October 22, 1999): 321–31.

Wright, Peter E., and H. Jane Dyson. "Linking Folding and Binding." *Current Opinion in Structural Biology* 19, no. 1 (February 2009): 31–38.

Zardecki, Christine. "Interesting Structures: Education and Outreach at the RSCB Protein Data Bank." *PLoS Biology* 6, no. 5 (May 2008).

Page numbers followed by *f* indicate illustrations.

235–38, 249n98; automation and, 3, 5, 13, 16, 20, 35–40, 54, 57, 74–75, 108, 112, 137, 143–45, 155, 211, 259n21; computer technology and, 3, 5, 10–13, 37, 40, 51–54, 68–75, 82–84, 114–15, 121, 127–29, 136–40; embodied knowledges and, 8, 15, 17–18, 22, 27–30, 32, 41, 75–76, 93, 99–104, 113, 204–21, 225–29, 236–38, 273n39; expert judgment of, 3–4, 20–22, 40, 93–96, 104–8, 136–40, 143–45, 189–91, 259n21; haptic creativity of, 17–18, 30, 74–78, 88, 98, 102, 123, 168–71, 216; intimacy and, xi, 32, 50–51, 65–68, 78–80, 93–96, 100–101, 112–17; ontological assumptions of, 130–35, 139–40, 148–55, 159–63, 191–97, 265n53; pedagogy and, x–xi, 2, 5, 20–25, 186–87, 204–9, 214f; proteins' liveliness and, 38–41, 99–100, 173–76; techniques of, x–xl, 2–3, 13–17, 21–22, 30, 41–49, 51–65, 74–76, 84–98, 242; vision's limitations and, x–xi, 13–15, 18–20, 35–36, 38–39, 54–58, 122–23, 126–27
Cullis, Anne, 257n39
Cummer, Clementine, 249n99
Curry, Stephen, 147

da Costa, Beatriz, 27
dance, 8, 22, 153–55, 206–10, 213, 218–29, 272n31, 273n39, 273n49
"Dance Your PhD Contest," 221–23, 225–27
"Dancing Dissertations" (Tierney), 221–22
DARPA (Defense Advanced Research Projects Agency), 25–26, 251n111
Daston, Lorraine, 125–27, 140–41, 259n21
Deleuze, Gilles, 28, 212, 224–25
Democritus, 122–23, 243n4
Descartes, René, 15
Despret, Vinciane, 212
diffraction, 14–15, 38, 51–58, 61–65, 115, 121, 123, 153. See also crystallographers; vision; X-rays
dilation (concept), 108–9
disciplining. See ecologies (moral); moral economy; objectivity; science

disenchantment, 5–6, 29, 185, 232. See also mechanism(s); objectivity
DNA, ix–x, 59, 68–69, 76, 81, 231–32, 234–37, 241, 251n114. See also genetics
docility, 46
Doyle, Richard, 231–32
Dumas, John, 151
Dumit, Joseph, 76
Durkheim, Émile, 260n29

E. coli, 12, 59, 188
ecologies (moral), 141–49, 155, 212, 234–37, 263n34
elasticity, 223–25, 274n49
electrical engineering, 177–79, 198–99
electron density maps, 56–57, 67, 69, 88, 89f, 89–91, 91f, 93–96, 112
electron microscopy, 7, 245n32
Elementary Forms of Religious Life (Durkheim), 260n29
Elizabeth II, 121, 122f, 123–24, 126–27, 133
El Lilly, 188
embodiment: dance and, 8, 22, 153–55, 206–10, 213, 218–29, 272n31, 273n39, 273n49; haptic creativity and, 18, 30, 74–78, 88, 98, 100–103, 123, 168–71, 216; kinesthetics and, 3–5, 99–100, 112, 116, 139–40; sensorium and, 14–15, 21–22, 41, 75–76, 93, 100–103, 113, 205, 210, 214. See also affect; epistemology; kinesthetics; pedagogy; vision
embryogenesis, 103–4, 133
emotions. See affect
emulation, 32, 113, 206–7, 209–11, 215–17, 224, 229
enchantment, 4, 29, 211. See also affect; disenchantment
entanglements. See affect; crystallographers; kinesthetics; objectivity; science
epigenetics, 234–35
epistemology: aesthetic anxieties and, 77, 82, 84, 124–29, 167–69, 182–87, 210–11, 224–25; affect and, 1–2, 112–17, 126–29, 244n4; automation and, 57, 108, 112,

liveliness. *See* biology; excitability; life; proteins

Lue, Robert A., 268n1

lure, the, 133–35, 165–67, 214–15, 237

lysozyme, 62, 115

The Machinery of Life (Goodsell), 169

Malcolm X, 49

Marcus, George, 249n99

Margulis, Lynn, 275n27

Martin, Emily, 42, 162

Mason, Frank, 194–97

matter and materiality: excitability and, 6, 9, 27–29, 38, 51, 65–68, 72, 140, 149–54, 184, 211–15, 229, 233–37; life's representations and, 4–5, 28–29, 159–65, 171–73, 197–203, 230–33; models and, 8–13, 18–19, 31–32, 71–104, 121–31, 141–44, 154, 159–62, 167–85, 198; proteins' liveliness and, xi, 3–5, 27–29, 38–41, 59–68, 112, 129–32, 148–55, 173–76, 182–203, 211–25, 232–37. *See also* life; mechanism(s)

Matter and Memory (Bergson), 213

Mauss, Marcel, 41, 63, 252n12

McClintock, Barbara, 100

McCullough, Malcolm, 75

mechanical engineering, 7, 159–63, 172–73, 176–79, 193–94

mechanism(s): atomic theories of matter and, x–xi, 167; biological research and, 25–27, 32, 159–69, 176–79, 197–203; fetishization and, 10, 25, 64, 116, 153, 195–97, 231; haptic aesthetics of, 77, 82, 84, 124–29, 169–71; life's representations and, 4–5, 28–29, 159–65, 171–73, 197–203, 230–33; materiality of, 171–76; neo-Darwinism and, 191–94; objectivity ideal and, 3–6, 27–29, 38–41, 59–61, 65–68, 112, 129–32, 148–55, 179–81, 194–97; pedagogy and, 166–67, 176–79, 191–94, 219–21; proteins' liveliness and, 173–76, 182–87, 194–97, 199–203, 233–37; temporality and, 187–91, 223–25, 273n49. *See also* disenchantment

Meinel, Christoph, 167

"A Ménage à Trois in Two Configurations" (Sadler), 150–51

Merleau-Ponty, Maurice, 15, 102, 108–9, 236

metabolomics, 10

microscopy, 14–15, 19, 55–57, 163–64, 186, 245n32, 247n56

Miller, Chris, 136–39, 143–45, 155

mimesis, 32, 124–27, 129–32, 206–7, 209–11. *See also* emulation; rendering; representation; verisimilitude

models: aesthetic anxieties and, 77, 82, 84, 124–29, 167–69, 182–87, 210–11, 224–25; affective dimensions of, 114–17; computer technologies and, 3, 5, 13, 16, 20, 35–40, 54, 57, 68–75, 92–93, 121, 137, 143–45, 155, 211, 259n21, 270n18; embodiment and, 206–9, 247n64; expert interpretation and, 93–96, 104–9, 111–12, 143–45, 259n21; fetishization and, 10, 25, 64, 116, 153, 195–97, 231; haptic creativity and, 17–18, 30, 74–78, 88, 98, 102, 123, 168, 216; kinesthetic limitations and, 4–5, 17–18, 92–93, 99–100, 103–4, 111–12, 114–17; as lures, 133–35; material theories and, 8–13, 18–19, 31–32, 71–104, 121–31, 141–44, 154, 159–65, 167–85, 198; media of, x–xi, 74–75, 80–96, 124–27, 173–74, 182–94, 206–9; pedagogy and, 20–22, 90–91, 93–96, 104–8, 111, 124–27, 247n64; performative aspects of, 129–40, 167–69, 189–94, 209–12; proteins' liveliness and, 38–41, 59–61, 65–68, 112, 127–32, 148–55, 173–76, 182–89, 199–203, 232–33; retractions of, 136–40; as snapshots, 66–67, 175, 217, 224, 270n19; temporality and, 187, 189–91, 223–25, 273n49; vision and, 13–15, 18–20, 35–37. *See also* crystallographers; epistemology; science

A Molecular Dance in the Blood, Observed (LiCata), 222, 222f

"Molecular Machinery" (Goodsell), 169

molecules. *See* atomic theory of matter;

performativity (*continued*)
129–35, 143–45, 167–69, 194–97, 209–12,
273n39; ontological implications of, 191–
94; pedagogy and, 20–22, 43–49, 186–87,
191–94, 204–11, 214f; rendering concept
and, 19–20, 132–35, 182–87, 211–12; scientists' explanations and, 5, 273n39; theories of, 262n26. *See also* dance; kinesthetics; pedagogy
perfumiers, 21–22
Perutz, Max, 11, 37, 58, 77–78, 80, 82–84,
85f, 113–14, 121–24, 133, 223, *plate 7*
pharmaceutical industry, ix–x, 3, 24–27,
137, 246n45, 262n9
phenomenology, 14–15. *See also* Merleau-
Ponty, Maurice
Philip, Kavita, 27
Pique, M. E., 87
Plasticine, 81–84, 128
Plexiglass, 90
Polanyi, Michael, 96–97, 108–9
practices. *See* craft; crystallographers;
ecologies (moral); epistemology; science
Proceedings of the National Academy of Sciences, 136
Project MAC, 85–87
Protein Data Bank (PDB), 10–13, 25–27, 62,
68–69, 106–10, 116, 126–27, 245n37
proteins: affective labor and, 50–51,
99–100; crystals of, x–xi, 13–15, 30, 40,
51–54, 58–65, 131, 153, 184; definitions of,
ix, 239–42; folding and, 7, 153–54, 159–
63, 172, 182–91, 204–9, 217–18, 224–25,
234–37, 241; liveliness of, xi, 3–5, 27–29,
38–41, 59–61, 65–68, 112, 129–32, 148–55,
173–76, 182–97, 199–203, 211–12, 215–21,
223–25, 232–37; mechanistic thinking
and, ix–xi, 4, 25–27, 159–67, 170–71,
182–87, 191–94, 197–203, 233–34; modeling of, x–xi, 74–80, 82–84f, 84–98, 126–
29, 211–12, 258n53, 270n18; pharmaceutical industry and, ix–x, 3, 24–27, 262n9;
sources of, 11–12, 58–59; structural focus
on, 8–13, 246n45; transduction and, ix,

32, 206, 213–15, 217–18, 229; visibility of,
13–15, 18–20, 51–58, 141–48, 245n32. *See
also* affect; crystallographers; matter and
materiality; pedagogy; vision
Protein Synthesis (Weiss), 219–21, 220–21f
proteomics, 10, 12, 246n45
protoplasm, 9, 163, 213
Puig de la Bellacasa, Maria, 51

race and racialization, 45–49, 251n114
Rancière, Jacques, 21
Rasmussen, Nicolas, 247n56
RCSB (Research Collaboratory for Structural Bioinformatics), 11–12, 245n37
reciprocal capture (term), 72–73, 100–101,
108–10, 175–76, 205–6, 210–11, 217–18,
229, 234–37, 255n68. *See also* affect; care;
crystallographers; intimacy
reflections, 52–53
rendering: capitalism and, 261n8; definitions of, 18–20, 29–32, 39, 114, 117,
123–24; kinesthetics and, 227–29; metaphor and, 159–65; mimesis and emulation and, 206–11; ontologies and, 139–
40; performativity and, 19–20, 132–35,
182–87, 211–12, 225–29; protein modeling and, 57–58; rendering and, 124, 130,
224–25; semiotic, 19, 124–33, 160–64,
172, 179–81, 200. *See also* affect; biology;
dance; models; performativity; science;
temporality
representation, 16–20, 22, 100–101, 122–27,
129–33, 148–49, 206. *See also* mimesis;
models
resemblance, 101, 123–24, 159–63, 209. *See
also* analogy; emulation; models; sympathy
retractions, 31, 136–40, 143–46
Rheinberger, Hans-Jörg, 78, 130
ribbon diagrams, 98f, 100f, 204, 217–18, 241,
plate 19
Richards, Frederic M., 89–90
Richardson, Jane and Dave, 153
Rifkind, Carole and Richard, 23–25, 41

Riskin, Jessica, 198
Röntgen, W. C., 51–52
Roosth, Sophia, 81
Rotman, Brian, 208
Roux, Wilhelm, 269n6
Royal College of Chemistry, 167
Royal Microscopical Society, 163
Ruggeri, Zaverio, 152

Saccharomyces cerevisiae, 12
Sadler, J. Evan, 150–51
Sagan, Dorion, 275n27
Schaffer, Simon, 14
Schrader, Astrid, 149, 265n53
science: aesthetic anxieties and, 77, 82,
    84, 124–29, 167–69, 182–87, 210–11,
    224–25; affect and, xi, 49–51, 58–65,
    68–73, 249n98; atlas tradition and, 12–13,
    15–17; body experiments and, 1–4, 18,
    32, 74–76, 102, 112, 170–71, 215–20; capi-
    talism and, ix–x, 25–27, 47–48, 186–87,
    237–38, 261n8; competition in, 115–16;
    ethnographic tradition and, 6–8; expert
    judgment in, 20–22, 40, 93–100, 104–8,
    136–45, 189–91, 259n21; fetishes and fe-
    tishization in, 10, 25, 64, 116, 153, 195–97,
    231; gender and, 121–22, 228–29, 257n39,
    260n26; information science and, 4–5,
    9–10, 168–69, 178–79, 231–36; kinesthe-
    tic dimensions of, 1–2, 15, 22, 41, 75–76,
    93, 100–103, 113, 205, 210, 214, 236; ma-
    teriality and, 8–13, 18–19, 31–32, 71–104,
    121–31, 141–44, 154, 159–62, 167–85, 198;
    mechanistic ontologies and, ix–xi, 5,
    159–67, 171–73, 179–80, 182–87, 230–33;
    models and, 15–17, 77–84, 123–24, 126,
    247n64; moral economy of, 141–49,
    154–55, 225–29, 263n34; narratives of,
    41–42, 230–33; objectivity and, 2, 127–
    29, 194–97; pedagogy in, 20–22, 38–49,
    63, 186–87, 247n64, 248n87; performa-
    tive culture and, 5–8, 18, 43–51, 182–87,
    206–9, 215–18; temporality and, 187–91,
    223–25, 273n49. *See also* biology

Science (journal), 10, 24, 51–52, 90, 136, 138,
    149, 170
Science Magazine, 8, 221–22
Scientific American, 82
The Scientist, 45
Scripps Research Institute, 169
sedimentation (of thinking), 19, 26, 124,
    129–33, 161, 171–73, 185, 197–203
Segal, Jerome, 86
self-fashioning, 43–49
sensorium, 14–15, 21–22, 41, 75–76, 93, 100–
    103, 113, 170–71, 204–15
Shapin, Steven, 14
Shapiro, James, 234–36
Shapiro, Larry, 23–25, 49, 63–65, 93–94,
    115–16, 170–71, 200–202
Sismondo, Sergio, 16
"Situated Knowledges" (Haraway), 164–65
Smith, Pamela, 75
snapshot (metaphor), 66–67, 175, 217, 224,
    270n19
"Some Analogies between Molecules and
    Crystals" (Caldwell), 51–52
Spinoza, Baruch, 28
Stacey, Jackie, 211
Steitz, Thomas, 146
Stengers, Isabelle, 72–73, 131, 133–34, 195–
    96, 255n68
Stent, Gunther, 9–10
stereoglasses, 69, 90, 92–93
sublime, 69–70, 182, 202
Suchman, Lucy, 211
sympathy, 100–102, 110, 113, 133, 209. *See
    also* emulation; resemblance
synesthesia, 17–18, 75–76, 86. *See also*
    sensorium
Szent-Gyorgyi, Albert, 214, 214*f*

tacit knowledges, 8, 20–22, 96–98, 103. *See
    also* epistemology
tangibility, 86–88, 109–10, 162, 169, 171–73.
    See also epistemology; kinesthetics; mat-
    ter and materiality; models; sensorium
Taussig, Michael, 209–10